CAMBRIDGE EARTH SCIENCE SERIES

Editors:
A.H. Cook, W.B. Harland, N.F. Hughes,
J.G. Sclater

General hydrogeology

General hydrogeology

Edited by
E.V. PINNEKER
*Institute for the Study of the Earth's Crust,
Novosibirsk, Siberia, USSR*

Translated by D.E. Howard and J.C. Harvey
Plymouth Polytechnic

CAMBRIDGE UNIVERSITY PRESS
Cambridge
London   New York   New Rochelle
Melbourne   Sydney

CAMBRIDGE UNIVERSITY PRESS
Cambridge, New York, Melbourne, Madrid, Cape Town, Singapore,
São Paulo, Delhi, Dubai, Tokyo, Mexico City

Cambridge University Press
The Edinburgh Building, Cambridge CB2 8RU, UK

Published in the United States of America by
Cambridge University Press, New York

www.cambridge.org
Information on this title: www.cambridge.org/9780521154833

Originally published in Russian as *Obshchaya gidrogeologiya*
by Izdatel'stvo 'Nauka' in 1980 and © Izdatel'stvo 'Nauka',
1980

First published in English by Cambridge University Press
as *General hydrogeology*
English edition © Cambridge University Press 1983

Paperback edition 2010

A catalogue record for this publication is available from the British Library

Library of Congress Catalogue Card Number: 82-9499

ISBN 978-0-521-24905-8 Hardback
ISBN 978-0-521-15483-3 Paperback

Contents

Contents

Preface

'General hydrogeology' is a name which is invested with a great variety of meanings. It is the name applied equally to a short introduction to hydrogeology and the whole complex of meanings which constitute the study of groundwater. Frequently it includes general information on geology, hydrology, and meteorology.

General hydrogeology as an inseparable part of hydrogeology is analogous, for example, to hydrogeochemistry or the dynamics of groundwater. It must therefore be invested with the same general, i.e. fundamental, hydrogeological considerations and, first and foremost, with the characteristics of the phenomenon and the laws which govern the distribution of water in the Earth's interior.

Among the fundamental hydrogeological considerations to which the attention of the reader of the present volume is drawn is data on the subject of hydrogeology, its definition as a science is given and the way is laid for its further development: there is a short review of the history of the subject and a discussion of the state of hydrogeological terminology. The question of the circulation of water in the Earth is examined from the modern standpoint and a description of the groundwater-bearing systems, distinguished according to the manner in which they are deposited, is given. The volume is concluded with chapters dealing with the features of hydrogeothermics and regional hydrogeological laws.

Just as in other sciences, hydrogeology did not develop uniformly: in some branches the changes in the concepts are great, in others they are preserved in almost the same form as they were formulated several decades ago. In particular, the bulk of the laws of distribution of water in the Earth's interior was established only in recent years and only a small number have been known for some time. Hence many chapters of the present volume are original and are based on new material. However, there are also sections where, in the past, information was very sparse and the authors have limited the content to present day concepts.

The authors of the various chapters and sections are shown in the Contents. Apart from the main contributors (E.V. Pinneker, B.I. Pisarskii, and S.L. Shvartsev) i.e. the authors of the different chapters, various sections were contributed by G.Ya. Bogdanov, V.N. Borisov, and K.P. Karavanov. V.A. Pleshevenkova and E.A. Rubinchik have done much in the work of preparing the manuscript. Valuable advice was received from the reviewers P.I. Trofimuka and I.S. Lomonosova and also from the members of the authors collective N.I. Plotnikov, O.N. Tolstikhin, and P.F. Shvetsov. All constructive remarks were valued and doubtless have improved the contents of this volume.

E.V. Pinneker

1

The subject matter of hydrogeology

1.1 The definition of hydrogeology, its content, development trends, and research methods

Hydrogeology is the study of water below the Earth's surface. This definition is found in almost all treatises, reference and textbooks published during the last century, and the term was used originally by Russian and Soviet (Gordeev, 1954) and then foreign hydrogeologists to cover the origin, movement, distribution and composition, and the regime of subsurface water. Examples of this treatment of hydrogeology include the works of P.N. Chirvinskii (1922), O.K. Lange (1931), F.P. Savarenskii (1935), D.W. Mead (1950), A.M. Ovchinnikov (1955), P. Fourmarier (1958), S.N. Davies & R.J.M. De Wiest (1967), A. Thurner (1967), P.P. Klimentov & G.Ya. Bogdanov (1977), as representative of the work by scientists of various countries, generations, and schools.

A detailed definition, which, in contrast to other early and recent definitions, most fully covers both the scientific and the engineering aspects, is given in the *Geological dictionary* (Anon., 1955, Vol. 1, p. 168): 'Hydrogeology is the study of subsurface water, its origins, manner of deposition, laws of motion, physical and chemical properties, interrelations with atmospheric and surface water and also the forms and conditions of man's activities on subsurface water . . . its economic value etc.'

Every scientific term must be absolutely specific, but unfortunately this cannot be said of subsurface water, which is now generally known as groundwater. It is by no means strange that different research workers still give this term very different meanings. As a result the hydrogeological content of their views suffers from lack of precision and needs clarification.*

*Strictly speaking the term subsurface water is not quite accurate. It has its origins in very ancient times when people mistakenly imagined that there was a subterranean water kingdom like the depths of Tartarus in Greek mythology, a watery abyss from which rivers and springs were supplied. The term has taken root in the literature both here and abroad.

Hydrogeology may be regarded as the study of groundwater in all its physical phases and various chemical compositions wherever it may be found below the surface of the Earth. However, not all hydrogeologists use the term groundwater in this sense (Meinzer, 1923; Tolstikhin, 1971). In much Soviet and foreign literature groundwater is often considered to be the 'free' liquid water that can move through voids in rock and soil. Exactly the same interpretation of the term is found in some well known groundwater classification schemes (Savarenskii, 1935; Ovchinnikov, 1955; Zaitsev, 1961a; Schoeller, 1962; and others). An intermediate point of view regards groundwater as separable into liquid and vapour phases, or water that is free, or physically bound in its surrounding rock material.

Groundwater is not pure water but is a natural solution of various amounts of several substances.

The definition given by F.P. Savarenskii (1935, p. 13) is probably still the best. Groundwater in the strict sense in his view is the finely divided water in the form of droplets which fills the voids (pores) in rock and which is capable of flowing through these voids. The English and American word 'groundwater' is synonymous with the Russian *'podzemnye vody'*, which translated literally means 'subterranean water'; the German equivalent is *'Grundwasser'* (Keilhack, 1935; Kühne, 1932; Todd, 1959; Gray, 1975; Richter & Lillich, 1975; Hähne & Jordan, 1978). It is used in the sense defined above when aspects of subsurface water, movement, composition, and economic use are being discussed. This definition has embedded itself deeply in science and for general practical purposes, and it is difficult to discuss the definition further without introducing controversial viewpoints.

Groundwater as defined above is a fundamental but not the only object of study within the compass of hydrogeology. Frozen groundwater, bound water of various kinds, and water/water vapour mixtures also play an important role, but the term 'water' is not really a suitable group name to include all of these. Therefore use of the word hydrogeology for the science of subsurface water in all its forms fails to reveal all the aspects concerned. The scope for research in the subject is vast. For example an understanding of the formation processes of groundwater or the laws which control its distribution demands that the hydrogeologist studies water in the ground in all its phases — ice, liquid, and vapour — and the physically and chemically bound forms of water in rock and soil.

Before defining hydrogeology precisely it is necessary to spend some time examining the reasons which make a precise definition necessary.

Hydrogeology is one of the Earth sciences. The word itself, from the Greek words ὑδραίνω (water), γεο (Earth), λόγος (discourse or study) shows that it is concerned with water inside the Earth. Studies develop from a basis of an analysis of the history of the formation of the Earth's crust, the outermost 30 km (approximately) zone of the Earth, because it is impossible to understand the history of water in the Earth without considering the rock surrounding it and

with which it was originally formed from the earliest times in Earth history. Hydrogeology has close links with many other branches of the Earth sciences (petrology, structural geology or tectonics, sedimentology, etc.). It is also a branch of hydrology in the broad sense because it is an inseparable part of a study of the whole of the Earth's hydrosphere (water above the ground). Geology and hydrology are the two fundamental bases of hydrogeology, and it was born at the interface of these two disciplines and its subsequent growth to the present day has been linked with both of them. The solution of hydrogeological problems has always required that other disciplines (physics, chemistry, hydraulics) be brought into account.

The word hydrogeology was introduced by G. Lamarck (in 1802) for the purpose of describing the processes of destruction, and deposition of sediments performed by water. It came into use in the 1880s for the study of water in the ground (D. Lucas in England, N.A. Golovkinskii in Russia, and others) a study which contributed to the economic use of groundwater in many countries. The direct consequence of the various uses of water was the establishment of hydrogeology as an independent scientific discipline.

The concept of hydrogeology requires that it remains an applied science concerned with the study of groundwater as a mineral. Its problems are mainly of a practical nature, to satisfy man's demand for fresh, and mineralized water for medicinal purposes, and to control it to avoid damaging and harmful effects. There are of course theoretical problems which need solving, but these are obviously of secondary importance.

The position of hydrogeology changed dramatically in the middle of the twentieth century when the scale of hydrogeological research increased sharply and the range of theoretical problems broadened. Figuratively speaking, hydrogeology entered a period of extended youth, and rapid growth. At that time in the same way as the other Earth sciences it was enriched by the achievements of the fundamental sciences — physics, chemistry, mathematics, and biology — and allied branches of knowledge (geology, geochemistry, geophysics, hydrology) and by the acquisition of a mass of new hydrogeological data. Hydrogeology became a complex science and independent schools of thought appeared, among which according to N.I. Plotnikov (1976) there were applied, methodological, and theoretical schools (Table 1.1). The majority of the new theoretical branches have been formed over the past 25–30 years — hydrogeochemistry, hydrogeothermics and palaeohydrogeology.

In the second half of the twentieth century hydrogeology entered an essentially new stage of development. According to Kedrov (1971), this stage may be described as transitional from the mere gathering and collation of facts, a *descriptive* stage, to the *explanatory* stage marked by the desire to explain the processes occurring in terms of physical laws inherent in the subject. At the same time a *prognostic* stage also came into being, which now plays a major part in forecasting the effects of man's activities within the interior of the Earth.

The transitional period and the crisis which arose during the rapid growth demands that the hydrogeological specialist master a vast mass of accumulated facts, and the re-examination of certain theoretical positions and the clarification of the future development of hydrogeology. The crisis was one of growth, not of decay — of trying to understand present day conditions, and looking for ways to make further progress. A fundamentally new approach to the definition of the scope of hydrogeology has begun to appear. In the course of discussions at conferences in the 1970s it was noted that, firstly, hydrogeology was changing from a study of the phenomena to a science concerned with processes and laws, and secondly its field of research had to be considered as a whole, just as in other sciences — a real material system, not merely water in the ground, the concept of which, as noted above, suffers from being rather vague.

Such a material system which forms the subject matter of hydrogeology may be regarded as an aqueous shell or membrane situated below the Earth's surface. It is closely connected with the rocks but differs because of its heterogeneity and it is therefore unsatisfactory to call it simply groundwater. This aqueous shell is a *subsurface hydrosphere* (Savarenskii, 1947; Ovchinnikov, 1955; Sydykov, 1973; Pinneker, 1975), or according to Plotnikov (1976), a *hydrosphere*. Meanwhile, between the actual content of hydrogeology (in so far as it is concerned with the free water in the Earth's crust) and its research content (the subsurface hydrosphere as a whole) there is still some divergence, but no doubt this will be reduced to a minimum in the future.

There are other opinions about hydrogeology. P.F. Shvetsov *et al.* (1973) regarded the totality of the subsurface water-bearing lithosphere (rock and soil) as the 'water-bearing system', and used the definition 'history of the formation of, and subsequent changes in the water-bearing system'. G.V. Bogomolov (1975) wrote that 'Hydrogeology is the science of the interactions of subsurface waters with solid and vapour phases of the materials in the Earth.' Neither definition can be considered exhaustive. However, the rational concepts which they contain (i.e. a systematic approach and the concept of interactions) can be accounted for if hydrogeology is recognized as the science of the subsurface hydrosphere — the water-bearing system of the interior of the Earth, in which different forms of water interact with rock, gases, and living things.

The subsurface hydrosphere is a fully defined material system containing all the water molecules in the Earth's interior. Water, both free and bound to other substances is the basic component, in the solid, liquid, and vapour states, and it changes from one state to another with changes in the hydrodynamic conditions. These components interact with one another, and they are so tightly bound up with the enclosing rock material that they effectively form part of the lithosphere. Moreover, these components naturally react upwards with the surface hydrosphere, atmosphere, and biosphere, and downwards to the Earth's mantle, and in fact the whole of the cosmos. Therefore the subsurface hydrosphere contains all the water situated below the surface of the Earth, penetrating the crust right down to the Earth's mantle.

Table 1.1. *The components of hydrogeology*

Nomenclature	Contents
Theoretical divisions	
General hydrogeology	Bases of the study of the subsurface hydrosphere, phenomena and the laws governing the distribution of water in the interior of the Earth
Hydrogeodynamics	Movement, regime, and resources of groundwater, hydrogeological modelling
Hydrogeochemistry	Laws governing the migration of chemical elements in the subsurface hydrosphere, the composition of groundwater and its formation
Hydrogeothermics	Thermal properties and characteristics of the subsurface hydrosphere
History of the subsurface hydrosphere (palaeohydrogeology)	Origins and evolution of the subsurface hydrosphere, the geological activity of water in the interior of the Earth and its role in various geological processes
Methodology and application divisions	
Methods of hydrogeology research (methodological hydrogeology)	Methods of carrying out hydrogeological research (surveying, fieldwork to establish limits of deposits, regime, experimental, laboratory, and test-rig work)
Exploitation of groundwater ('prospecting' hydrogeology)	Study of the deposits of groundwater and the use of groundwater for water supplies, irrigation, and for medicinal, industrial, and thermal energy purposes
Engineering problems with groundwater ('engineering' hydrogeology)	Influence of groundwater on the degree of saturation of mineral deposits and effects on mining, irrigation conditions for agricultural land, erection of industrial and other types of construction
Conservation of the subsurface hydrosphere (technogenic hydrogeology)	Pollution and exhaustion of resources of the subsurface hydrosphere, preventive measures, and control of its regime
Regional hydrogeology	Regional studies and the description of groundwater and other components of the subsurface hydrosphere

We will now formulate a precise definition. *Hydrogeology* is the science of the subsurface hydrosphere. It is concerned with the *history of the underground hydrosphere, its resources and composition, the laws governing the distribution of its components, the processes which take place within it, and its interaction with the surrounding Earth envelope, and also the economic value of the components of the underground hydrosphere and man's influence on them.* Both the theoretical and practical aspects of hydrogeology are accounted for in this broad definition. The essence of hydrogeology may be defined briefly not as underground water, but as the water-bearing system of the Earth's crust and the processes which take place within it, i.e. the life of the subsurface hydrosphere.

Subsurface water is an important component of the subsurface hydrosphere both in the cognitive and economic sense, and it will remain the basic subject of study for the hydrogeologists of the future. Only the directions of research will vary, and these, for the purpose of studying subsurface water will include unbreakable links with the other components of the subsurface hydrosphere, and will take into account water exchange and mass transfer for the subsurface hydrosphere as a whole, and not just for some individual component no matter how important it may be.

The further development of hydrogeology requires the optimal combination of the 'descriptive', 'explanatory', and 'prognostic' elements. In the future, just as in the past, the descriptive studies of the subsurface hydrosphere will receive careful attention, i.e. the gathering of new information chiefly about the qualitative and quantitative evaluation of groundwater. The 'explanatory' stage will be at a higher level and will have to formulate new laws, study processes, and delve more deeply in discovering the life of the subsurface hydrosphere. The important difference from early thoughts on hydrogeology will be the leading role taken by scientific forecasting. The practical problems facing hydrogeologists before the middle of the twentieth century amounted mainly to the use of subsurface water for economic purposes and a struggle with related problems, and further problems which subsurface water posed for the construction industry. Other problems have now arisen, one of which is the role played by water in the formation and decomposition of mineral deposits.

The other major problem will be the control of the subsurface hydrosphere. This is bound up with a whole complex of theoretical problems concerning the laws of distribution, the history of all the water in the Earth's interior, and the processes that govern the formation of the various components in the subsurface hydrosphere. There remains the question of the controlled interference by man in the affairs of the subsurface hydrosphere, mainly in order to exploit the water resources contained there. Such interference is now occurring on an ever increasing scale.

Hydrogeologists use research methods employed in the other Earth sciences, and adapt them to the study of groundwater. These methods are borrowed to a considerable extent from geology (surveying, mapping, palaeohydrological analysis), hydraulics (hydrogeodynamic methods), and geo-

chemistry (hydrogeochemical methods) being examples. The methods of the fundamental sciences, physical, chemical, physical-chemical, and mathematical, are widely employed.

In evaluating the various components of the subsurface hydrosphere hydrogeologists are guided by the methods of the natural sciences, prospecting techniques, and geophysical methods. Furthermore, specific local observations lead to the formulation of separate and individual parameters for the area concerned. Calculations, laboratory experiments, and modelling of natural processes began to be introduced into hydrogeology earlier than into its allied sciences.

The list of methods employed in the study of the subsurface hydrosphere can be continued, but even the most complete list would not give a good picture of the methodology of scientific research. The methods enumerated form *in toto* a complex set of hydrogeological methods which generally form an outline of the subject matter. As for the methodology of hydrogeological research, here, as in other Earth sciences, it relies on the *principle of actualism* — 'the present is the key to the past'. Therefore a comparison of the geological processes of the present and the past epochs is undertaken not mechanistically but dialectically. In general the modern methodology of hydrogeological research amounts to the following:

First the elucidation of the local hydrogeological regime is unsatisfactory without an *historical approach*. The subsurface hydrosphere is the product of a long period of geological history. Therefore its components (groundwater, voids, etc.) must be studied from the natural-historical aspect. Here reconstructions of the hydrogeological past and prognoses about the future come to the aid of the research worker. To be able to see the subsurface hydrosphere in historical perspective, to be able to follow its history from the past to the present, and from the present to the future is the essence of theoretical generalizations and scientific hydrogeological forecasting.

Secondly the water in the Earth's interior is in a state of continual movement, change, and renewal. It changes itself and changes its surroundings by interaction. The ways by which it moves are very varied (mechanical, thermal, etc.), but they may be reduced to a single complex form, a geological form (Kedrov, 1964), concerned with the water exchange and mass transfer processes which take place in the subsurface hydrosphere. In so far as changes in this environment have no simple expression even for quite large regions of the Earth's surface, studies of the laws governing the development of this environment must be based not only on the consideration of the features of separate regions, but also general laws. Recognition of the interrelationship of the general and the particular in the subsurface hydrosphere demands a *comparative approach* together with the use of analysis (the division of a process or phenomenon into its component parts or features), and synthesis (the uniting of components which are the results of analysis). The comparative approach is widely used for the elucidation of local hydrogeological regimes involved with hydrogeodynamics and hydrogeochemistry.

Thirdly the subsurface hydrosphere is closely connected

with and is constantly reacting with the other internal zones or shells of the Earth. It cannot exist in isolation from the solid material of the lithosphere, the surface hydrosphere, the atmosphere, the biosphere, even the mantle and the cosmos generally. More comprehensive studies of the internal links between the components of the subsurface hydrosphere and these other entities require a thorough knowledge of the physiogeographical and the geological structure of the Earth's interior and the thermodynamic conditions of the waters therein. The study of the subsurface hydrosphere demands a *sophisticated approach*. Such an approach assumes the employment of different research methods, which will guarantee that the results obtained are unbiased and confirm the laws as revealed by a large number of facts. The processes by which the water in the Earth's interior participate can only be studied against the background of their interrelationships with the components of the environment and with the aid of various research methods.

The three approaches outlined above may be called the *comparative-historical method*. This method is the very basis of the Earth sciences, including hydrogeology. In its relationship to the subsurface hydrosphere it reflects the objective nature of hydrogeology and the laws of natural development.

The solution of the theoretical problems and practical questions facing hydrogeologists demands the mastery of an ever increasing stream of information and the putting forward of new ideas. These are indispensable conditions without which there is no scientific and technical progress. Facts need effective summary and clear comprehension; they are able to call to life other views and to disprove traditional concepts. In hydrogeology, just as in any other science, ideas that repudiate the traditional point of view and therefore at first sight seem, using Einstein's expression, 'stupid', are of especial interest: the question is − are they sufficiently paradoxical to be true?

The large-scale introduction into hydrogeology of the laws of the exact sciences and the progressive methods used in the allied Earth sciences offer great prospects. This path has already shown that it gives hopes of results, and is far from exhausted. An example of this is the mathematical formulation of the sophisticated comparative-historical method. In hydrogeology at present comparative description rules, on the principle of 'larger−smaller', or 'a little or a lot'. The comprehensive introduction of quantitative indicators into hydrogeology, i.e. the use of numbers and measurements in the characterization of geological processes is the most important task of future research into the subsurface hydrosphere (Sidorenko, 1964).

Finally, the traditional methods of hydrogeological research must also be developed in breadth, by taking in more and more, and newer, territory (in the end the whole Earth and then the planets), also probing deeply to obtain information about the composition and the properties of the water at great depths. While the hydrogeology of the land masses has been very thoroughly studied, we have only just started to study the hydrogeology of the ocean basins. Information on the water in the deepest layers of the Earth's crust has appeared. The contours of the deep-seated hydrogeology are

being more clearly revealed to the hydrogeological research workers of the subsurface hydrosphere; it is becoming quite clear that the geological processes that are associated with the water in the upper levels of the crust reflect the processes occurring in the depths, in the mantle. In order to obtain an understanding of the depths of the subsurface hydrosphere some well-established methods for the study of the crust (geophysical, geochemical, etc.) will play an important role alongside the new research methods.

1.2 A short history of hydrogeological thought

The history of the development of hydrogeology as a science is very much shorter than the history of ideas about origins, movement, or more especially the use of water from the interior of the Earth. Although hydrogeology is only one hundred years old, views on the nature of the subsurface hydrosphere began to appear in ancient times, and the conflict of opinions about different concepts began to appear in antiquity.

The birth of hydrogeological ideas. Even at the dawn of civilization man began to be interested in groundwater, which played an important part in water supply and irrigation. The medicinal properties of mineral waters were known long ago. In the Near East, in Central Asia, India, and China 3000−5000 years ago they knew how to construct water wells, adits, water-raising machines and pumps, capping of sources, and conduits. The use for economic purposes of 'Nature's miracle', as they called underground water in ancient times, was based naturally on the laws of distribution of water and its behaviour.

The Sumerians, the first inhabitants of Mesopotamia, knew how and where to find water. The pumping technology which existed in ancient Egypt was the best of its age. Mining activities would have been impossible without a knowledge of the theory of water inflow. To build the great *qanat* irrigation systems in ancient Persia, which are huge even by present day standards, the builders must have had a clear understanding of the way in which water is precipitated and moves underground. Man learned early to use what we now call 'hydrogeological information' in order to forecast earthquakes and volcanic eruptions.

The first attempts to systematize knowledge of the subsurface hydrosphere were made by the ancient philosophers. At the turn of the fifth and sixth centuries BC Thales of Miletus taught that the basis of all existence was water, upon which the land 'floats'. According to his views sea water is driven by the wind into the interior of the Earth, from whence, under the pressure of the rocks, it rises to the surface and emerges as springs. Plato held similar ideas. His pupil Aristotle (fourth century BC) went further. According to his view sea water falls into rivers after evaporation, followed by condensation on the surface, but underground water is formed from cooling air in caves and spaces within the rocks. He did not reject the idea of infiltration by rain water, but this is assigned a subsidiary role in the feeding of springs. Aristotle considered the rocks to be the source of the minerals in underground water. 'Water is essentially of the same quality as the ground through which it flows' – his proposition remains fundamental to the present day.

The subsurface hydrosphere was also a subject of constant study for the natural philosophers of ancient Rome. They expressed their views more precisely. Some of them clarified Plato's view about the penetration of sea water under the Earth and its desalination in the filtration process (Titus Lucretius Carus); others attempted to reconcile the views of Plato and Aristotle, considering that the primary source of groundwater was sea water in combination with the condensation of moisture (Annaeus Seneca). Pliny the Elder was of the latter school; it was he who classified mineral sources, and had interesting thoughts on their origins, which were based on Aristotle's idea of the interdependence of rock composition and groundwater quality.

Marcus Vitruvius Pollio in the first century BC expressed opinions on the penetration of rain and melting snow which are close to those held today. In his well known work *De architectura*, in which he also laid down the base of natural science, he was one of the first to explain correctly the circulation of water in nature and the emergence of springs as being the result of infiltration of atmospheric precipitation into the ground, while allowing the possibility of condensation in the ground of water vapour in the air, and hot vapours which rise from the depths of the Earth. Vitruvius Pollio may be considered to be the father of the *infiltration theory* of groundwater.

However, it was not the ideas of Vitruvius Pollio but those of Seneca, who had rejected the possibility that groundwater could be fed by the penetration of atmospheric precipitation into the ground, which were accepted without question in Europe for several hundred years in the Middle Ages. The thinkers of the Near East and Central Asia did not subscribe to these views. One of these was the scholar and scientist Al-Buruni (973–1048), a native of Khiva (Khwarizm) in the Uzbek SSR, who, six or seven centuries before the Europeans, had already formed an accurate conception of the 'nature of spouting springs' and the cause of hydrostatic pressure or head. This is what he wrote in the year 1001 in his work *al-Athar al-Baqiya* (*The chronology of ancient nations*): 'when water comes from underground reservoirs lying higher than ground level, it rises and gushes upwards. If however the reservoir is lower, then the water is not able to rise to the surface of the ground.' He also had a clear idea of the role of water in the formation of minerals.

The first work to systematize the knowledge of groundwater was that of the Persian naturalist M. Karadi (died 1016). This work was called *The search for waters hidden under the Earth*. In it a wide range of problems was illuminated, which would have done credit to a modern textbook on hydrogeology: (1) water circulation, (2) pressure and non-pressure water, and surface water, (3) the use of phreatophytes in the search for underground water, (4) a description of some field experiments, (5) the quality of groundwater from the point of view of its suitability as a potable water supply. Unfortunately, this work became known to Europeans only in the second half of the twentieth century.

Drilling for water started in China. The cable tool drilling rig was invented there several thousands of years ago for drilling wells. In Europe drilling began in the seventeenth century. Since 1126 in the province of Artois in the north of France the drilling of water wells has released gushing water. Groundwater under a head of pressure was then called artesian from the Roman name for the province (Artesia). Brine-producing wells have been drilled in Russia since 437, when the technique of rotary drilling and the lining of wells with wooden pipes reached a high level of technology.

Interest in groundwater only reappeared in the sixteenth century, after a long period of stagnation of scientific thought. Relying on concrete facts – observations of the inflow of water into mines, the Saxon naturalist Georg Agricola (Bauer) expressed his views on the phenomenon of underground water partly as being caused by the infiltration of surface water, and partly by the condensation of water vapour rising from below. A 'magic wand' was widely used in the search for water. This water divining method, based on the twisting of a willow twig or rod held in the hand when placed above groundwater, has been used since ancient times. In the Middle Ages it was surrounded by an aura of secrecy and was considered to be a gift of the 'chosen'.

Tibetan medicine at this time had attained considerable success in the understanding of the medicinal properties of mineral waters. There exist documents in which it is clear that in the twelfth to sixteenth centuries the Tibetan lamas knew about the geological conditions for the emergence of mineral waters and used this knowledge for prospecting, particularly in that area now known as the Mongolian People's Republic.

The Frenchman B. Palissy actively championed the theory of water infiltration into the Earth's interior. His views, which exist in the form of a lively dialogue between theory and practice, are supported by a multitude of observations on springs and wells in connection with various natural phenomena. 'I never had any books – only the Heaven and the Earth, which supplied everything' he loved to say. But Palissy's studies could not stand up to the authority of Kepler and Descartes, the foremost minds of the time, who likened the ground to a living organism which was situated in the ocean and which digested it: 'water moves in the rivers and in underground streams like the blood in the veins and arteries of an animal', declared Descartes. As for springs, according to Kepler, they are none other than the final results of the exchange of materials, which takes place within the interior of the Earth.

The views of the Jesuit A. Khirker of Wurzburg, unfettered in their imaginative sweep and adapted to the dogmas of the Church, enjoyed great popularity in the eighteenth century. His work *The subterranean world* was widely used as a geological textbook. Khirker connected the formation of the subsurface hydrosphere with the infiltration of water from the oceans into the interior of the Earth. This water, reaching the red-hot magma, turns into steam, liberates the salts from the rocks and then rises to the surface, either in the form of hot springs, or as cool water after passing through caverns. He rejected Aristotle's view on the changing of air into water, and as far as rain water and melt water are concerned, they, according to Khirker, can only produce temporary currents of short duration. The composition of groundwater was explained as being the result of interaction with dissolved material from the rocks. Thus, in spite of the fantastic nature of the majority of his views, Khirker's work also contained rational ideas.

In the second half of the seventeenth century the infiltration theory of the formation of groundwater received quantitative support from the French naturalist E. Mariotte, based on research by his compatriot P. Perrot. Thus the scientific basis for the study of the water balance was laid down; this permitted the currently held views on the infiltration of sea water into the interior of the Earth and opposed any contradiction with the results of measurements. Perrot's book *The origin of springs*, published anonymously in 1674 is regarded as the first work in the field of hydrogeology, three centuries of which were widely celebrated in 1974 on the initiative of UNESCO. This book considers the River Seine as an example, and shows that atmospheric precipitation is sufficient to maintain 'an uninterrupted flow of water in the rivers and springs'.

The works of Mariotte appeared later, but they became widely disseminated. He demonstrated that groundwater is fed from atmospheric precipitation. The idea of the pluvial origin of groundwater began to be associated with his name. The mechanism of the infiltration of atmospheric precipitation through the ground, the presence of water-bearing and impermeable horizons, the debit factor of springs depending on the amount of rainfall, these and other questions were brilliantly answered by Mariotte, an inquisitive scientist, who, in the opinion of Meinzer, deserved more than anyone else the title of founder of the science of the study of groundwater.

Views on the atmospheric origin of groundwater received support in the treatise of the Italian A. Vallisieri, in which the pressure of artesian water as observed in natural occurrences is explained by the existence of water-bearing horizons (aquifers) that are lower than the region (catchment area) feeding them. The infiltration theory was confirmed by the great Russian scientist M.V. Lomonosov. In his work *The layers of the Earth* a clear picture is sketched of the interrelationship between surface and subsurface waters. Citing the experience of miners he confirmed that: 'in dry years the inflow of water into mines is not such a pressing problem as in wet years, when very much more water penetrates into the interior of the Earth'. According to Lomonosov the melting of buried ice has a great influence on the feeding of groundwater. The Frenchman G. de Lametri at the end of the eighteenth century performed a great service by disseminating Mariotte's ideas. With the help of experiments he succeeded in studying the permeability of rocks. He assigned equal roles in the formation of groundwater resources to rain water infiltration and to the precipitation of water from mists over mountains.

Despite the wide recognition of the atmospheric origin of groundwater, doubts were many times cast on this idea by, for the most part, followers of Descartes and Kepler (R. Plot in England, H. Kefferstein in Germany, E. Babine in France,

and others). But their belief that rain water could not penetrate deeply did not stand up to criticism.

The fierce controversy which broke out at the end of the eighteenth century and the start of the nineteenth century between the 'plutonists' who believed in the magmatic origin of rock (the Hutton school), and the 'neptunists' (the Werner school) who ascribed a marine sedimentary origin to all rocks, had a favourable effect on the development of theories concerning the appearance of water in the Earth's interior. The same must be said of the work of Charles Lyell in the 1830s who, as Engels said, brought common sense to geology.

By the end of the first half of the nineteenth century atmospheric precipitation was regarded as the only source of groundwater. The French geologist Elie de Beaumont gave the most comprehensive account of the new ideas. His account added to the external water circulation an internal circulation which was associated with it, in which water percolated under the Earth and sea water participated. Elie de Beaumont also pointed out the possibility of replenishment of the subsurface hydrosphere by water which was expelled by cooling and crystallizing magmas. His concept may be regarded as a complete theory of water movement within the Earth.

At the same time the main aspects of the distribution and formation of groundwater began to take shape. Geographers, hydrologists, and geologists occupied themselves in the description of fresh water and mineral water springs. The classification of the mineral waters of Russia by V.M. Severgin and G. de Thiury's investigations into the origin of the artesian wells of France are some examples. A particularly large amount of data was obtained from experiments carried out over vast areas of Russia. William Smith, the great English civil engineer and geologist, gave in 1827 the first account of groundwater in Great Britain, when he solved the problem of increasing the underground water supplies for the town of Scarborough.

Thus in the middle of the nineteenth century there existed quite accurate ideas on the origin, distribution, and composition of groundwater. It is true that the information was by no means complete, but these ideas were no longer speculatory, because they had been confirmed by experiment and observations.

The founding of hydrogeology. The emergence of hydrogeology as a science dates from the second half of the nineteenth century, because it was in this period that there was a fundamental breakthrough in the understanding of the laws that govern local hydrogeology, and the formulation of hydrogeological theory.

The rapid economic development of the countries of Western Europe, North America, and then Russia, brought about the widespread use of groundwater. The Paris and Moscow artesian basins, groundwater in Germany, and water in mineral deposits were investigated by drilling. Great cities like Paris, Vienna, Berlin, and Chicago changed over to underground water supplies. At the same time the mineral waters of Southern Europe, the Caucasus, and the Yellowstone National

Park, the karst geology of the Balkans and Alsace, and the artesian waters of the USA were all under investigation.

There were vital achievements in research into infiltration processes. While supervising the water supply to Dijon in France, the engineer A. Darcy in a report on experiments which he had conducted (1856) described in mathematical terms the mechanism of infiltration of water through a porous medium. This established the fundamental laws of hydrogeology. Another Frenchman, the scientist G. Dupuit, using Darcy's law as a basis, derived a formula for the determination of water inflow into a borehole, and the German hydraulics engineer A. Tim and the Austrian F. Forchheimer used mathematical methods in studies of groundwater movement.

Experimental methods were perfected. Scientific methods replaced water divining in the search for groundwater.* In 1856 the Abbé Paramell described in detail the methods for discovering springs and underground water courses using geological data. The first hydrogeological maps appeared, of which the 1867 map of the depth of groundwater bodies of St Petersburg is a good example. Russian geologists (I.V. Mushketov, N.A. Sokolov, S.N. Nikitin, and others) started carrying out large-scale studies of groundwater simultaneously with geological surveys. I.V. Mushketov referred to the phenomena of groundwater in his textbook *Physical geology* published in 1888. The Frenchman A. Deles in 1861 made a quantitative determination of the volume of all the water in the ground – 1175 million km^3 – i.e. comparable with that in the Pacific Ocean. At the same time the German B. Lerch, and the Canadian T. Hunt carried out the first experiments in what we would now call hydrogeochemistry. Analyses of natural waters (including groundwater) began to be issued from the laboratory of K. Fresenius in Wiesbaden.

There gradually emerged a branch of the geological sciences concerned with groundwater alone. Its foundations were laid in the 1860s by, for example, the Russian Academician G.P. Gel'mersen in his paper 'Artesian wells in general and in Russia in particular', and by the above-mentioned Abbé Paramell in his book *Studies on springs*.

Russian geologists who realized that this science differed from the ordinary 'study of springs' because of its deep geological roots began to use the new word 'hydrogeology' widely. Russia was the first nation to appoint a government hydrologist; H.A. Golovkinskii, who organized the drilling of artesian wells in the Crimea and Black Sea region, was the first to hold the post.

A systematic exposition of the state of knowledge about groundwater was published at the end of the nineteenth century in French by A. Daubrée (1887) and in German by H.I. Haas (1895). These were hydrogeological textbooks, although they did not include the term 'hydrogeology' itself. It is more-

over not mentioned as such in much later works (Slichter, 1902; Ototskii, 1906; Keilhack, 1912, 1917; Prinz, 1919). A. Daubrée put the question about studying the subsurface hydrosphere from an historical point of view, proposing that the dynamics, composition, and regime of the underground waters of former epochs can be reconstructed from the minerals which they deposited. Interesting views on the origin and distribution of groundwater can also be found in the works of Haas.

At this time the theory of the atmospheric origin of groundwater ruled supreme. People tried to apply it not only to near-surface but also to deep-lying water. The German geologist B. Kotta proposed that atmospheric precipitation could penetrate to great depths, was heated there, and subsequently emerged at the surface as hot springs. 'All other theories, often formulated with great skill and wit, turn out to be groundless – a game of the imagination' (Kotta, 1859, p. 39). Nevertheless, fierce arguments about the penetration of water into the interior of the Earth flared up periodically. The Austrian engineer O. Folger came out decisively against the proposition that 'All underground water is derived from rainfall.' He considered that 'There was no underground water from rain.' In his opinion, moist air entered the interior of the Earth and this moisture condensed and added to the underground water. The rational ideas in Folger's hypothesis on the condensation of water vapour in the Earth's interior were shared to some extent by various other scientists (Keilhack, 1912), but his ideas had a great number of opponents (H.I. Haas, E. Vol'ni, and others) who rejected it completely.

At the beginning of the twentieth century the argument was resolved by the Russian scientist A.F. Lebedev. By accurate measurement he established that in comparison with percolation, condensation yielded a much smaller amount of groundwater.

In the nineteenth and twentieth centuries a great mass of data on deep-lying groundwater was accumulated outside Russia. Whilst the formation of groundwater in the upper horizons was successfully explained by the infiltration of water or the condensation of vapour, such an explanation in relation to geysers, thermal springs, brine, oil, and salt deposits met serious objections. New ideas arose. For geysers and thermal waters, the famous Austrian geologist E. Suess (1902) put forward the hypothesis of *juvenile* water. This was the name he gave to water that was formed in the Earth's interior from oxygen and hydrogen. Juvenile water (i.e. newly created water), in contrast to vadose water which is moving about in the ground, comes from magma as steam and first enters the hydrologic cycle via thermal springs and volcanoes. In *Das Antlitz der Erde (The face of the Earth)* published in 1909 Suess's hypothesis evoked a lively discussion, in the course of which the author himself introduced substantial corrections to his original ideas. In his work water vapour was not considered to be juvenile; only the hydrogen given out by magma which formed water on combining with oxygen of the atmosphere was considered juvenile. But the essence of the idea remained the same.

The hypothesis of *fossil* water of marine origin also

---

*From the very beginnings of hydrogeology and up to the middle of the twentieth century the leading minds in the subject regarded the searching for water with a 'magic wand' with scepticism. However, latterly attempts have been made to put this on a scientific basis (N.N. Sochevanov and others), calling it a biophysical method. It is proposed that people with sensitive nervous systems can use the 'magic wand'.

became popular at the start of the twentieth century. The idea that ancient water could be buried and preserved in sedimentary basins was expressed independently by the Austrian geologist G. Gofer and the Russian Academician N.I. Andrusov (1908) and the American hydrologist A.C. Lane (1908). Fossil water is formed in the crust of the Earth in sedimentary basins in silt and other porous sediments of former geological periods. Some water may escape upwards during diagenesis of sediments, but some may be trapped. Conditions in the deeper parts of structural depressions between continents, and in oil and salt deposits are very favourable for the trapping of water.

The hypothesis of juvenile and fossil water had from the very beginning both protagonists and opponents. Because of lack of supporting data the original proposition underwent considerable changes. Arguments on the subject have not ceased even today.

The Swiss geologist L. Brun opposed Suess's elegant concept which so cleverly linked endogenetic processes with the nature of water in the ground; he expressed the opinion that 'volcanoes are dry' — i.e. he believed the water of volcanic eruptions to be of secondary origin. G. Gofer (in 1925), following B. Kotta, proposed that the water itself is not juvenile, but is mobilized by the heat of the eruption. It is true that these points of view, and those of Suess, were fully supported, but those who supported the juvenile water hypothesis were forced to make a careful evaluation of the quality of the water which was liberated from the magma. Suess's supporters (A.P. Gerasimov, G. Göthe, N.A. Ovil'vi, and others) no longer regarded every mineral spring as the appearance of juvenile water, but believed that during their journeys to the surface these waters became mixed with vadose water; therefore, strictly speaking, only the dissolved salts and gases are juvenile. Nevertheless, the existence of water of magmatic origin remains a cornerstone of this hypothesis, which in the first half of the twentieth century divided many geologists and hydrogeologists.

Groundwater of high chloride and other mineral content was considered to be of marine origin. The clear incompatibility between the salinity of these waters and present day sea water caused two schools of thought to emerge (Smirnov, 1976).

1. The proposal of the former existence of marine basins in which the water in terms of salinity and composition was analogous to that found in oil deposits, i.e. oilfield brines (R. Mills, R. Wells, V.A. Sulin, and others). According to this concept saline water and brines are a relic of ancient 'calcium chloride' seas.

2. If however it is considered that the salt content of the water of sedimentary basins does not differ from that of present day oceans, then the composition of groundwater must be regarded as the result of the metamorphism of the saline water on the seabed in sedimentary environments. D. Rogers, L. Mrazek, R. Nil, A.D. Archangelskii, N.S. Kurnakov, and others support this theory of the genesis of brines.

Logical conjectures play an important part in this concept of 'calcium chloride' seas and sea water metamorphism.

Hence the fossil water concept is treated cautiously and critically by oilfield and salt deposit geologists, among whom it has become widely disseminated.

The end of the nineteenth century was marked by great advances in the theory of groundwater movement (C.S. Slichter, J. Boussinesq). The direction of water infiltration into fissured rock, deviating from the linear law of Darcy, was deduced and observed by A.A. Krasnopolskii (1912). Later N.N. Pavlovskii worked out the theory of groundwater movement below hydraulic engineering works. The latter is the author of the theory of anomalous movement of groundwater and the method based on electrohydrodynamic analogy (EHDA), which has been widely used in the laboratory for modelling infiltration processes.

Russian scientists have made great advances in the understanding of the laws which govern local hydrogeological regimes. S.N. Nikitin (1900) who became the founder of regional groundwater studies and the designation of hydrogeological regions devised methods for area-based studies of groundwater and of hydrogeological regions.* The ideas of V.V. Dokuchaev on the zonal nature of soils and the role of forests in the water balance became the basis of the theory of zonality and regime of groundwater (P.V. Ototskii, E.V. Oppokov, and others). As has already been noted A.F. Lebedev's research work was of fundamental significance. He not only demonstrated the true role of condensation in groundwater recharge, but he also outlined the mechanism of moisture movement in soils, and distinguished between the different types of water in rocks. A.D. Stopnevich, at the start of the twentieth century mentioned the need to preserve groundwater. Data on groundwater in permafrost zones began to appear. This aspect of groundwater was summarized by Professor A.V. L'vov of Irkutsk (1916). Another Siberian scientist, M.G. Kurlov put forward a description of the chemical composition of water in the form of pseudo-droplets; this 'formulation' has now been accepted throughout the world and deservedly bears the name of its author.

The French (A. Mage, L. Poshe, R. Chalo, E.A. Martel, and others), the Germans (K. Keilhack. E. Luger, E. Prinz, and others), and the Americans (C.S. Slichter, D.W. Mead, O.E. Meinzer, and others) have made major contributions to groundwater science. Their ideas have formed the basis of original textbooks which have become widely known. The book by Keilhack, the greatest authority on groundwater in Central Europe, which was reprinted three times (1912, 1917, 1935) was especially popular; also popular were the works by O.E. Meinzer (1923, 1935) a specialist in the field of hydrogeological terminology, theoretical, and applied hydrogeology. These and other works by foreign authors (C.S. Slichter, I. Richert, G. Gofer, E. Prinz, and others) have been translated into Russian.

In the USSR immediately after the Civil War (1917–22), hydrogeological research became part of a national plan. By the 1920s and 1930s, relying on a rich national background of

---

*In S.N. Nikitin's original definition of hydrogeology the word was defined as the science of the 'underground branch of the total natural circulation of the Earth's water'.

experience and completely unique factual material, Soviet scientists put forward new theoretical works in literally every branch of hydrogeology, i.e. the matter of groundwater classification, hydrogeological zonation principles, problems of the formation and distribution of water resources in the Earth's interior, experimental methods of hydrogeological research, and many other topics. The first All-Union Hydrogeological Congress, which was held in Leningrad in 1931, became a review of the achievements of groundwater research workers. It summed up the state of groundwater studies and outlined future tasks.

During these years a school of Soviet hydrogeologists arose under the leadership of the talented scientist and great scientific organizer F.P. Savarenskii. He was responsible for the profound theoretical summary about the zonality of groundwater, its formation in the arid regions, and chemical breakdown. The first hydrogeology textbook in Russian was written by P.N. Chirvinskii (1922), who may be regarded as the founder of the historical school in groundwater science. In the field of regional and local hydrogeology much work was done by P.I. Butov, V.S. Il'in, O.K. Lange, N.F. Pogrebov, A.N. Semikhatov, and B.K. Terletskii. Thanks to the work of A.I. Dzens-Litovskii, N.N. Slavyanov, and N.I. Tolstikhin the study of mineral waters and their geochemistry was set up. A large number of instruction manuals were published, in which the fundamentals of the study of the methods of hydrogeological research were laid down.

In the 1930s the dynamics of groundwater were recognized. In this field great credit is due to the Soviet scientist L.S. Leibenson who deduced the equations of motion of aerated water, to P.Ya. Polubarinova-Kochina the creator of fundamental ideas on the dynamics of groundwater, and to G.N. Kamenskii, the author of the theory of unsteady flow of groundwater. The laws that govern the motion of groundwater in deep-lying horizons were established by A.I. Silin-Bekchurin. Among the foreign scientists the Americans C.V. Theis, M. Muskat, and M.K. Hubbert must be mentioned. The last used the concept of 'force potential' in his investigations into the flow of immiscible liquids in multiphase systems.

Of the great variety of scientific works which have contributed to the development of hydrogeology *The history of natural waters* by Academician V.I. Vernadskii, which was published in the years 1933–35, is of special significance. In it he summarized the many centuries of work on the Earth's water, its geological and its geochemical activity.

Vernadskii paid particular attention to the subsurface part of the hydrosphere, rigorously analysing its various parts. He considered magma to be a solution of water, silicates, and aluminosilicates, which is supersaturated with gases. In his view part of the water vapour, coming from below, is actually juvenile in the sense of the word as used by Suess, but the bulk of the water penetrates into the deeper layers of the lithosphere from above and is of atmospheric origin. Furthermore, renascent (i.e. born again) water liberated from minerals also plays a part in groundwater recharge. Water synthesis occurs not only in magma but elsewhere in the crust (e.g. water of organic origin). Vernadskii suggested that instead of vadose

and juvenile waters they should be divided into stratal bodies, or veins of water. The last, which are thermal and ascending, are connected with the deep zones of the crust and sometimes with magmatic masses.

Vernadskii attached great importance to the transition of water from one state to another. Water as a liquid is represented mainly by solutions, in the deeper zones of the Earth it turns into water vapour, part of which is captured by magma. According to Vernadskii there exists a single united water balance between the various components of the subsurface hydrosphere (and the hydrosphere in general). He regarded groundwater as a solution which interacts with rocks, gases, and living organisms, in which the gaseous composition is determined both by biochemical and metamorphic processes.

Vernadskii's ideas became widely accepted in the USSR and abroad and his *History of natural waters*, described as an 'encyclopaedia of water' essentially completed the establishment of hydrogeology as a science.

Apart from Vernadskii's work, there appeared at this time a whole series of textbooks which have become reference works for hydrogeologists. The standard works and monographs of R.R. Vyrzhikovskii, R. Dachler, G.N. Kamenskii, V. Kühne, O.K. Lange, A.F. Lebedev, M. Muskat, F.P. Savarenskii, A.I. Silin-Bekchurin, V.A. Sulin, C.F. Tolman, N.I. Tolstikhin, D.I. Shchegolev, E. Embo, and J. Stiny are representative. N.I. Tolstikhin laid down the basis for the study of groundwater in the frozen regions of the lithosphere. The first compendiums on regional groundwater appeared in the USSR (M.M. Vasil'evskii, A.N. Semikhatov, N.N. Slavyanov, and others), and *Hydrogeology in the USSR* began to be published, and the hydrogeological group VSEGEI (All-Union Geological Institute) directed by N.F. Pogrebov made the first hydrogeological map of the USSR. Unfortunately the Great Patriotic War (1941–45) interfered with the completion of this work.

Modern hydrogeology. If hydrogeology was created as a modern science towards the end of the Second World War, then the following years may be called its heyday. This period of flourishing owed its rapid development to the demands put on it by the situation in which on the one hand there was a colossal demand for groundwater as a form of mineral resource, and on the other hand there was the global interference by man with the subsurface hydrosphere. These demands stimulated fundamental applied scientific thinking both to eliminate the harmful effects and problems associated with groundwater during civil engineering construction works, and mainly to prevent the pollution and exhaustion of groundwater. Hydrogeologists faced the new problem of the rational use and conservation of the subsurface hydrosphere.

The rapid development of industry and agriculture, the unprecedented growth of population, and the progressive pollution of surface sources of water were the basic reasons for the sharply rising demand for pure water. The fraction of groundwater in the water used for general purposes, including irrigation, is increasing yearly. In the USSR for example, after the Great Patriotic War this was 5–6%, at the end of the

1960s it was 13%, and by the end of the 1970s it had reached 20%. This trend will no doubt continue in the future. Moreover, for several cities (Kiev, Minsk, Vitebsk, Tumen, Tomsk, Yakutsk, etc.) and arid regions (Turkman SSR, South Kazakhstan, etc.) the fraction of groundwater exceeds 50% of the total water used. In East Germany the equivalent amount as an average over the country is 60–80%. There are countries where the demand for water is completely (Saudi Arabia), or almost completely (Denmark, Tunisia), met from groundwater. Groundwater is practically the only source of water for general use and irrigation in the arid zones and these comprise 60% of the Earth's land surface.

Groundwater is a valuable mineral. Naturally, nowadays, fresh water is of the greatest importance. But other types of groundwater, mineral and thermal, which are used for medicinal and industrial purposes, or in the production of thermal energy are of no less interest. From this 'liquid' mineral resource are extracted cooking salt, soda, bromine, iodine, lithium, and other elements.

The hydrogeological conditions of a region determine the course of construction of cities, factories, hydraulic engineering projects, mines, and railways. A knowledge of hydrogeology is vitally necessary even in the design stage, and then in the construction stage. For example, during the construction of the great Baikal Amur Magistral research carried out in the 1930s to 1970s enabled complex problems to be solved: to find underground sources of water within deeply frozen ground, to forecast water inflow into tunnels, to outline practical measures for drainage along the line, to avoid difficult ground, to discover mineral and water deposits, etc.

Progress has been made towards solving the problem of rational use and conservation of groundwater. The solution of this problem is closely bound up with a complex of other problems concerning the exploitation of natural resources and the conservation of the environment as a whole: the question of man's interference deep into the regime of the subsurface hydrosphere, planned redistribution of its resources, and its general control.

'If society is faced with a technical necessity', wrote Engels,* 'then this forces science to move forward more quickly than ten universities.' This is what happened to hydrogeology in the middle of the twentieth century. The colossal and ever-growing demand for groundwater, the question of its rational exploitation and use, and its conservation, made hydrogeological research a matter of urgency on a global scale.

The scope of the theoretical and methodological problems changed. As has been noted in Section 1.1, the development of hydrogeology had arrived at a new stage which was marked by the mastering of a huge mass of accumulated factual material, the re-examination of certain theoretical positions and the appearance of several new methods of exploitation. All this led to the formulation of new, original theoretical concepts concerning geological processes and the laws governing them.

Marx, K., Engels, F. Selected letters. Moscow: Gospolizdat, 1947, p. 469.

Firstly we will examine the general state of development of hydrogeological research in the post-war years. Regional studies of groundwater are being made on an unprecedented scale in the USSR and embrace a wide range of problems. This has been made possible by planned hydrogeological mapping and exploratory work carried out over the whole of the territory of the USSR, and is supported by drilling for groundwater and other minerals. In conjunction with the search for oil the deep-lying water-bearing horizons of artesian basins began to be explored. The effect of these explorations, carried out on a strictly scientific basis is difficult to overstate. They have made possible the discovery of new artesian basins and groundwater deposits, enabled the reserves of fresh and mineral water to be evaluated, and finally enabled the use of groundwater in the national economy to be sharply increased.

The All-Union Research Institute for Hydrogeology and Engineering Geology, one of the older sections of the Ministry of Geology of the USSR, having been founded in 1939, and the F.P. Savarenskii Laboratory for Hydrogeological Problems which operated under the auspices of the Soviet Academy of Sciences from 1944 to 1961, became centres for theoretical hydrogeology. A network of theoretical research was started in a number of institutes of higher education (the Moscow Geological Prospecting Institute, the Leningrad Mining Institute, the Tomsk Polytechnical Institute, etc.), and universities (e.g. Moscow University). Later the Ministry of Geology of the Uzbek SSR set up the scientific–industrial unit 'Uzbekgidro-geologiya', and the Institute of Hydrogeology and Hydrophysics of the Academy of Sciences of the Kazakh SSR, the section of the Academy of Sciences of the Georgian SSR devoted to Hydrogeology and Engineering Geology, and a Commission for the Study of the Groundwater of Siberia and the Far East under the Siberian Branch of the Academy of Sciences of the USSR were created.

Hydrogeological research is now being carried out by huge divisions of industrial, design, and research organizations, the Ministry of Geology and the Ministry of Irrigation and Water Supply of the USSR. The training of teams of geologists was started in many other cities such as Kiev, Odessa, Rostov, Gomel, Tbilisi, Alma-Ata, Tumen, Vilnius, Voronezh, Saratov, Chita, and Yakutsk, apart from Moscow, Leningrad, Tashkent, Dnepropetrovsk, Novocherkassk, Kharkov, Perm, and Tomsk, where such facilities had existed before the war. U.M. Akmetsafin, M.E. Altovskii, A.E. Babinets, B.A. Beder, G.V. Bogomolov, I.M. Buachidze, V.M. Fomin, N.K. Ignatovich, V.V. Ivanov, G.N. Kamenskii, N.A. Kenessarin, N.N. Khodzhibaev, B.I. Kudelin, V.N. Kunin, O.K. Lange, F.A. Makarenko, K.I. Makov, N.A. Marinov, A.M. Ovchinnikov, N.I. Plotnikov, P.F. Shvetsov, N.N. Slavyanov, M.V. Syrovatko, V.G. Tkachuk, N.I. Tolstikhin, P.A. Udodov, I.K. Zaitsev, and many others have made enormous contributions to the founding, organization and the undertaking of hydrogeological research in the post-war years.

Abroad, beginning in the 1950s, hydrological research of differing degrees of detail has been undertaken in many countries. On the initiative of UNESCO a comprehensive study of groundwater has been carried out in the arid zones of the

Earth, thanks to which considerable quantities of groundwater have been found in the developing countries of Africa (in particular those in the waterless deserts of the Sahara), Asia, and Latin America. The hydrogeological conditions in Western Europe and the USA have been studied in detail, where groundwater resources have been almost completely recorded. The hydrogeologists of the socialist countries (O. Ginie, Ya. Etel, A. Klechkovskii, L. Luchner, Ya. Dovgiallo, R. Kadere, B. Stepanovič, I. Tsishang, E. Schmidt, K. Shcherev) have achieved great success. The scientists of the USA (D.K. Todd, D.E. White, R.J.M. De Wiest, and others) and France (G. Castany, G. Marga, H. Schoeller) have made considerable contributions to the understanding of the laws of hydrogeology. Interesting papers have been published in West Germany (W. Richter, K. Frikke, G. Matthess), England (S. Buchan), Belgium (P. Fourmarier), Austria (J.G. Zötl), New Zealand (A.J. Ellis), and Canada (R.O. Van Everdingen).

In some countries (USA, France, etc.) hydrogeology is not divided into separate branches. The study of groundwater more often than not comes within a complex of hydrogeological research (the evaluation of water resources for example), or within the geology of mineral deposits (for example the exploration for oil and gas). Sometimes research is carried out only in the 'chemistry of groundwater' or 'the hydraulics of groundwater'.* The narrowness of such approaches is obvious. However, in some branches (mainly the applied branches) the scientists of the developed western countries have achieved great successes: remote methods for the study of groundwater, the use of programmed recording instruments and computers in hydrogeodynamic calculations, the automation of regional observations, and the introduction of accurate quantitative methods in the evaluation of interactions within the 'water–rock' system.

In spite of the differences in approach to the problems of hydrogeology, in recent years contacts between the scientists of different countries have become closer, in respect of the understanding of hydrogeological processes and the laws governing them. These contacts arise, in the first place, in discussions at symposia and congresses at which the problems of hydrogeology are discussed; in the second place, in branches of international bodies engaged in the study of groundwater (e.g. the International Association of Hydrogeologists and the International Association of Hydrogeological Sciences); and finally during joint research by scientists from different countries. The hydrogeological work undertaken in the International Hydrogeological Decade (1965–74), or the economic cooperation which exists within the framework of Comecon are further examples.

The results of joint research are published yearly. In the works of the 23rd Session of the International Geological Congress (Prague, 1968), the description of mineral and thermal waters of the Earth was published in two volumes under the editorship of G. Kachur and Ya. Etel. Since the end

*The book of the famous hydrologist D.K. Todd (Todd, 1959, 1980) *Ground water hydrology.* The hydraulic or chemical approach can be found in many articles in the journal *Groundwater*, which covers similar topics and which has been published in the USA since 1963.

of the 1960s a number of countries have worked on the preparation of a series of maps of Europe. The question of a hydrogeological map of the world is still on the agenda.

The flow of abundant and varied information obtained as a result of theoretical and experimental research, and the introduction of progressive methods of the allied sciences (modelling, the quantitative evaluation of processes, analysis by using isotopes, etc.) have resulted in the birth of new ideas. It must be noted that in the middle of the twentieth century the scientific content in hydrogeology grew quite rapidly; there was no feeling that there were deficiencies in the hypotheses of the subsurface hydrosphere, although quite often there were no convincing facts to support many of them.

It is impossible here to cover or even estimate the basic ideas of modern hydrogeology, because they are so many and varied. As has already been noted at the beginning of the chapter what is fundamentally new is the transition of hydrogeology from the 'collating' stage to the 'explanatory' and 'prognostic' stages – its transformation from a science of phenomena to the science of processes and laws, the science of the subsurface hydrosphere or, using the terminology of P.F. Shvetsov *et al.* (1973), the science of water-bearing systems. An understanding of the processes of water exchange and mass transfer in the subsurface hydrosphere, a study of the interactions of the subsurface hydrosphere with the environment, the establishing of the laws which govern the formation and distribution of its components – these are theoretical fields with some purpose.

A.M. Ovchinnikov very accurately observed in 1955 that hydrogeology in the USSR differs from the study of groundwater that is done abroad in its inclusion of the scope of natural phenomena connected with the formation of groundwater. This difference can be seen in a complex study of the component parts of the subsurface hydrosphere, not only of groundwater, but also water bound up with rock, and water in the vapour and solid states. It can also be seen in the desire to understand the laws governing the formation and occurrence of groundwater in the Earth's interior as broadly and as deeply as possible.

Modern hydrogeology is a sophisticated science. It consists of several independent branches (see Table 1.1); many of these put into practice new ideas and are based on the methodological and applied work of the post-war years.

General hydrology. G.V. Bogomolov, P.P. Klimentov, V.F. Derpgol'ts, O.K. Lange, A.M. Ovchinnikov, W. Richter, N.I. Tolstikhin, D. Todd, R.J.M. De Wiest, P. Fourmarier, P.F. Shvetsov, and A. Schoeller have written works on the subject of hydrogeology and the water found in the interior of the Earth. In the works of I.K. Zaitsev, N.K. Ignatovich, B.L. Lichkova, F.A. Makarenko, N.A. Marinov, N.V. Rogovskaya, D.S. Sokolov, and others, the laws governing the distribution of groundwater and other components of the subsurface hydrosphere are revealed. New factual material has been obtained on the circulation of water, the genetic aspect of groundwater, and the origins of the water resources of the Earth's interior.

Hydrogeodynamics. This is the study of the movement,

regime, and resources of groundwater, and of hydrogeological modelling. V.D. Babushkin, N.N. Bindeman, N.N. Verigin, F.M. Bochever, N.K. Girinskii, S.N. Davies, G. Castany, B.I. Kudelin, A.A. Konoplyantsev, G. Marga, L. Luchner, P.Ya. Polubarinova-Kochina, A.I. Silin-Bekchurin, M. Hantush, I.A. Charny, V.M. Shestakov, V.N. Shchelkachev, and others have made great contributions to the development of this branch. The movement of groundwater has been successfully studied by computer modelling on various computers (I.K. Gavich, I.E. Zhernov, V.S. Luk'yanov, V.I. Lyalko, and others). Calculations on the unsteady flow of groundwater in conditions of turbulent flow have been devised. A theory for the prognosis of the regime of groundwater has been built up, and regional and local evaluations have been carried out.

Hydrogeochemistry. The founders of this branch were V.I. Vernadskii and A.M. Ovchinnikov. It arose as a science in the middle of the twentieth century as a result of the work of A.A. Brodskii, A.N. Buneev, M.G. Valyashko, A.P. Vinogradov, G.A. Goleva, I.K. Zaitsev, V.V. Ivanov, A.I. Perel'man, E.V. Posokhov, V.S. Samarina, S.I. Smirnov, V.A. Sulin, N.I. Tolstikhin, D. White, P.A. Udodov, V.M. Shvets, and others. It is concerned with the laws governing the migration of the chemical elements in aqueous solutions, the formation of the composition of groundwater and the character of the interaction in the 'water–rock–gas–liquid' system.

It has been successfully applied to the search for mineral deposits, especially oil, salts, and metals. In recent years radiohydrogeology and isotope hydrogeochemistry, etc. have begun to be recognized as separate branches of hydrogeochemistry. There is a special division devoted to the study of the composition of mineral waters.

Hydrogeothermics. This is a young branch of hydrogeology, concerned with the thermal properties of the subsurface hydrosphere. B.F. Mavritskii, F.A. Makarenko, and N.M. Frolov were among its founders. Special geophysical methods are mainly used to study the hydrogeothermal regime of the Earth's crust.

History of the subsurface hydrosphere. The historical approach to the life of the subsurface hydrosphere is the basis of hydrogeological research. Its successful application enabled the specialized section known as palaeohydrogeology to develop. Work was done by E.A. Baskov, G.V. Bogomolov, S.A. Vagin, A.A. Kartsev, A.M. Ovchinnikov, Ya.A. Khodzhakuliev, and S.A. Shagoyants. The role of water in geological history is enormous. Without a knowledge of it one cannot understand the exogenic and endogenic processes, the laws which govern the formation and decomposition of mineral deposits.

Hydrogeological research methods. The methods of carrying out field, experimental and laboratory work, the processing of hydrogeological information and its systematization are the concern of this section. New methods (for example remote sensing and the use of isotopes) have been added to the armoury of the traditional methods in recent years, and completely different methods are applied to the dumping of industrial waste, the replenishment of groundwater resources, etc. In the field of systematic hydrogeology M.E. Al'tovskii, G.N.

Kamenskii, P.P. Klimentov, N.I. Plotnikov, and A.I. Silin-Bekchurin have done much useful work.

Applied hydrogeology. This may be provisionally divided into 'prospecting' and 'engineering'. The first is the economic exploitation of groundwater, the second is the struggle with problems caused by groundwater during the construction of various civil engineering projects. Both are old branches of hydrogeology; in recent years they have undergone further development.

'Prospecting' hydrogeology is the study of the deposits and the exploitation of groundwater, and is concerned with groundwater as a complex mineral: fresh water is used as a general source of supply and for irrigation, mineral and thermal waters are used for medicinal and therapeutic purposes, and as a thermal energy source.

In the field of 'engineering' hydrogeology (apart from the working out of measures to combat groundwater during the construction of civil and hydraulic engineering projects, railway construction, etc.) the hydrology of mineral deposits is divided into independent branches dealing with solids (A.I. Dzens-Litovskii, P.P. Klimentov, N.I. Plotnikov, M.V. Syrovatko, S.V. Troyanskii, and D.I. Shchegolev), oil and gas (M.A. Gatal'skii, A.A. Kartsev, V.N. Kortsenshtein, V.A. Krotova, G.M. Sukharev, and V.A. Sulin), and into irrigation hydrogeology (A.G. Vladimirov, D.M. Kats, and M.M. Krylov).

Preservation of the subsurface hydrosphere. This became established as an independent branch of hydrogeology quite recently. The aims of the study are to preserve the subsurface hydrosphere from exhaustion and pollution. This problem is completely new, work on it has only just begun and will be solved by devising suitable preventive measures, right up to means of controlling the whole regime of the subsurface hydrosphere.

Regional hydrogeology. This branch is engaged in the study of the groundwater of a defined area. The following problems arise: (1) the establishing of the range of distribution of groundwater and qualitative and quantitative aspects, (2) the origins and history of the water in the part of the Earth's crust under investigation, and (3) the study of the laws governing the formation of and the composition of groundwater.

Regional groundwater reports have resulted in the production of hydrogeological maps (for example maps of scales 1 : 2 500 000 and 1 : 5 000 000 edited by I.K. Zaitsev, B.I. Kudelin, N.A. Marinov, and others), and in the production of descriptive monographs (Kamenskii *et al.*, 1959; Lange, 1959, 1963); *The hydrogeology of Asia* (Anon., 1974); and *The hydrogeology of Africa* (Anon., 1978). The multi-volume publication *The hydrogeology of the USSR* is a great achievement (Table 1.2).

From the above review one can see that over the last hundred years hydrogeology has developed by way of a change of principles, the putting forward of new ideas, the formation of new schools, and the defining of new problems. Moreover, concepts have changed, and this has forced a special examination of the problem of terminology.

Table 1.2. *Multivolume publication 'Hydrogeology of the USSR' (VSEGINGEO, Editor-in-chief A.V. Sidorenko, Deputy Chief Editors N.V. Rogovskaya, N.I. Tolstikhin & V.M. Fomin)*

Volume No.	Title	Editor-in-chief	Year of publication
I	Moscow and neighbouring district	D.S. Sokolov	1965
II	Byelorussian SSR	G.V. Bogomolov	–
III	Leningrad, Pskov, and Novgorod districts	I.K. Zaitsev	1967
IV	Voronezh, Kursk, Belgorod, Bryansk, Orlov, and Tambov districts	D.S. Sokolov M.R. Nikitin	1972
V	Ukraine SSR	F.A. Rudenko	1971
VI	Donbas (Don Basin)	D.I. Shegolev	1971
VII	Moldavian SSR	N.A. Plotnikov	–
VIII	Crimea	V.G. Tkachuk	1970
IX	North Caucasus	N.A. Grigor'ev	1968
X	Georgia SSR	I.M. Buachidze	1970
XI	Armenia SSR	A.M. Ovchinnikov	1968
XII	Azerbaijan SSR	N.V. Rogovskaya	1969
XIII	Lower Volga and Kama region	T.P. Afanas'ev	1970
XIV	Ural	V.F. Preis	1972
XV	Bashkir ASSR	E.A. Zubrova	1972
XVI	Western Siberian Plain	V.A. Nudner	1970
XVII	Kemerovo district and Altai territory	M.A. Kuznetsova O.V. Postnikova	1972
XVIII	Krasnoyarsk territory and Tuva ASSR	I.K. Zaitsev	1972
XIX	Irkutsk district	V.G. Tkachuk	1968
XX	Yakut ASSR	A.I. Efimov I.K. Zaitsev	1970
XXI	Chitin district	N.I. Tolstikhin	1969
XXII	Buryat ASSR	A.I. Efimov	1970
XXIII	Khabarovsk territory, and Amur district	N.A. Marinov	1971
XXIV	Sakhalin Island	O.V. Ravdonikas E.G. Chapovskii	1971
XXV	Coastal region	N.A. Marinov	–
XXVI	North-east USSR	O.N. Tolstikhin	1972
XXVII	Murmansk and Karelia ASSR	I.K. Zaitsev	1971
XXVIII	Lower Don, and North-east Azov	V.N. Vasil'eva	1970
XXIX	Kamchatka, Kurile and Komandorskie Islands	G.A. Goleva	1972
XXX	Estonian SSR	B.N. Archangel'skii	1966
XXXI	Latvian SSR	A.I. Dzens-Litovskii	1967
XXXII	Lithuanian SSR	A.R. Kondratas	1969
XXXIII	North Kazakhstan	N.M. Frolov	1966
XXXIV	Karagandansk district (Central Khazakhstan)	N.E. Falevich	1970
XXXV	West Kazakhstan	A.V. Sotnikov	1971
XXXVI	South Kazakhstan	V.I. Dmitropovskii	1970
XXXVII	East Kazakhstan	A.P. Kuznetsov	1971
XXXVIII	Turkmen SSR	B.B. Mitgarts	1972
XXXIX	Uzbek SSR	G.A. Mavlyanov	1971
XL	Kirghiz SSR	V.S. Tuytyukin	1971
XLI	Tadzhik SSR	V.S. Samarina	1972
XLII	Komi ASSR	V.G. Chernyi	1970
XLIII	Orenburg district	E.I. Tokmachev	1972
XLIV	Arkhangelsk and Vologda district	A.A. Makkaveev	1969
XLV	Kaliningrad district	A.R. Kondratas	1970

Summary

Part I	The basic laws of the distribution of groundwater in the territory of the USSR	N.V. Rogovskaya	1976
Part II	Basic laws of groundwater formation in the territory of the USSR	L.A. Yarotskii	1978
Part III	Groundwater resources of the USSR and prospects of their exploitation	L.S. Yazvin	1977
Part IV	Influence of industrial activity on hydro-geological and engineering geology conditions	I.V. Garmonov	1973
Part V	The engineering zonation and the laws governing the formation of the engineering geology conditions in the USSR	G.G. Skvortsov	1975

1.3 Hydrogeological terminology

The number of terms used in hydrogeology grows continuously. The information 'explosion' in hydrogeology, accompanied by the establishment of a mass of new facts and laws, unavoidably causes a demand for a periodic reconsideration of the terminology, its systematization and improvement. This work is all the more important because the traditional and the newly-coined terms are not always beyond criticism from the point of view of the correspondence of their meanings with other terms.

Several attempts have been made to put the terminology of hydrogeology in order. The compendium of concepts, definitions, and terms published in 1923 by the great American groundwater specialist O.E. Meinzer and translated into Russian ten years later was of very great help to the science. In the USSR N.N. Slavyanov, F.P. Savarenskii, M.M. Vasil'evskii, O.K. Lange, N.I. Tolstikhin, and others have devoted considerable efforts to hydrogeological terminology. Nevertheless the trouble with the terminology which is observed today is beginning to retard the development of hydrogeology as a science.

Vagueness of terminology, as is well known, is to be found in all branches of science, especially in the geological sciences (Vassoevich, 1974). However, if it can be said that in other branches of geology, for example in tectonics and lithology, notable work has been done in the improvement of terminology, hydrogeologists cannot make the same claim. In any case, neither the *Dictionary of hydrogeology and engineering geology* (Anon., 1971), nor the *Classification of groundwater* have solved the problem. Moreover, in the last 25–30 years, apart from the rare exception of Hähne & Jordan (1978), a single-minded investigation into the state and prospects of the improvement of the terminology used has been entirely absent.

Without proposing the task of a re-examination of the hydrogeological terminology, we will try to explain the causes of the above-mentioned difficulty and indicate how it may be eliminated (Pinneker, 1977).

Terms should concisely reflect the sense of the concepts with which they are invested. An elementary requirement of any science consists of the achievement of one-to-one correspondence between the terms used and the concepts. The terminology used in hydrogeology does not always satisfy this criterion: the terms often have more than one meaning and are sometimes inconsistent.

As an illustration let us consider the various meanings of the term 'hydrogeological structure'. It is usually used (Zaitsev & Tolstikhin, 1963) to designate a geological body which contains groundwater (an artesian basin or a hydrogeological massif). It was used in a different sense by N.K. Ignatovich (1944): by 'hydrogeological structure' he understood a combination of definite indications which enabled geological–structural forms to be classified according to the degree of permeability.

To which should preference be given? It appears that the word 'structure' is translated from the Latin as construction, or arrangement. Hence 'hydrogeological structure' must give a picture of the spatial distribution of groundwater and its interrelationship with rocks. Then the different meanings of the term will be eliminated and a correspondence to the meaning which is peculiar to the term 'structure' in other sciences will be attained. As for the most widely used meaning of the term 'hydrogeological structure' (Zaitsev & Tolstikhin, 1963), it is better to reject it altogether because it is not at all correct.

Lack of precision is also to be found in the terms which are used to designate fundamental concepts, i.e. the most general terms which are consequently used to form new concepts and terms. The majority of them are traditional and are embedded in the course of the historical development of ideas on the subsurface hydrosphere. In their modern meanings the traditional terms do not always reflect the sense with which they were earlier invested. After all, are the meanings of the term 'juvenile' as used by Suess (1902) and, for instance, Kapchenko (1966) really identical? In precisely the same way the definition of the term 'artesian' applied to a basin or to groundwater obviously cannot be considered to be comprehensive, since in the general understanding of the term it is explained only in terms of the hydrostatic head and the geostatic pressure, but other forces which are observed in 'young' basins are not considered at all. Now consider the term 'hydrogeological region' of some order (first, second, etc.). This in no way reduces the imprecision, and is in fact in error, because a region is already a taxonomic unit of zonation.

The cases of the use of inaccurate terminology can be cited at length. Thus the term 'circulation' implies motion round a circle. We apply this to natural water but inaccurately talk of the 'circulation of groundwater', for groundwater participates in only a part of the hydrologic cycle as a whole. The expression 'the exit of a source' is similarly inaccurate, because by the term 'source' the emergence of groundwater is understood.

Together with the multiplicity of meanings and the imprecision of certain terms in hydrogeology there is also the reverse phenomenon, i.e. the multitude of terms with almost identical meanings. The aforementioned terms 'hydrogeological structure', 'water pressure (or geohydrodynamic) system', 'body of groundwater', and 'groundwater reservoir (hydrogeological reservoir)' are used in more or less the same sense (as a geological body, containing groundwater). So far none of these has been accorded preference.

There is great disagreement about the word 'ion-salt' (i.e. the macrocomponent) composition of groundwater.* Some authors use the word as a single compound adjective, others use two different forms for anions and cations. There is not even unanimity about which order to write the compound adjective: in increasing order (A.M. Ovchinnikov, N.I.

*In order not to refer to this further, we note that the authors of this work, when it is not specially specified, use the term ion-salt composition in the form of two independent adjectives: the first for anions, the second for cations. In the compound adjectival form the components are distributed according to the principle of 'shade and colour', i.e. according to the increase in the content. The dominant ion stands at the end of the compound adjective and determines the name of the chemical type of the water. In the absence of special specifications the mineralization and the content of microcomponents (before and after the pseudo-fraction in Kurlov's formula) is given in g/litre.

Tolstikhin, I.K. Zaitsev, E.V. Posokhov, and others), or in descending order (M.E. Al'tovskii, O.K. Lange, and others) of the percentage of ions. Not all of these nomenclature systems are of fundamental significance, there are discrepancies in the hydrogeological terminology here. The result is that without special explanation it is impossible to understand which ion-salt is being considered by different authors from the name itself.

The examples given show that among the reasons for the unsatisfactory state of hydrogeological terminology one must distinguish the absence of clear and logical correspondence between the terms and the phenomena to which they refer. From this absence stems not only the fact that the terms have several meanings and are imprecise, but also alternative meanings, and sometimes there are mistakes of a terminological character.

Therefore the primary task is to make better definitions of the concepts that are represented by contradictory or debatable terms, and as far as possible achieve simplicity of the latter. If the deficiencies in hydrogeological terminology originated from imprecision of concepts and lack of coordination in the terms, then it would be a comparatively easy task to make them more precise: in order to do this it would be sufficient to discuss the meaning of the terms and their use. In fact the problem is very much more complex. There are many problems which require discussion, in which it is not the hydrogeological terms themselves, but the concepts which they reflect which give rise to disagreements. It is no longer an argument about words and linguistic tastes. The question is about theoretical ideas, or about the relation to actual topics, phenomena, and processes which are being studied.

The causes of the disagreements which result in the terminological discussions of this kind may, in general, be explained in the first place by the limited nature of our knowledge, and in the second place by the different methods of approach to research. There are probably other reasons too. However, the nomenclature is fundamental.

There are in fact still a great number of hydrogeological concepts which at a given stage of study do not lend themselves to simple interpretation. There are processes and phenomena, the explanation of which is still in the realms of hypothesis, or is, in general, beyond the frontiers of our knowledge. It is impossible to give a precise definition to these. Take as an example the dynamics of groundwater of the deep horizons of artesian basins. What determines the dynamics — the hydrostatic head, or the geostatic pressure? How does groundwater move — downwards or upwards? The answers to these questions are most contradictory. But after all, the formulation of hydrogeodynamic concepts, and consequently the ideas too, depend upon them. There can be no doubts about the difficulties of perfecting terms relating to such ideas.

The different notions on one or any other phenomenon may arise not only from the absence of sufficient information. The method of approach to a phenomenon is also of great importance, as are the point of view and the concepts of the researcher studying it. As is well known, the completeness of

the characterization of an object depends upon the position from which it is viewed. Therefore different researchers in one and the same situation often invest the very same set of circumstances with different contents and use for them far from the same nomenclature.

We will analyse from two different positions the terms 'hydrogeochemistry' and 'hydrochemistry'. Which should be preferred? O.A. Alekin (1972) considers that to isolate from hydrochemistry (a branch of which is concerned with the chemistry of all natural waters) that part, hydrogeochemistry (a branch in which only the chemistry of groundwater is studied), would be to contradict the idea of the unity of natural waters in the Earth. Hydrogeologists oppose him: according to the views of V.I. Vernadskii it is necessary to study not the chemistry but the geochemistry of groundwater and the natural waters in general. The argument, as can be seen, is convincing. Therefore, without rejecting the term 'hydrochemistry', it is obviously more correct to use the terms 'hydrogeochemistry' and 'hydrogeochemical' (in application to the research method).

In an analogous fashion, when the matter of groundwater arises, the terms 'hydrogeothermal' (the geothermics of groundwater), 'hydrogeodynamic' or 'geohydrodynamic' (the geodynamics of groundwater), and 'hydrogeophysical' (the geophysics of groundwater), are to be preferred to 'hydrothermic', 'hydrodynamic', and 'hydrophysical'.

It has been recently proposed (Shvetsov et al., 1973) that the term 'groundwater' should be replaced by the term 'subsoil water'. It would seem that such a change is logical: in certain circumstances, when talking of water which is lying below the soil, the expression 'subsoil water' would be more accurate. But what then to call the water lying lower than the Earth's surface where there is no soil layer (in mountains or on a shelf) — internal Earth water? This equivalent expression is more accurate but the attempt to introduce it was not successful. Probably it is necessary to recognize as correct and best the word 'groundwater', which as shown above, is imprecise, but is nevertheless very universal and has come firmly into current use by scientists and engineers.

How different the definition of one and the same concept can be we saw too in the example of such a fundamental term as 'hydrogeology'. One could continue the list of such definitions that are affected by the different approach to exactly the same set of circumstances. The question in this case is one of fundamental concepts.

And so, together with the disarray in the terminology, which, with goodwill and the desire to do so can be eliminated, the imperfection of hydrogeological terminology for the most part is caused by a large number of disagreements of principle. So for the latter we will hardly be able to negotiate round the table. The elimination of such disagreements inevitably calls for a reconsideration of the facts and principles which lie at the base of the concepts. In other words, one should not apply the same requirements to all concepts.

From the above-mentioned causes of the terminological problem, the way to improve the hydrogeological terminology is as follows:

(1) It seems that the process should begin with the devising of a system of concepts in order to obtain the foundation of scientific hydrogeological language. Kosygin (1974) has experience in creating such systems. This experience should be used to the utmost, leaning in particular on the systematic approach and the distinguishing of the basic types of geological branches (statistical, dynamic, and historical).

(2) Simultaneously with the devising of the system of concepts the terminology used to designate well-studied objects, phenomena, and processes demands an urgent and critical analysis. This analysis will include a clarification of the concepts, the definitions of which do not involve disagreements of principle. Precisely those terms which are named by them suffer from a multitude of meanings, indeterminacy, or alternative versions. The introduction of precision more than anything must be carried out in certain fundamental concepts, because it is precisely these which are used for the construction of new terms.

(3) Finally, after establishment of a new system of concepts it will be possible to pass to a review of the facts and principles which lie at the basis of the concepts, for which precise or exhaustive definitions are absent because of the limited nature of our knowledge or disagreements in principle. This work is most difficult; it will demand a great deal of time, laborious research, good preparation, and thoughtful consideration.

The achievement of precise and perfect hydrogeological terminology implies the formalization of the concepts and definitions. This process is already taking place, especially in connection with the use of computer processing of hydrogeological information. As experience shows, formalization simplifies the construction of new terms, the precision of concepts, and reduces discrepancies of all kinds. There is no doubt that a terminology, faultless from the point of view of formal logic, must be as simple as possible and available to a wide circle of specialists.

It is impossible to ban the use of any particular word. This must not be forgotten. However in order to carry on a discussion, it must be clearly and firmly known what stands behind the terms being used. Referring to this R. Descartes put it well when he said: 'Define the meaning of words and you will save humanity from half of its errors' (quoted from Vassoevich, 1974). Bearing in mind that a review of the geological terminology is connected with a prolonged process of perfecting the concepts and definitions, we are forced at the present time to use the existing terms. This necessity remains, but as the founder of cybernetics N. Wiener has said 'Working with inexact concepts gradually clarifies them and makes them more precise.'

The perfecting of the terminology is a complicated process which is influenced by the various opinions of different researchers. A concerted effort is required in order to eliminate a subjective approach to the evaluation of and the naming of one and the same set of circumstances. In principle it is not an important matter in which establishment they are employed, but of course it would be best if they were members of the Scientific Committee for Hydrogeological Problems of the Academy of Sciences of the USSR. It is important that the commission on hydrogeological terminology should work systematically. It should as far as possible be sufficiently representative to make use of not only the native but also the foreign store of hydrogeological concepts and terms, to conduct the investigation in liaison with the Committee of Scientific and Technical Terminology of the Academy of Sciences of the USSR. Recommendations on hydrogeological terminology should be published periodically. After thorough discussion they will come into force and be adopted by the All-Union Hydrogeological Congress, or the International Hydrogeological Conference.

Thus, hydrogeologists face an important and difficult problem. It is impossible to cut the terminological Gordian knot: the task is patiently to untie it.

References

Alekin, O.A. (1972). The achievements of hydrochemistry in the 50 years of the existence of the USSR and its tasks in the near future. *Vodnye resursi*, No. 3, 25–32.

Anon. (1955). *Geological dictionary*, 2 vols. Moscow: Gosgeoltekhizdat. Vol. 1, 403 pp., Vol. 2, 445 pp. (2nd edn (1973). Moscow: Nedra. Vol. 1, 486 pp., Vol. 2, 456 pp.).

Anon. (1971). *Dictionary of hydrogeology and engineering geology*, 2nd edn. Moscow: Nedra. 216 pp.

Anon. (1974). *The hydrogeology of Asia.* Moscow: Nedra. 575 pp.

Anon. (1978). *The hydrogeology of Africa.* Moscow: Nedra. 372 pp.

Bogomolov, G.V. (1975). *Hydrogeology and the fundamentals of engineering geology.* Moscow: Vyshaya shkola. 320 pp.

Chirvinskii, P.N. (1922). *Textbook of hydrogeology.* Rostov-on-Don: Gosizdat. 74 pp.

Davies, S.N. & R.J.M. De Wiest (1967). *Hydrogeology*, 2nd edn. New York: Wiley. 463 pp.

Fourmarier, P. (1958). *Hydrogéologie.* Paris: Masson. 294 pp.

Gordeev, D.I. (1954). Main stages in the history of Russian hydrogeology. *Trudy Laboratorii gidrogeol. problem.* Moscow: Izd-vo AN SSSR, Vol. 7, 383 pp.

Göthe, G. (1925). *Groundwater and springs.* Leningrad, Moscow: Gosizdat. 304 pp. (Translated from the German.)

Gray, D.A. (1975). The scope of hydrogeology. *Q. J. Engng. Geol.*, 8, 177–91.

Hähne, R. & H. Jordan (1978). Bemerkungen zur internationalen hydrogeologischen Terminologie. *Zeitschr. f. angewand. Geologie*, No. 4, 181–6.

Ignatovich, N.K. (1944). The laws governing the occurrence and formation of groundwater. *Dokl. AN SSSR*, 45 (3), 133–6.

Kamenskii, G.N. (1947). *Prospecting for groundwater.* Moscow, Leningrad: Gosgeolizdat. 313 pp.

Kamenskii, G.N., M.M. Tolstikhina & N.I. Tolstikhin (1959). *Hydrogeology of the USSR.* Moscow: Gosgeoltekhizdat. 366 pp.

Kapchenko, L.N. (1966). The present state of the problem of the age of deep juvenile groundwater. *Litologiya i polezn. iskopaemye*, No. 4, 75–87.

Kedrov, B.M. (1964). On the geological forms of movement in relation to other forms of movement. In *Interaction of sciences in the study of the Earth*, pp. 129–51. Moscow: Nauka.

Kedrov, B.M. (1971). History of science and the principles of its study. *Philosophical questions*, No. 9, 78–89.

Keilhack, K. (1912). *Lehrbuch der Grundwasser- und Quellenkunde*, 1st edn.

Keilhack, K. (1917). *Lehrbuch der Grundwasser- und Quellenkunde*, 2nd edn.

Keilhack, K. (1935). *Lehrbuch der Grundwasser- und Quellenkunde*, 3rd edn. Berlin: Borntraeger. 575 pp.

Klimentov, P.P. & G.Ya. Bogdanov (1977). *General hydrogeology.* Moscow: Nedra. 357 pp.

Kosygin, Yu.A. (1974). *Fundamentals of tectonics.* Moscow: Nedra. 215 pp.

Kotta, B. (1859). *Geological maps.* SPb. 267 pp. (Translated from the German.)

Kühne, V. (1932). *Groundwater studies.* Moscow, Leningrad: Gostroiizdat. 196 pp. (Translated from the German.)

Lange, O.K. (1931). *Short course in general hydrogeology.* Moscow, Leningrad: ONTI. 161 pp.

Lange, O.K. (1959). Groundwater of the European part of the USSR, Part 1, 280 pp. of *Groundwater of the USSR.* Moscow: Izd-vo MGU.

Lange, O.K. (1963). Groundwater of Siberia and Central Asia, Part 2, 248 pp. of *Groundwater of the USSR.* Moscow: Izd-vo MGU.

L'vov, A.V. (1916). *Finding and proving of water supply resources in the western part of the Amur railroad in permafrost soil conditions.* Irkutsk. 881 pp.

Mead, D.W. (1950). *Hydrology, the fundamental basis of hydraulic engineering,* 2nd edn. New York: McGraw-Hill. 728 pp.

Meinzer, O.E. (1923). Outline of groundwater hydrology with definitions. *US Geol. Surv. Water Supply Paper,* No. 494, 1−71. (Translated into Russian in 1933 as *Hydrogeological concepts, definitions, and terms,* 120 pp.)

Meinzer, O.E. (1935). *Groundwater studies.* 240 pp.

Nikitin, S.N. (1900). *Groundwater and artesian water in the Russian plain.* SPb. 71 pp.

Ototskii, P.V. (1906). Groundwater and forests, primarily in middle latitudes. In Groundwater, its origin, life, and distribution. *Trudy opytnykh lesnichestr.,* No. 4, 1−300.

Ovchinnikov, A.M. (1955). *General hydrogeology,* 2nd edn. Moscow: Gosgeoltekhizdat. 383 pp. (1st edn, 1949, 356 pp.)

Pinneker, E.V. (1975). Hydrogeology − the science of the subsurface hydrosphere. *Vodnye resursi,* No. 4, 130−3.

Pinneker, E.V. (1977). Hydrogeological terminology. *Izv. AN SSSR Ser. geol.,* No. 7, 119−24.

Plotnikov, N.I. (1976). *Groundwater − our treasure.* Moscow: Nedra. 208 pp.

Pošepny, F. (1893). *The genesis of ore deposits.* Am. Inst. Min. Eng. 149 pp.

Prinz, E. (1919). *Handbuch der Hydrologie.* Berlin: Springer-Verlag. 381 pp.

Richter, W. & W. Lillich (1975). *Abriss der Hydrogeologie.* Stuttgart: E. Schweizerbartsche Verl. 281 pp.

Savarenskii, F.P. (1935). *Hydrogeology,* 2nd edn. Moscow, Leningrad: ONTI. 336 pp. (Also, 1939, 212 pp.)

Savarenskii, F.P. (1947). The principles of hydrogeological zonation. *Sov. geologiya,* No. 19, 19−23.

Schoeller, H. (1962). *Les eaux souterraines.* Paris: Masson. 642 pp.

Shvetsov, P.F., A.A. Konoplyantsev & V.M. Shvets (1973). The modern content, main branches, and the organizational forms of development of hydrogeology in the USSR. *Izv. AN SSSR. Ser. geol.,* No. 2, 56−66.

Sidorenko, A.V. (1964). *Geology, the science of the future.* Moscow: Znanie. 64 pp.

Slichter, C.S. (1902). The motions of groundwater. *US Geol. Surv. Water Supply Paper,* No. 67, 1−106.

Smirnov, S.I. (1976). Evolution of the subsurface hydrosphere: kinetic and palaeohydrogeological aspects of its study. *Byul. MOIP. otd. geol.,* No. 6, 46−59.

Suess, E. (1902). Ueber heisse Quellen. *Verhandl. Gesell. deutsch. Naturforsch. und Aerzte (Leipzig),* 71, 133−51.

Sydykov, Zh.S. (1973). Classification scheme for natural sciences, including hydrogeology. *Vestn. AN KazSSR,* No. 9, 15−20.

Thurner, A. (1967). *Hydrogeologie.* Vienna, New York: Springer-Verlag. 350 pp.

Todd, D.K. (1959). *Ground water hydrology.* New York: Wiley. 334 pp.

Todd, D.K. (1980). *Ground water hydrology,* 2nd edn. New York: Wiley. 463 pp.

Tolstikhin, N.I. (1971). Classification of groundwater. *Zap. LGI,* 12 (2), 3−15.

Vassoevich, N.B. (1974). All-Union conference on the problems of terminology. *Izv. AN SSR Ser. geol.,* No. 10, 161−4.

Vernadskii, V.I. (1933−36). The history of natural waters (Vol. 2 of *History of the minerals of the Earth's crust*). No. 1, (1933). Leningrad: Goskhimizdat. 403 pp. No. 2, (1934). Leningrad: ONTI. 202 pp. No. 3, (1936). Leningrad: Khimteoret. pp. 403−562.

Vernadskii, V.I. (1960). History of natural waters (Vol. 2 of *History of the minerals of the Earth's crust*). In *Vernadskii, V.I. Collected works,* Vol. 4, Book 2. Moscow. 651 pp.

Zaitsev. I.K. (1961a). Some problems of the terminology and classification of groundwater. In *Material on regional and prospecting hydrogeology,* pp. 111−60. (Leningrad: Trudy VSEGEI, nov. ser. No. 46).

Zaitsev, I.K. (1961b). The methods of constructing review hydrogeological maps. In *Material on regional and prospecting hydrogeology,* pp. 7−48. (Leningrad: Trudy VSEGEI, nov. ser. Vol. 61).

Zaitsev. I.K. & N.I. Tolstikhin (1963). Fundamentals of the structural−hydrogeological zonation of the USSR. In *Material on regional and prospecting hydrogeology,* pp. 5−35. (Leningrad: Trudy VSEGEI, nov. ser. Vol. 101).

2

The subsurface hydrosphere

The Earth is spherical in shape. The division of the external parts of the Earth into geospheres which was proposed by E. Suess and subsequently developed by D. Merry and V.I. Vernadskii, implies the separation of atmosphere, hydrosphere, lithosphere, and biosphere.

The hydrosphere is the aqueous envelope of the Earth. Water occupies and permeates the three other geospheres. Surrounding our planet the substance H_2O is found throughout the atmosphere and the lithosphere, and more than three quarters of the biosphere consists of water.

F.P. Savarenskii (1947, p. 21) distinguished a surface and a subsurface hydrosphere. The latter, as we have seen in Section 1.1 occupies that part of the aqueous envelope which is situated below the surface of the Earth. 'It is impossible to consider the subsurface hydrosphere separately from the surface hydrosphere.' Neither did he recommend that the upper and lower zones of the subsurface hydrosphere be isolated the one from the other.

2.1 The extent of the Earth's hydrosphere

Natural water is a single entity. The masses of the water in the geospheres are closely interdependent and a balance is maintained between them. 'Any phenomenon involving water in nature', wrote V.I. Vernadskii (1960, p. 24), 'glacier ice, the immense ocean, rivers, soil solution, geysers and mineral springs, constitutes a single whole which is directly or indirectly, but nevertheless deeply, interconnected'. In the above quotation is formulated the popular idea of the unity of the Earth's waters, their inseparability, and their mutually interacting states.

Water is ubiquitous and therefore the hydrosphere does not possess any clear boundaries and the concept of it suffers from indefiniteness.

In early definitions of the hydrosphere it was conceived of as an aqueous envelope situated between the atmosphere and the lithosphere. In such a concept, which is shared even today by some scientists (Tolstoi & Shvetsov 1977), it is the world ocean and the surface waters of the land. Sometimes atmospheric moisture and glaciers are assigned to the hydrosphere, i.e. its lower boundary runs over the Earth's surface, the bottoms of rivers, seas, and oceans, but the upper boundary is lost in the stratosphere. The hydrosphere is considered to be well marked up to the tropopause, higher than this the water molecules suffer photodissociation (Fedoseev, 1974).

The exclusion from the hydrosphere of groundwater and other interior waters artificially divides the subsurface links of the water cycle from the surface, which is not in accordance with the opinions of modern hydrologists. For example, according to M.I. L'vovich (1974), the concept of the 'hydrosphere' has the same meaning as the concept of all free water of the Earth. He includes in the composition of the hydrosphere vapour in the atmosphere, glaciers, soil moisture, and groundwater apart from the world ocean and the surface waters of the land (see Table 2.1).

G.P. Kalinin (1968) and A.I. Chebotarev (1975) hold similar points of view on the extent of the hydrosphere. The authors of the book *The world water balance and the water resources of the Earth* (Anon., 1974) which was prepared in the USSR from the results of the hydrogeological decade (1965–74), include in the hydrosphere the water contained in living organisms. Their ideas on the extent of the Earth's water resources are shown schematically in Table 2.2.

The extent of the hydrosphere is defined sufficiently precisely and its boundaries are more or less clear, which cannot be said about the water situated in the Earth's crust and possibly the mantle.

As shown in Tables 2.1 and 2.2, there is on Earth, in a static condition about 1.4×10^9 km^3 of water. The volume of the world ocean, which contains almost the whole of the mass of the hydrosphere, has been determined with great accuracy. There are insignificant divergences (of opinion) among modern scientists concerning the volume of other parts of the surface

hydrosphere – water contained in glaciers, lakes, rivers, and the atmosphere.

The static volume at any instant gives only a general picture of water resources. The hydrosphere is a dynamic system in which the water cycle and the rate of exchange play a great role. The circulation of water in nature is a grandiose process involving the whole hydrosphere: the atmosphere, surface, and subsurface waters, and also those of the biosphere, the mantle, and the cosmos (Fig. 2.1). Incidentally, water passes from one phase or condition to another, and furthermore in the subsurface hydrosphere the interdependence between free and bound water is established. It is because of the circulation, which will be discussed in Chapter 4 that the unity of natural waters is established.

There is very much more divergence of opinion among scientists on the volume of groundwater and the subsurface hydrosphere. The reason for this is the different estimation of the parameters which are taken into account, mainly the surrounding rocks and the water which they contain.

In the sedimentary envelope of the Earth, that is on average to a depth of 5 km, water everywhere, apart from the regions of recent volcanic activity and permafrost, is in a liquid state. From approximately 8 to 16 km the temperature is close to the critical temperature for water (374 °C) and for aqueous solutions (425–540 °C). At higher temperatures water vapour cannot change to liquid no matter how high the pressure. In different states, water, in the form of molecules of H_2O, can be found in all layers of the Earth's crust (Makarenko *et al.*, 1972). According to A.M. Ovchinnikov (1955), V.F. Derpgol'ts (1962), and I.A. Fedoseev (1974), the lower boundary of the hydrosphere reaches the Moho, close to which, due to the high temperature and pressure, there is continuous synthesis of water molecules, together with dissociation. There is no doubt that there is aqueous fluid in the mantle; usually the hydrogen bonds are broken: there exists 'potential' water in the form of hydrogen and oxygen.

M.I. L'vovich (see Table 2.1) considers groundwater to

Table 2.1. *Mass of the hydrosphere and the rate of water exchange (L'vovich, 1974)*

Hydrosphere type	Actual volume $\times 10^3$ km^3	Same volume as %	Complete renewal of the water, recycling time (years)
World ocean	1 370 323	93.96	2 600
Groundwater	(60 000)	(4.12)	(5 000)
Including zones of active exchange	4 000	0.27	330
Glaciers	24 000	1.65	(10 000)
Lakes	280[a]	0.019	–
Soil moisture	(85)[b]	(0.006)	(0.9)
Atmospheric vapour	14	0.001	0.027
River water	1.2	0.0001	0.033
Total hydrosphere	1 454 193	100	2 800

[a] Including about 5000 km^3 in reservoirs.
[b] Including about 2000 km^3 of irrigation water.
Parentheses indicate estimated values.

include free (gravitational) water and physically bound water which may participate in the cycle (down to a depth of about 5 km). His determination of the quantity of groundwater is very close to that of F.A. Makarenko (1948, 1966). According to the latter, in the upper 5 km layer of the Earth's crust there are 86.4 million km³ of water, consisting of 13.7 million km³ (16.2%) of free water, 35.8 million km³ (42.4%) of physically bound water, and 34.9 million km³ (41%) chemically bound water. The quantity of free and physically bound water in the Earth's interior as calculated by M.I. L'vovich is 60 million km³, and according to F.A. Makarenko it is 49.5 million km³. The volume of groundwater stated in Table 2.2 is about half

this amount; this is because the thickness of the surrounding rock is taken as 2.5–3 km (with absolute limit of 2 km).

In the above definitions only a small part of the water contained in the interior of the Earth is considered. V.I. Vernadskii (1934) attempted to determine its quantity correctly. He approached the original parameters used for the calculation from a completely different position. Because the quantity of hydrogen in the Earth's crust is about 1% and is almost completely found incorporated in water, Vernadskii obtained for the 16 km layer of the lithosphere a figure of 1800–2000 million km³ of water. Subtracting from this the volume contained in the ocean he calculated the volume of the subsurface hydrosphere to be 450–600 million km³. Thus, the mass of the water in the world ocean and the crust is calculated to be of the same order of magnitude, or if not, then close to it. In this calculation the average quantity of water in the rocks is 8.5%, in which, as Vernadskii proposed, the liquid phase is found down to a depth of 3.10 km, but the water is mostly in the vapour phase.

A similar method, but using an incomparably greater amount of data, was used by A. Poldervaart (1951–55) and V.F. Derpgol'ts (1962–65) to calculate the volume of the subsurface hydrosphere. These researchers (Poldervaart, 1957; Gavrilenko & Derpgol'ts, 1971) evaluated the water in an Earth's crust of thickness 35 km below the continents and 4.7 km under the oceans (the calculated volume of the lithosphere is 8240×10^6 km³). Taking the value of the water content of the lithosphere as being 10.2 and 12.8 respectively, Poldervaart found the volume of the subsurface hydrosphere

Table 2.2. *The Earth's water resources (Sokolov, 1974)*

Water types	Volume × 10³ km³	%
Ocean	1 338 000	96.52
Ice and snow	24 012.1	1.74
Groundwater	23 400[a]	1.69
River channels	2.12	0.00015
Lakes	176.4	0.013
Marshes	10.3	0.0007
Atmosphere	12.9	0.0009
In living organisms	1.12	0.00007
Total (rounded off)	1 385 600	100

[a]Not including groundwater in the Antarctic, which is estimated to be 2000 km³.

Figure 2.1. General circulation of water in nature (after Z.S. Abramov).
1, free marine water; 2, sedimentary cover; 3, crystalline rocks of the crust; 4, magma chamber; 5, mantle; 6, upper and lower boundaries of intensive water exchange.

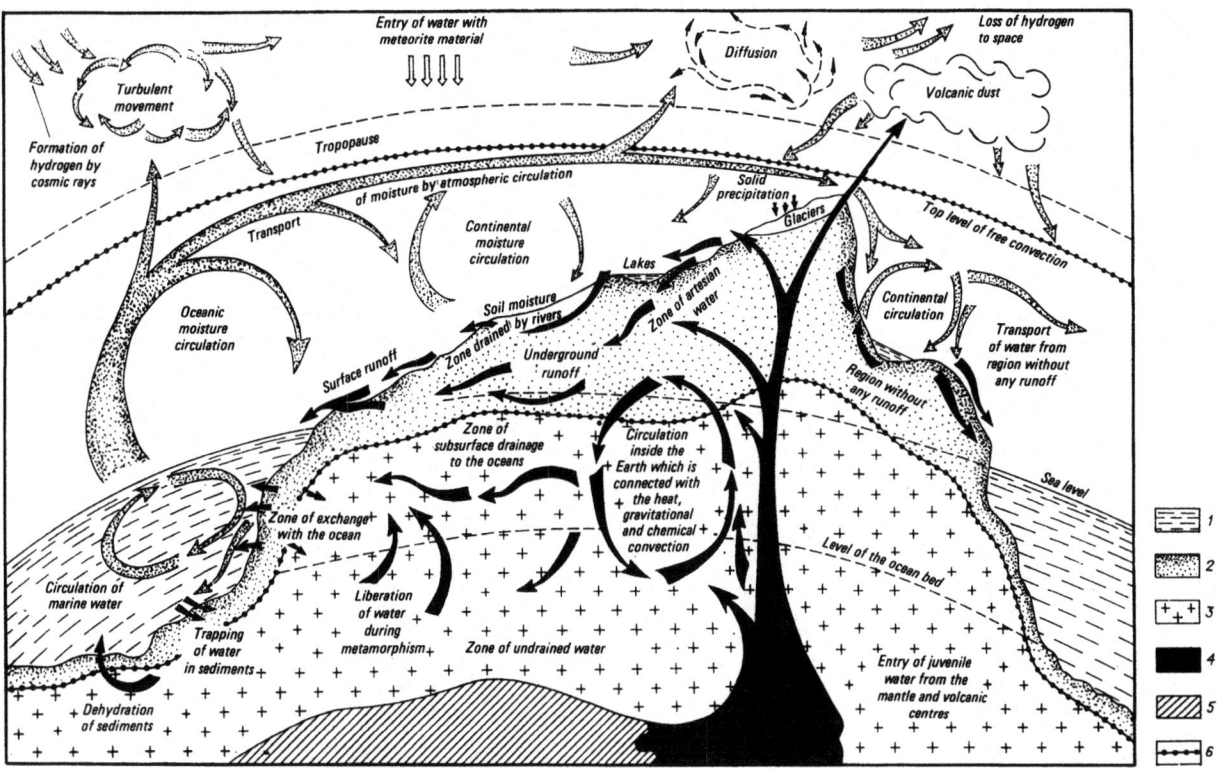

to be 840 million km^3 and Derpgol'ts calculated it to be 1050 million km^3 (see Table 2.3). The total quantity of water in the Earth, as calculated by both investigators coincides and comes out to be 2500 million km^3.

On comparing Tables 2.1 and 2.2 with 2.3 it is not difficult to see that for the total volume of the hydrosphere Derpgol'ts uses figures which are 1.5−2 times greater than those used by M.I. L'vovich or A.A. Sokolov. Such a difference relates entirely to the volume of subsurface hydrosphere. Sometimes on the other hand, a very high figure has been calculated (Makarenko *et al.*, 1972), in comparison with which the figures of Poldervaart and Derpgol'ts appear to be reduced by at least a factor of three. The exact determination of the quantity of the Earth's water resources, mainly the volume of the subsurface hydrosphere, will remain the most important task of geological and geographical scientists for the near future.

It follows from the above that the term 'hydrosphere' is now used in several senses. In the broad sense (Fedoseev, 1974), the hydrosphere is a complex aqueous envelope which stretches from the tropopause down to the upper layer of the mantle. We too will adhere to such a semantic meaning.

2.2 The components of the subsurface hydrosphere

Water in the Earth's crust is bound to some or other degree to the rocks and is found in different states. In passing we will point out that the water of the interior of the Earth is closely bound not only to the rocks, but also to the gases, which virtually saturate the subsurface hydrosphere. In the upper layers nitrogen is the main constituent, but as the depth increases so does the amount of methane and carbon dioxide, and at the level of the upper mantle the main constituent is probably hydrogen.

V.N. Kortsenshtein (1977), taking the volume of gas in the subsurface hydrosphere to be 10 km^3/km^3 (although in reality it is higher, up to 30−60 km^3/km^3) obtained a minimum value of the gaseous content of the subsurface hydrosphere of about 10 billion km^3, i.e. twice the volume of the Earth's atmosphere.

In the light of the process which we have just explained, the subsurface hydrosphere is the most important regulator of

the degasification of the Earth and at the same time is a trap which catches the gases emanating from the mantle.

Types of water in rock. A.F. Lebedev (1936), on the basis of well conducted experimental investigations established that several types of water existed in rock: (1) in the form of vapour, (2) hygroscopic, (3) pellicular, (4) gravitational (capillary, suspended, and in a state of descending), (5) in the solid state, (6) water of crystallization, and (7) chemically bound.

Lebedev's idea of the types of water in the rocks was developed by V.A. Priklonskii, A.A. Rode, B.V. Deryagin, A.M. Vasil'eva, E.M. Sergeev, and others.

N.I. Tolstikhin (1971) approached the division of the waters of the Earth's interior from rather different, but on the whole similar, standpoints. In comparison with the 'genetic' classification proposed by this author in 1956 (Maksimov, 1959) the unwieldiness of the system examined was removed and the previously-noted inaccuracies were eliminated (Vel'mina, 1970). There is in both classifications the question of groundwater in the broad sense, which embraces practically all the components of the subsurface hydrosphere.

Tolstikhin (1971) distinguishes the following types and subtypes of water in the Earth's interior:

 (I) gaseous state (steam): (1) free and (2) vacuolar;
 (II) liquid state (water): (1) free and (2) vacuolar;
(III) solid state (ice): (1) holocryogenic and (2) cryophoric;
(IV) physically bound: (1) pellicular and (2) hygroscopic;
 (V) chemically bound: (1) weakly chemically bound and (2) strongly chemically bound.

Neither of these classifications mentions water in the supercritical state. Considering this omission and bearing in mind the requirements of hydrology, then water contained in rocks ought to be divided into the following:

(1) in the form of vapour;
(2) in the solid state;
(3) physically bound and hygroscopic (tightly bound, pellicular, and osmotic);
(4) free (vacuolar), capillary, and gravitational;
(5) chemically bound (water of crystallization, zeolites, and in minerals;
(6) in a supercritical condition.

The three states of water are shown on the pressure−temperature graph in Fig. 2.2 (after A.I. Tugarinov, 1973).

Table 2.3. *Quantitative characteristics of the hydrosphere (Gavrilenko & Derpgol'ts, 1971)*

Location of water	Volume × 10^6 km^3	Mass × 10^{15} t	%
Hydrosphere	2460	2530	100
Oceanic crust	1500	1550	61
Continental crust	960	980	39
Surface water	1410	1460	58
World ocean	1370	1420	−
Ice	30	30	−
Rivers and lakes	10	10	−
Water inside the Earth	1050	1070	42
Sedimentary rocks	190	190	−
Crystalline rocks	860	880	−

Figure 2.2. Phase diagram of water with respect to temperature and pressure (after K. Krauskopf).

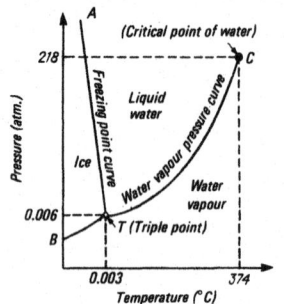

Along the boundaries TA, TC, or TB there are two phases and at the point (T) three phases can be present provided that the system is stable, i.e. if neither the temperature nor the pressure in the system changes.

With pressure and temperature values higher than the critical point (C), differences between the liquid and the vapour disappear. The diagram given in Fig. 2.2 is for pure water. The system of aqueous solutions behaves in a considerably more complex manner depending on the composition and concentrations of the dissolved materials.

Water in the form of vapour is extremely mobile, filling the pores and other spaces in the rocks which are not occupied by the liquid phase. Near the Earth's surface the water vapour content varies from a fraction of a per cent to several per cent of the total amount contained in the spaces. Water in the gaseous state and aqueous solutions pass through the rock even in the temperature range 100–450 °C. Hot steam in the form of gaseous mixtures, usually referred to as steam *hydrothermals* is found, for example, in the regions of recent volcanic activity, where the gas phase H_2O fills all the interstices completely and is under great pressure – up to 218 atmospheres for pure water (see Fig. 2.2).

Between the Earth's surface and the surface of the groundwater, water vapour moves from layers with greater elasticity to layers with less elasticity, or from more humid regions to those in which the humidity is less. The transfer may take place in different directions. When water vapour is in a saturated condition and has maximum elasticity at a given temperature, its movement is determined only by the value of the temperature and will be directed from places with high temperature to places with lower temperature.

Water vapour penetrates downwards from above, from the atmosphere, or is formed indirectly within the Earth by condensation. As a result of the condensation of vapour on the surface of mineral particles, free and bound forms of water are formed. Evaporating water is always in equilibrium with other forms of water and with water vapour in the atmosphere, therefore evaporation in some places and condensation in others demonstrate the fundamental influence of the redistribution of moisture in the soil and rocks.

Water vapour as a gas possesses a number of anomalies. In particular it cannot take part in the nutrition of plants. E.M. Sergeev *et al.* (1971) consider that part of the molecules of water vapour are associated with complex forms of $(H_2O)_n$, in which *n* varies from 2 to 5 molecules. Such complexes are adsorbed on the surfaces of mineral particles together with individual molecules of H_2O.

Water in the solid state, that is in the form of ice, is characteristic of the subsurface hydrosphere in permafrost regions, and in the zone outside the permafrost which is seasonally frozen. It also crystallizes inside rocks and forms streaks and veins which sometimes reach a thickness of several tens of metres – ice-filled rock fissures.

In general the following varieties of subsurface ice can be distinguished: (1) buried ice of surface origin; (2) fossil ice of subsurface origin; (3) ice-inclusions, which enter the composition of rocks; (Anon., 1970).

It is necessary to say a few words about fossil ice of subsurface origin. It may be seasonal and many years old. Lode and stratal ice bodies are widely distributed in the alluvial and lake deposits of the Baikal region, Yakutsk, and the north-east of the USSR. With underground flow the ice bodies can grow, which leads to the formation of pingos or *hydrolaccoliths* which contain a growing ice nucleus.

Ice which forms as inclusions in rocks is subdivided into ice-cement and segregational ice. The former variety appears between pieces of rock when water freezes in rocks; the latter separates in the form of little crystals in wet rocks (usually with humidity of more than 25%).

Gaseous hydrates are unique phenomena; they are solid crystalline substances which are compounds of H_2O molecules with natural gases, more often than not hydrocarbons. Superficially they resemble compressed snow. In the layer of perennially frozen rocks even industrially viable deposits of gaseous hydrates are found (the Messoyakhskoe gas deposits in the north of the Krasnoyarsk region). Under high pressure and at favourable temperatures the formation of hydrates occurs, in particular on the walls of wells and gas pipelines. The hydrates evaporate rapidly on reaching the surface, leaving pools of water and emitting gas copiously.

Physically bound water is found to some extent or other as an interaction with the surface of rock particles. On this basis it is divided into tightly bound (hygroscopic) and loosely bound (pellicular and osmotic). These varieties are easily distinguished from free water by their properties: their average density is much higher ($1.2–2.4$ g/cm^3), their mobility is extremely low, freezing point considerably lower than 0 °C (in kaolinite about −20 °C, and in montmorillonite even lower: −100 °C). There are many types of physically bound water in clays. They are held to the surface of the crystals with forces which are many times greater than that of gravity.

Hygroscopic moisture is formed when molecules of evaporating water cover the surface of the mineral particles. This can be seen in the classical scheme of A.F. Lebedev (Fig. 2.3). Water vapour is adsorbed mainly on the surfaces of finely dispersed rock particles and is very firmly held by electrostatic forces. Hence it is called 'tightly bound' or 'adsorbed'.

Several categories of hygroscopic water are distinguished according to the level of the energy bond. Moisture filling the corners or defects in the crystal lattice has the least mobility and forms a so-called 'immobile' layer. The thickness of this is from one to three molecules. Such moisture does not form a continuous layer around particles. It is possible to drive it off by heating to a temperature of 150–300 °C. The next layer covers the surface of the rock particles directly and has a thickness of 10–20 molecules. This is called the 'water of the basal surface of the clay particles'. The water of this layer, which corresponds to the maximum degree of hygroscopicity, i.e. to that moisture content which is the result of the adsorption of water vapour by particles under an elasticity of 100%, is characterized by a certain mobility and can be driven off at temperatures of 90–120 °C. However, the transfer of water with such stable bonds is possible only when it is in the vapour phase. Other varieties of hygroscopic water differ in the level of the energy bond within the mineral particles (Sergeev *et al.*, 1971).

Under maximum hygroscopic conditions, the tightly bound water completely fills the space between fine-grained particles. Its content is determined by the mineralogical and granular composition of the rock, and by the composition of the exchangeable cations. In clayey soils the hygroscopic moisture comprises 15–20% and more, but when the particles are larger it does not exceed 5% of the mass of the mineral material. The influence of the mineralogical composition is even greater. Thus for fine-grained particles the maximum hygroscopicity may reach 0.9% in quartz, 8–17% in feldspars, and 36–48% in micas.

The binding energy of pellicular water is less than that of hygroscopic water and it freezes at a temperature of about $-1.5\,^\circ C$ depending on its quantity and the duration of the freezing. It forms as it were a second film over the hygroscopic water and moves from places with the thickest films to those where the film is thinner (see Fig. 2.3). The film is maintained by molecular forces. The influence of the intermolecular bonds reduces rapidly as the film thickness increases and at the surface of the film it is insignificant. The water is therefore called 'weakly bound'. The speed with which it moves is a linear function of the temperature. The external layer of the pellicular water is available to plants and may serve as a medium for the growth of micro-organisms.

E.M. Sergeev *et al.* (1971) distinguish between two varieties of weakly bound water. The first is secondarily oriented water in multilayers and surrounds the mineral particles and adsorbed ions. It is held by molecular forces which arise between the molecules of tightly bound water and the molecules of the newly formed film. It is also properly called pellicular water in the sense used by Lebedev. The maximum moisture content forms the total content of hygroscopic and pellicular water. This parameter gives the amount of physically bound water which is in the rock under the action of the forces of molecular attraction. The maximum molecular moisture content is: sand 1–7%, sandy loam 9–13%, loams 15–23%, and clays 25–40%. Another variety distinguished by Sergeev *et al.* is osmotic water. This is formed by the penetration of molecules of aqueous solvent into the diffuse layer of colloidal particles where the ion concentration is greater than that of the solution. This has the lowest level of the energy bond. Osmotic water is not pellicular but is rather a transitional variety towards free water. It causes swelling to colloids and forms around the colloidal particles a stable film of solvent – a solvent layer.

The maximum amount of physically bound water in the rock corresponds to the moisture of maximum swelling. The molecular and osmotic forces which hold the weakly bound water exceed the force of gravity. However, with increasing film thickness a point is reached when they cannot hold it any longer and the physically bound water separates and becomes free.

Physically bound water can be liberated from rocks by drying (at a temperature of $105–110\,^\circ C$), centrifuging, or placing the rocks in a vacuum. Pressing out is often used. As the experiments of P.A. Kryukov, V.D. Lomtadze, A.E. Babintsa, V. Engelhardt, and others have shown, at a pressure of 3000–5000 kg/cm^2 weakly and strongly bound water of clays is able to go to the free condition. Because it is squeezed out from the microcapillary pores (of diameter less than 0.0002 mm) of the rocks or wet sediments on the sea-bed or elsewhere, such water is called *pore solution* (Anon., 1968). Under natural conditions pore solutions are squeezed out as the sediments become denser under their own weight.

Free water includes water contained in minerals, capillary, and gravitational. The last is also groundwater.

Water inside inclusions fills spaces of different sizes. Sometimes it is found together with gas, forming gas–liquid inclusions within minerals. Quite often the solution is highly mineralized (several hundreds of grams/litre concentration) as, for example, the trapped water of Iceland spar. Essentially it is the trapped remains of the medium in which the mineral formed, mechanically included in the mineral in the form of free water. The water in inclusions may be freed by heating or by the mechanical destruction of the minerals.

Capillary water occupies the corners of the pores, and with increasing moisture content it completely fills the capillary pores. It is held and moves in the rock mainly under the action of capillary (meniscus) forces, which arise at the boundary of the water and air contained in the rocks.

Capillary-junction water, distributed in the corners of pores, is immobile. Water filling the capillary pores is divided into capillary-rise water and capillary-suspended water according to whether it is connected with the groundwater or not. In contrast to capillary-junction water these two varieties are free to migrate under the influence of surface tension and transmit hydrostatic pressure.

Capillary-rise water (capillary water proper) is found in the zone known as the capillary fringe rising above the groundwater connected to the infiltration water (Fig. 2.4). This is

Figure 2.3. Diagram of the different types of water in rocks (after A.F. Lebedev).
The large empty circles are rock particles and the little circles round the rock particles represent water molecules.
1, sites not all filled; 2, maximum hygroscopicity; 3 and 4, rock particles with pellicular water which moves to particles with thinner pellicular water layer (the dashed line shows equal thickness of the water sheath); 5, particles of rock with free water.

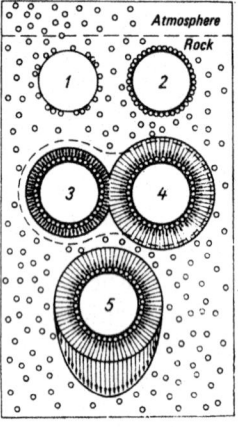

maintained by the rise of groundwater through the capillaries above the groundwater zone. The amount of water in the rock corresponding to complete saturation of all the capillary pores is called the capillary moisture content. Its magnitude depends on the porosity and the structure of the rock. The height of the capillary rise is a minimum (or even zero) in gravels, and other coarse-grained rocks, and reaches a maximum in clays (Table 2.4).

Capillary-suspended water does not reach the level of the groundwater (see Fig. 2.4). It arises in both homogeneous and in layered rocks, more often than not in sandy loams because of the penetration of atmospheric precipitation into the ground. This form of capillary water is held by surface tension and does not reach the capillary fringe. The degree to which the capillary pores are filled by suspended water varies greatly. The amount of suspended water held in rocks by capillaries is customarily called its field capacity. When all the capillaries are full the state is known as saturation moisture content. The extent to which the pores are filled with moisture gives a measure of the water-bearing capacity. According to A.A. Rode, the extent to which the pores are full when moisture is present varies from 40–100% depending on the composition and structure of the rocks. If evaporation continues for a long period this water disperses, sometimes completely.

The chief property of gravitational water is its movement under the action of gravity and the pressure gradient. Gravitational water may be divided into *infiltration water* of the zone of aeration, which percolates downwards from above, and *moving groundwater* in the saturated zone. This moves in the form of a current along the subsurface horizons (see Fig. 2.4). The amount of gravitational water depends on the grain size of the material, the porosity of the rock, and the amount of fissuring in it. In clays gravitational water occurs in very small quantities and as the clay layers thicken it is practically absent. In the case of coarse materials (sands and gravels), or with increasing degree of fissuring of the rocks the amount of gravitational water exceeds all the other forms.

Saturation moisture content corresponds to the amount of water in the rock when all the pores and fissures are completely filled with water, i.e. it is the maximum possible quantity of physically bound capillary and gravitational water. It is observed for example in water-bearing strata along which there is a moving current of groundwater.

In clays the saturated and capillary moisture content are not usually differentiated because the porosity is due almost exclusively to capillary pores. For clays and colloids, the volume of which increases when wetted, the total moisture content will be higher than the moisture content for normal porous materials only when approaching the moisture content for maximum swelling. As for granular crystalline rocks, which do not swell when wetted, their total saturation moisture content is equal to the maximum moisture at the given porosity of the rock.

The ability of water-bearing rock to yield water by means of free flow under gravity contains an element of efficiency based on the difference between full and maximum moisture content. This is expressed numerically in the form of a coefficient of water yield, that is the ratio of the quantity of water which the rock can yield to the total volume of the rock. The greatest value of this coefficient is 10–30% in coarse-grained and fissured rock, and it is insignificant in clays and in massive sedimentary rocks (less than 1%).

Infiltration water is found in the *zone of aeration*. This zone is understood to be the space between the surface of the Earth and the top level of the groundwater. In the zone of aeration the pores are filled with air, water vapour, or physically bound water. Infiltration water flows periodically, in spring when the snow melts, and after rain has fallen. Reduction of the volume of the zone of aeration occurs, for example, in the strip bordering on artificial reservoirs (see Fig. 2.4). Vertical downward movement continues until the water meets a restraining layer of low permeability. Groundwater currents are formed above such an impermeable layer. The zone of aeration changes to the zone of saturation (see Fig. 2.4), where the pores and fissures are filled with water. Only oil and gas deposits and other strata where the pores are also filled with oil and gas are exceptions to this.

The direction and velocity of the infiltration water flow is determined by the pressure gradient. In the upper layers the transfer of groundwater from high hypsometric levels to lower is caused by the hydrostatic pressure. At great depths the pressure gradient also increases as a result of the geostatic pressure (i.e. under the action of the weight of the rock above), and a current of water is produced, which is squeezed out of the compacting layer as a result of the presence of endogenic forces, mainly tectonic in character.

When the question of the zone of saturation arises, one has to bear in mind water in the liquid state. The lower bound-

Figure 2.4. The effect of an artificial reservoir on the thickness of the zone of aeration.
I, zone of aeration; II, confining bed.
1, capillary water; 2, dry rock; 3, saturated rock; 4, confining bed; 5, that part of the zone of aeration which becomes saturated as a result of the pressure of the reservoir; 6, direction of water movement.

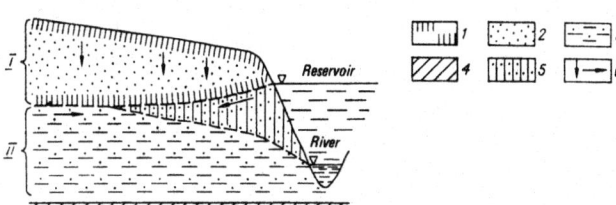

Table 2.4. *Maximum height of the capillary fringe*

Rock material	Height (cm)
Coarse-grained sands	2–3.5
Medium-grained sands	12–35
Fine-grained sands	35–120
Sandy loam	120–350
Clayey loam	350–650
Clay	650–1200

ary of this zone extends to the depth at which the rock is at critical temperature (374–450 °C); deeper than this is the zone in which water is in a supercritical state (Fig. 2.5). Within the limits of the *zone of saturation*, starting at depths between 1.5 and 2 km the physically bound water, under the influence of temperature and pressure goes to the free condition. At the lower boundary, i.e. at depths of greater than 12–20 km on the continents and deeper by 2–3 km in regions of recent volcanic activity, water is found mainly as water occluded in the crystal lattices of minerals.

Chemically bound water enters into the composition of minerals, entering into the crystal lattice as H_2O, OH^-, H^+, and H_3O^+. When it is included in the form of H_2O molecules it is known as water of crystallization. The hydrogen and oxygen found in hydroxyl-containing minerals is referred to, not quite correctly, as water of composition, because it changes to H_2O only after it has separated from the mineral. There is also the water in the zeolite mineral group, which is essentially a form of water of crystallization. Water of crystallization is a property possessed by minerals which are formed in conditions of relatively low values of temperature and pressure. Natron ($Na_2CO_3.10H_2O$) contains more than 50% water (64% in fact), mirabilite ($Na_2SO_410H_2O$) contains 55% water, bischofite ($MgCl_2.6H_2O$) contains 53% water, etc. The hydration of minerals occurs in the zone of supergene mineralization or during metasomatism. Separation of water of crystallization is caused by the destruction of the crystal lattice and the formation of anhydrous compounds. In the majority of cases this requires heating to a temperature of not greater than 300–400 °C. Sometimes hydrated minerals (natron in particular) release water at normal surface temperatures. Others, like polygallite ($NaCl.2H_2O$) are in general extremely unstable and are only temporary products of mineral formation. However, more often during initial heating intermediate crystallohydrates with smaller water content are formed, and these subsequently release water at higher temperatures. The transformation of gypsum to anhydrite:

$$CaSO_4.2H_2O \rightarrow CaSO_4.H_2O\ (107\ ^\circ C) \rightarrow CaSO_4\ (170\ ^\circ C),$$

is an example of this.

The water of hydration in zeolites is similar to water of crystallization; however, it separates from these minerals over a considerable temperature range and without the destruction of the crystal lattice. In zeolites the water is weakly bound, almost like hygroscopic moisture. Zeolites comprise a large group of minerals, for example analcime ($Na_2Al_2Si_4O_{12}.H_2O$) and natrolite ($Na_2Al_2Si_3O_{10}.2H_2O$). The separation of water from zeolites can occur even without heating in a medium of low water vapour pressure. Because of the structural peculiarities of zeolites the water which separates is given off with a change in the thermodynamic conditions and is easily restored later.

Minerals containing hydroxyl and hydrogen are formed under conditions of high pressure. OH^- and H^+ are incorporated in the mineral molecule with such a strong bond that they can only be separated by complete destruction of the mineral, as a rule at temperatures from 400 up to 1300 °C (Table 2.5). A.S. Povarennykh (1966) has proposed the following scheme which depends upon the position of the hydrogen atom relative to the oxygen atom for the transition from the hydroxyl bond, which is essentially ionic, to the hydrogen bond, which is essentially covalent:
  (1) hydroxyl bond, distance OH–OH = 5–3 Å (NaOH, $Ca(OH)_2$, $Mg(OH)_2$, etc.);
  (2) hydroxyl–hydrogen bond, distance OH–OH = 3.0–2.7 Å ($Zn(OH)_2$, $Al(OH)_3$, H_2O are all crystallohydrates);
  (3) the hydrogen bond, distance O–H–O = 2.7–2.5 Å ($NaHCO_3$, $KHSO_4$, $HAlO_2$, etc.).
Oxonium ions H_3O^+ enter into the structure of a mineral as an analogue of monovalent ions, often replacing them isomorphically. Among these minerals there are alunite, several silicates, and other minerals.

Water in a supercritical state. A pressure of 218 atmospheres and temperature of 374–450 °C are considered to be critical for aqueous solutions. For temperature and pressure values greater than these the physical state of H_2O is such that there is no difference between liquid and gas: the molecules have the speed of the gas molecules and the density approaches unity. Sometimes this gas–liquid form of H_2O is not quite accurately called a fluid (Makarenko *et al.*, 1972). Because it is accepted that a fluid is a multicomponent mixture of volatile materials which also includes, apart from H_2O other products of the degassing of the mantle (see Section

Figure 2.5. Diagram that shows the principles of phase zonation of the subsurface hydrosphere (scale does not permit the zone of aeration to be shown).
1, cryolithic zone; 2, zone of saturation; 3, zone of water in the supercritical condition; 4, upper mantle; 5, the boundary between the sedimentary and granite layers; 6, Conrad discontinuity (separates the granite layer from the basalt); 7, Moho surface.

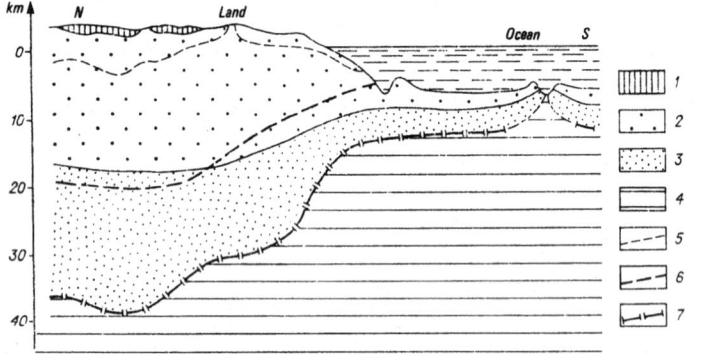

Table 2.5. *Temperature of dehydration of some minerals*

Mineral	Formula	Temperature of dehydration (°C)
Phlogopite	$KMg_3(OH,F)_2\ [AlSi_3O_{10}]$	1090
Talc	$Mg_3(OH)_2\ [Si_4O_{10}]$	930
Muscovite	$KAl_2(OH)_2\ [AlSi_3O_{10}]$	820
Serpentine	$Mg_6(OH)_8\ [Si_4O_{10}]$	670
Kaolinite	$Al_4(OH)_8\ [Si_4O_{10}]$	560

3.2), it is more accurate in such a case to talk of water in the supercritical state.

Water in this state differs in many of its properties from the vapour and liquid states. Its hydrogen bond is very unstable and the H_2O molecule splits into H^+ and OH^-. The viscosity is reduced and consequently the migratory ability is much greater than that of 'liquid' water. With increase of temperature and pressure its power to dissolve, which simultaneously depends on the concentration of solution, increases. The latter also determines the parameters of the critical point. For example if the critical values for pure water are 374 $^\circ$C and 218 atmospheres (see Fig. 2.2), then for NaCl solution of strength 50 g/l, these parameters will increase to 430 $^\circ$C and 340 atmospheres (Gavrilenko & Derpgol'ts, 1971).

The supercritical state is characteristic of water in magma. The ability of different magmas to hold water, depending on composition, reaches 4–10%. In the interior of the Earth the transition of H_2O from a supercritical state to steam or liquid proceeds continuously. There are no clear boundaries between fluid, pneumatolytic, and hydrothermal mineral-forming systems. However, this transition is accompanied by an increase in volume of H_2O (up to 1.5–2 times) and the precipitation of ore minerals from the solution is of prime geological importance.

2.3 The laws governing the localization of groundwater in rocks

The water-collecting properties of rocks. Groundwater is contained in porous or fissured permeable rocks (Fig. 2.6), which allow water to move if there is an excess pressure. These rocks are called water-bearing, or aquifers, in contrast to impermeable rocks (aquicludes). The latter are impermeable, or allow water to move only very slowly, and divide or underlie rocks containing groundwater.

According to the hydrogeological classification of the rock shown in Fig. 2.7, water-bearing rocks, depending on their permeability, form hydrogeological collectors (aquifers), and the impermeable rocks act as hydrogeological isolators. In the interior of the Earth the laws governing the localization of groundwater are predetermined first and foremost by such water-collecting properties as *porosity* and *permeability*.

The volume of all the spaces in rock is correctly called the void volume, but it is more often called porosity (see Fig. 2.6). Reservoir rocks containing water are divided into three categories according to the character of the spaces; these are

(1) granular with pore-like voids, (2) fissured with slit-like discontinuities (fissured slates and shales, magmatic and metamorphic rocks), and (3) containing mixed pores and discontinuities (for example fissured sandstones and porous limestones, the volume of voids of which is determined by the pores, and the paths of movement of groundwater in the discontinuities). The greater the porosity, the more the rock can hold water, i.e. the degree of porosity determines the capacity of the aquifer. It is not just the total magnitude of the porosity, but the size of the voids and the extent to which they interconnect which is of importance in determining the movement of groundwater. For instance, clays, the porosity of which reaches 50–60%, are practically impermeable to groundwater, but sandstones and fissured rocks with a porosity of only 10% are good aquifers. In fact one feature of clay is the vast number of subcapillary pores (of diameter less than 0.0002 mm), in which water cannot move and which for the most part are isolated from one another.

Rocks which fall into the category of aquifers are characterized by capillary (0.0002–0.1 mm) and supercapillary (more than 0.1 mm) channels of interconnected pores. Groundwater moves mainly through the supercapillary pores; however, the hydrostatic pressure is also capable of being transmitted through capillary pores, in which the movement of groundwater takes place under the influence of surface tension. In voids, which are as a rule free of capillary water, the groundwater moves under the influence of gravity and difference in the head.

Pores in rock may be interconnected (open) or isolated (closed). The total volume of open and closed pores, independent of their size, shape, and spatial relationship one to another is called the total (absolute) porosity. This is expressed as the ratio of the volume of all the pores to the total volume of the rock.

For hydrogeological purposes the important factor is not the total but the dynamic or *effective* porosity, which is the volume of the open pores through which the transfer of liquid takes place under the influence of the pressure gradients which occur in nature, relative to the total volume of the rock. The effective porosity should not be confused with the *open* porosity which characterizes all open pores. The movement of liquid is not through the total volume of the open pores, because part of the pore space, for example at the junctions between particles, is always occupied by capillary weakly or

Figure 2.6. Basic types of rock porosity (after Meinzer, 1923). 1, unconsolidated well-sorted grains and high porosity; 2, well-sorted porous pebbles and high porosity; 3, unconsolidated poorly sorted grains and low porosity; 4, reduced porosity as a result of cement between grains; 5, cavernous, the porosity increases as the result of solution; 6, porosity caused by fracturing.

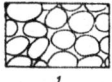 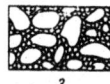  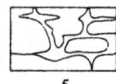 

Figure 2.7. Hydrogeological classification of rocks (after Stepanović, 1962).

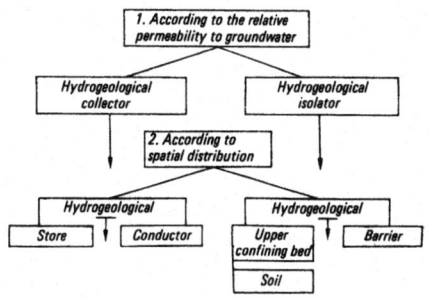

strongly bound varieties of water. The open porosity is always greater than the effective porosity.

Permeability is the ability of the rock to let liquids and gases move when a pressure gradient exists. Hydrogeologists are mainly interested in hydraulic conductivity because this governs the movement of water through aquifers. Quantitatively the flow is determined by the hydraulic conductivity, which is a function of porosity and the viscosity of the fluid concerned.

The darcy (d) is the unit of intrinsic permeability. This refers to a specimen of rock of 1 cm length and cross-sectional area 1 cm^2, through which flows 1 cm^3/s of a liquid of viscosity 0.001 Pa s, and across which there is a pressure difference of 1 atmosphere. The hydraulic conductivity, which has the dimension of metres/day or metres/second, is the flow velocity per unit hydraulic gradient. The following relationship exists between the coefficient of intrinsic permeability (K_p) and the hydraulic conductivity (K):

$$K = K_p \frac{\gamma}{\mu},$$

in which γ is the specific weight in g/cm^3 and μ is the coefficient of dynamic viscosity in Pa s.

As can be seen in Table 2.6, for fresh groundwater at a temperature of 20 °C, the intrinsic permeability in darcys roughly corresponds to the hydraulic conductivity of 0.85–0.9 m/day.

Rock can be extremely *permeable* (with an intrinsic permeability from 10 to several hundred thousand millidarcys (md), *semipermeable* (10–0.1 md), and practically *impermeable* (less than 0.1 md). A more detailed division of values is given in Table 2.6.

Sandstones, well sorted sands, carbonate rocks with solution cavities, and fissured crystalline rocks are designated permeable rocks. The pore space, which occupies 10–40% usually, is uniformly distributed throughout the volume of the rock and consists mainly of supercapillary or large capillary pores. Pores of other dimensions have subordinate significance, only a small part of the volume of the pores is occupied by physically bound water.

Unsorted semipermeable rocks (sandy clay sediments, sandstones with clay cement, chalk-like limestones, etc.) have pores of different sizes (pores for the most part of fine capillary and subcapillary dimensions).

Rocks with subcapillary or closed pores (clays, argillites, unweathered shales/slates, compact crystalline rocks) are practically impermeable. In nature however, there are no completely impermeable rocks. As the experiments and calculations of Myatiev (1947) showed, at considerable excess pressure even layers of clay up to several tens of metres thick may be permeable to water. Moreover, cases are known in which, together with water, dissolved substances are 'squeezed' through such layers. Under 'normal' hydraulic gradients the migration of dissolved substances through aquifuges is possible by molecular diffusion (Smirnov, 1971).

Thus in certain thermodynamic or hydrogeological conditions, taking geological time scales into account, clay formations are also permeable. However with the temperatures and pressures that are the rule in the upper part of the Earth's crust (down to depths of 2–5 km), rocks with permeability values of less than 0.1–0.001 md may be regarded as reliable aquicludes. According to G.Ya. Bogdanov & V.M. Kononov (1975), clay layers of thicknesses of greater than 10 m must be regarded as impermeable. Frozen rocks make good aquicludes.

As the pressure increases at deeper levels, the porosity and the permeability decrease. It is true that there are exceptions, for example when sandstones or limestones are found at depths of 3–6 km and their porosity ranges from 1.5–20%. However, on the whole, it is an order of magnitude less at such depths than in the upper horizons.

Sedimentary rocks (and sedimentary-volcanic or pyroclastic rocks) differ sharply from crystalline rocks (magmatic and metamorphic). We will examine these separately.

The hydrogeological stratification of sedimentary rocks is the result of the alternation of aquifers consisting of layered, porous-layered, fissured-layered, or more complex types over a wide area. As yet there is no well-founded classification system for these layers into grades of aquifer rocks.

Since the end of the last century the water-bearing horizon (English and French, *aquifer*, German, *Grundwasserleiter*) has been accepted as a fundamental type of water-bearing rock. So far, very different definitions of this concept have been given by various research workers.

According to the available definitions (M.E. Al'tovskii, A.M. Ovchinnikov, A.S. Ryabchenkov, R.J.M. De Wiest, and others) what must be understood by the term aquifer is a

Table 2.6. *Average values of the coefficients of filtration and permeability of some rocks for the conditions of movement of water at 20 °C (after N.A. Plotnikov)*

Rock characteristics	K (m/day)	K_p (d)
Very well permeable (pebble gravels with coarse-grained sand, highly karsted limestones and highly fissured rocks)	100–1000 and more	<116–1160
Well permeable (pebbles and gravels partly packed with fine sand, coarse-grained sand, pure medium-grained sand, fissured and other rocks)	10–100	11.6–116
Permeable (pebbles and gravels mixed with fine sand and partly with clay, medium-grained and fine-grained sands, poorly karsted, not very well fissured, and other rocks)	1–10	1.16–11.6
Semi-permeable (fine-grained sands, sandy loams, poorly fissured rocks)	0.1–1	0.12–1.16
Slightly permeable (loams, very poorly fissured rocks)	0.001–0.1	0.0012–0.12
Almost impermeable (clays, dense marls, and other massive rocks with insignificant permeability)	0.001 and less	<0.0012

relatively consistent and unified (in the hydraulic sense) thickness or stratum of permeable rock, the pores, fissures, or voids of which are filled with groundwater.

An aquifer is usually composed of homogeneous or nearly homogeneous rock from the lithofacies point of view. Its assignment to a definite unit of the stratigraphic sequence is by no means compulsory. It may consist of one or several layers of water-saturated rocks; in the first instance it will be a single rock bed, in the second it will be complex, double or multi-layered.

The aquifer rocks that make up the aquifer as a whole must have a single hydraulic surface which is free, or piezometric. It is isolated from higher or lower lying aquifers by aquicludes. This does not of course exclude the possibility of hydraulic connection of separate aquifers among themselves, for example in places where the wedging out of aquifer rocks occurs, or in the presence of hydrogeologic 'windows' (faults, tectonically weakened zones, etc.).

The following fundamental varieties are distinguished on a basis of the nature of the bedding and the hydrogeodynamic features:

(1) Groundwater horizons starting from ground level — the first layer of rock constantly fed with infiltration water, and containing water with a free piezometric surface, lying above an aquiclude, or above rocks of a different degree of permeability.

(2) Groundwater horizons in interstratified deposits enclosed between aquicludes — confined aquifers; they can be under no pressure, or under pressure (artesian).

Groundwater occurring in the different parts of the section must be assigned to different water-bearing horizons. However, regarding this point, opinions vary about the necessity to divide, for example, water-bearing alluvium and the groundwater zone of intensively fissured bedrock which underlies it into two distinct bodies of groundwater. A.S. Ryabchenkov (1968) and I.K. Zaitsev (1971) recommend that this should be done, but G.Ya. Bogdanov & V.M. Kononov (1975) combine them into a single groundwater zone if they form a single hydrologic unit. The opinions of research workers on the maximum thickness of a water-bearing layer differ in precisely the same way.

A *water-bearing complex* must be considered as a much larger unit of the hydrogeological section. It is a formation of water-bearing rock in which it is impossible to designate independent water-bearing horizons because of rapid changes in the composition and properties of the rocks with depth in consequence of the complex structural features of the geology, and often because it is difficult to study the formation (Anon., 1971).

According to G.Ya. Bogdanov & V.M. Kononov (1975), a water-bearing complex is a consistent thickness of rocks of the same or different ages and the same or different types, which is bounded by similar or other confining beds. These hinder or exclude hydraulic connection with adjacent water-bearing complexes and thereby furnish the given water-bearing complex with a hydrogeodynamic and partly hydrogeochemical individuality. In the opinion of these authors, a

water-bearing complex consists of a number of water-bearing horizons which are to differing degrees self-contained, and in which the head of the groundwater changes significantly throughout the section and thus predetermines the degree of hydraulic connection of its separate and individual horizons.

There are also different opinions concerning the volume of a water-bearing complex. It is considered that its thickness must not exceed that of the stratigraphic sub-stages or stage (Pinneker, 1966), sometimes it rises to a series, or even a system (Bogdanov & Kononov, 1975). It is important that the water-bearing complexes should be distinguished according to lithofacies features and the related hydrogeological features.

The largest subdivisions of the hydrogeological stratification of sedimentary and sedimentary-volcanic deposits are known by various names — a water-bearing series, hydrogeological horizon, horizontal aquifer, etc.

Of all the names proposed for the largest subdivision of hydrogeological stratification, the most successful is that due to N.A. Marinov (1961), a *water-bearing* or *hydrogeological formation*. This combines water-bearing rocks which, although they are of different ages and have their own individual laws governing the accumulation, distribution, and formation of groundwater, are lithologically, genetically, and in the infiltration sense, homogeneous.

Water-bearing formations are often divided regionally by confining beds or stratigraphic breaks; they include several similar water-bearing complexes. Each such formation is distinguished from the others by its palaeohydrogeological development, by the general hydrogeodynamic and hydrogeochemical features which are peculiar to it alone.

The following water-bearing formations are considered to be fundamental in a section through a cover of epi-Hercynian platform formations:

(1) sandy-clayey friable rocks with pore-stratal water;

(2) sedimentary-volcanic cover, or layers with fissure-stratal and stratofissure water;

(3) indurated sandy-clayey rocks with fissure-stratal water;

(4) carbonate formations with fissure-stratal or karst-stratal water;

(5) salt deposits containing stratified water-bearing horizons between saliferous beds.

Sometimes several similar water-bearing layers alternate in a section and are divided regionally by confining beds. In particular, such a situation is characteristic of a section through the Western Siberian artesian basin, which is the largest in the world.

The water-bearing capacity of crystalline rocks is characterized by great variety, which arises because of the type of fissuring and the degree to which they are fissured. Magmatic and metamorphic aquifers can in the first place be assigned to the upper weathered horizons of these rocks, the zone of intensive fissuring or macroporosity formed by the exogenic processes (weathering, leaching, etc.) and characterized by water in fissures (Zaitsev, 1971): this is the zone of regional rock fissuring. In the second place there is the system of differentiated pores, the zone of tectonic disturbances (faults), of intrusive igneous bodies, and the degree of fissuring which accompanies them, which can be traced down from the surface to the depths and contain water of the vein type (I.K. Zaitsev calls this the zone of localized rock fissuring).

As a consequence of the irregular fissuring of crystalline rocks, the above-mentioned capacities of the water-bearing rocks are usually distributed irregularly throughout the section. Sometimes there are no hydraulic connections between them. Because of this some hydrogeologists consider the division of water-bearing sections instead of groundwater zones to be more expedient. On the other hand, other research workers by analogy with sedimentary deposits propose calling both kinds of groundwater bodies water-bearing horizons, which for the zone of intensive fissuring is in many cases not without foundation, but for the faulted zone or the regular confining zone this is, more often than not, unacceptable. Referring to the water infiltration into the crystalline rocks being considered, the most well established term in hydrogeological research is the *water-bearing* or *permeated zone*.

The permeated zone consists of magmatic or metamorphic rocks which are quite closely fissured to various degrees. Its shape is also different and its attitude ranges from the horizontal to the vertical. Containing beds are usually compact and massive varieties of these same rocks which have not undergone fissuring.

Groundwater in the zone of intensive exogenic fissuring is of wide distribution. Individual accumulations to some extent connect with each other and as a rule are characterized by the absence of a head of pressure. The permeated thickness varies enormously — from several metres up to several tens of metres, sometimes more than 100 m.

By way of contrast, groundwater in the zones of tectonic disturbance and intrusions is characterized by its localized and usually linear distribution, forming vein-type aquifers. The width of such 'veins' may reach 0.5–1 km. Fault zones in shallow deposits (up to 100–200 m) contain fresh water, but the deeper deposits contain mineral, thermal, carbonated, and saline waters, etc. These are under high pressure; when the aquifers are penetrated water gushes from the boreholes.

Water-bearing fault zones, besides occurring in crystalline rocks, are found in indurated sedimentary rocks. For example they are widely distributed throughout the sedimentary cover of the Siberian shield.

Thus the laws governing the localization of groundwater in sedimentary rocks depend mainly on the lithofacies of the rocks. In crystalline rocks the lithology is relatively insignificant and of second order importance, whilst fissuring is the major factor. There are, of course, other factors, in particular the location of the water-bearing rocks (see Fig. 2.7). One could very tentatively divide this groundwater realm into conductors of water as channels of supply, and storage or accumulator aquifers. However, it is possible for one and the same water-bearing rock formation to act both as a conductor and as an accumulator, as in places where outcropping catchment area rocks dip inwards as elements of a basin structure below an impermeable confining bed.

References

Anon. (1968). *Pore solutions and methods of studying them.* Minsk: Nauka i tekhnika. 231 pp.

Anon. (1970). *The hydrogeology of the USSR. Vol. 20, Yakutsk ASSR.* Moscow: Nedra. 383 pp.

Anon. (1971). *Dictionary of hydrogeology and engineering geology.* Moscow: Nedra. 216 pp.

Anon. (1974). *The world water balance and the water resources of the Earth.* Leningrad: Gidrometeoizdat. 683 pp.

Bogdanov, G.Ya. & V.M. Kononov (1975). Some problems of hydrogeological stratification. *Izv. vuzov. geologiya i razvedka*, No. 2, 99–104.

Chebotarev, A.I. (1975). *General hydrogeology.* Leningrad: Gidrometeoizdat. 524 pp.

Derpgol'ts, V.F. (1962). Major sources of the natural waters of the Earth. *Izv. AN SSSR. Ser. geol.*, No. 11, 18–31.

Fedoseev, I.A. (1974). Hydrosphere: its boundaries and water masses. *Izv. AN SSSR. Ser. geograph.*, No. 2, 24–32.

Gavrilenko, E.S. & V.F. Derpgol'ts (1971). *The deep hydrosphere of the Earth.* Kiev: Naukova dumka. 272 pp.

Kalinin, G.P. (1968). *Problems of global hydrogeology.* Leningrad: Gidrometeoizdat. 377 pp.

Klimentov, P.P. & G.Ya. Bogdanov (1977). *General hydrogeology.* Moscow: Nedra. 357 pp.

Kortsenshtein, V.N. (1977). Evaluation of the global resources of dissolved gases in the subsurface hydrosphere. *Dokl. AN SSSR*, 235 (2), 448–9.

Lebedev, A.F. (1936). *Soil- and groundwater*. 4th edn. Moscow, Leningrad: Izd-vo AN SSSR. 312 pp.

L'vovich, M.I. (1968). The scientific basis for the general use and preservation of water resources. In *Water resources and their general use*, pp. 3–32. Moscow: Mysl'.

L'vovich, M.I. (1974). *World water resources and their future.* Moscow: Mysl'. 447 pp.

Makarenko, F.A. (1948). Results of the study of the underground runoff. *Trudy Laboratorii gidrogeol. problem im. F.P. Savarenskogo* Vol. 1, pp. 51–66.

Makarenko, F.A. (1966). Water in the Earth. In *The water cycle*, pp. 86–104. Moscow: Znanie.

Makarenko, F.A., V.A. Il'in, & V.N. Kononov (1972). Physical model of the subsurface hydrosphere. In *Hydrogeology and engineering geology*, pp. 15–25. (Moscow: Internat. Geol. Congress 24th Session. Dokl. Sov. geologov).

Maksimov, V.M. (ed.) (1959). *Reference manual of hydrogeology*, 1st edn. Leningrad: Gostoptekhizdat. 836 pp. (2nd edn in 2 vols, Leningrad: Nedra. Vol. 1, 592 pp., Vol. 2, 360 pp.)

Marinov, N.A. (1961). Hydrogeological formations. *Razvedka i okh. nedr.* No. 8, 40–3.

Meinzer, O.E. (1923). Outline of groundwater hydrology with definitions. *US Geol. Surv. Water Supply Paper*, No. 494, 1–71.

Meinzer, O.E. (1935). *Groundwater studies.* Leningrad, Moscow: ONTI. 240 pp.

Myatiev, A.N. (1947). Pressure environment of groundwater and wells. *Izv. AN SSSR, otd. tekhn. nauk.*, No. 9, 1069–88.

Ovchinnikov, A.M. (1955). *General hydrogeology.* Moscow: Gosgeoltekhizdat. 383 pp.

Pinneker, E.V. (1966). *Brines of the Angara–Lena artesian basin.* Moscow: Nauka. 332 pp.

Poldervaart, A. (1957). The chemistry of the Earth's crust. In *The Earth's crust*, pp. 130–57. Moscow.

Povarennykh, A.S. (1966). *Crystallochemical classification of minerals.* Kiev: Naukova dumka. 547 pp.

Ryabchenkov, A.S. (1968). The construction of hydrogeological field maps. In *Methodological instructions for hydrogeological survey on scales 1 : 500 000, 1 : 200 000, and 1 : 50 000.* pp. 150–4. Moscow: Nedra.

Savarenskii, F.P. (1935). *Hydrogeology.* Moscow, Leningrad: ONTI. 336 pp.

Savarenskii, F.P. (1947). The principles of hydrogeological zonation. *Sov. geologiya*, No. 19, 19–23.

Sergeev, E.M. et al. (1971). *Ground management.* Moscow: Izd-vo MGU. 596 pp.

Smirnov, S.I. (1971). On the relationship of molecular and filtration diffusion of minerals in groundwater. *Trudy VNII gidrogeol. i inzh. geol.*, collection 21, 12–36.

Sokolov, A.A. (1975). How much water is there in the Earth? In *Man and the elements.* Leningrad. 63 pp.

Stepanovič, B. (1962). *Principi opšte hidrogeologije*. Beograd: Rad. 144 pp.

Tolstikhin, N.I. (1971). Classification of groundwater. *Zap. LGI*, 12 (2), 3–15.

Tolstikhin, O.N. (1974). *Icings and groundwater of the north-east USSR*. Novosibirsk: Nauka. 164 pp.

Tolstoi, M.P. & P.F. Shvetsov (1977). The hydrosphere and its limits – another appraisal. *Izv. AN SSSR Ser. geograph.*, No. 3, 90–3.

Tugarinov, A.I. (1973). *General geochemistry*. Moscow: Atomizdat. 288 pp.

Vel'mina, N.A. (1970). *Features of the hydrogeology of the frozen zone of the lithosphere*. Moscow: Nedra. 326 pp.

Vernadskii, V.I. (1934). *Geochemical essays*, 4th edn. Moscow, Leningrad, Georgia, Novosibirsk: Gorgeonefteizdat. 380 pp.

Vernadskii, V.I. (1934). History of natural waters No. 2 (Vol. 2 of *History of the minerals of the Earth's crust*). Leningrad: ONTI. 202 pp.

Vernadskii, V.I. (1960). History of natural waters (Vol. 2 of History of the minerals of the Earth's crust). In *Vernadskii, V.I. Collected works*, Vol. 4, Book 2. Moscow. 651 pp.

Zaitsev, I.K. (1971). The principles of hydrogeological stratification. In *Problems of hydrogeological mapping and zonation*, pp. 45–55. Leningrad.

3

The origin of the water in the Earth's interior

3.1 The evolution of the Earth's hydrosphere

Primary sources of water. The problem of the evolution
of the hydrosphere occupies scientists of the most varied
branches of knowledge. Modern concepts of the origins of the
water in the Earth's interior were formed in the middle of the
twentieth century thanks to the work of A.C. Lane, B.L.
Lichkov, W.W. Rubey, N.M. Strakhov, J.L. Kulp, L.A.
Lenkevich, Ph.H. Kuenen, A.B. Ronov, V.F. Derpgol'ts, H.H.
Hess, and many others. A more comprehensive basis belongs to
A.P. Vinogradov (1959, 1962, 1967), whose ideas are now
shared by Soviet and foreign scientists.

The origin of the hydrosphere is indissolubly linked with
the development of the Earth itself as a planet. The generally
accepted hypothesis of O.Yu. Schmidt and others is the formation of our planet from dispersed material followed by
gradual heating of the initially cold Earth, and the differentiation of it into outer shells or envelopes.* The hydrosphere
is one such shell. It arose simultaneously with the hot rocks as
a result of melting and irreversible degassing of the material of
the mantle which A.P. Vinogradov likened to the mechanism
of zone refining, which is a process widely used in industry for
the removal of impurities during the purification of metals.

The separating out of the shells of the Earth began at the
dawn of geological history 4–5 billion years ago. As A.P.
Vinogradov has proposed, during the heating of the materials
of the mantle as the result of energy from the decay of radioactive elements it became divided into two phases: refractory
peridotite (which forms the substrate of the lithosphere) and

---

*It is the most widely held viewpoint, but not the only one. V.I.
Ferronskii (1974), following V.M. Gol'dshmidt and others considers that the hydrosphere was formed from the initial hot
Earth at the last stage of its formation during the gravitational
compression of the gas cloud. From these points of view, the
mantle must be, in the opinion of some petrologists, 'dry'. However, such a concept has not gained wide acceptance.

easily fusible basalt (which forms the Earth's crust). During this process not only the light but the most volatile components of the basaltic magma, water vapour and gases, rose to the surface of the Earth. The initial crust of the Earth was formed from the basalt. As it approached the surface of the Earth the fusible fraction continued to be enriched with water. When the supercritical conditions no longer existed the bulk of the water gradually became transformed to the liquid state, taking dissolved substances with it to the surface.

The picture of the appearance of the hydrosphere is sketched as follows. The original liquid water was given the name *juvenile* (i.e. virginal), being the first water created from hydrogen and oxygen.

The mantle is the source of all natural water. In it is contained about 20×10^{18} tonnes of water in all states (Vinogradov, 1959), of which from 7.5 to 12.5% ($1.5-2.5 \times 10^{18}$ tonnes), and according to some scientists, 17–24% ($3.8-4.8 \times 10^{18}$ tonnes) has migrated to the Earth's crust and the world ocean, i.e. has formed the hydrosphere. It can be provisionally accepted that 3.4×10^{18} tonnes of water have evaporated from the mantle.

Another source is the cosmos, or more accurately asteroids from the solar system which fall on to the Earth. The cosmic source comes via meteorites which contain on average 0.5% by weight of water. In all about 1.0×10^{14} tonnes of water have fallen on to the Earth (Derpgol'ts, 1962), which, with respect to geological time, is four orders of magnitude less than the water that has come from the mantle.

The upper layers of the atmosphere may be another potential source of water for the Earth. V.I. Vernadskii has pointed out the possibility of the formation of aqueous accumulations in the so-called 'silvery clouds' at a height of more than 80 km. Now research workers with the help of artificial satellites in the upper layers of the atmosphere have established that at a height of 230–250 km there are atoms of hydrogen and oxygen and the formation of molecules of water is not excluded.

In comparison with the mantle, the cosmos and the upper layers of the atmosphere have given very little water. It is literally 'a drop in the ocean'. A vastly greater amount of water is lost from the Earth to interstellar space. Thus the mantle is practically the only source of water on the Earth.

The evolution of the hydrosphere. When the evolution of the hydrosphere is discussed, one has in mind primarily the history of the world ocean, its birth and development, and its establishment.

The bulk of the products of the degassing of the mantle was ejected, probably at the dawn of geological history of the Earth in the first hundreds of millions of years. These products began to arrive at the surface in the form of gases and water through the forming crust. As a result of the flat relief of the initial basalt crust the water covered almost the whole of the Earth's surface, i.e. the initial ocean was truly a world ocean, but it was distinguished by being much less deep than today.

According to the calculations of P.N. Kropotkin (1964), about 90% of the total volume of the hydrosphere appeared in the Archaean era, and only 10% in the subsequent period of geological time. From the above figures, which are usually put forward as evidence of the practically single event of the degassing of the interior of the Earth, the inevitable conclusion which comes to mind is that the world ocean is very ancient indeed.

From these points of view the development of the Earth's crust went through the sequence of the building up of dry land and the transforming of the crust from oceanic to continental masses. However, the floor of the world ocean is a comparatively young formation. The recent deep-water depressions date only from the Jurassic and there are no rocks of a greater age on the floor of the ocean.

After the Early Proterozoic, contemporaneously with the end of the nuclear, i.e. the early stage of the development of the crust and the beginning of the formation of shield areas, the output of water from the mantle decreased noticeably. Water passed through the crust in overwhelming quantities and reached the surface, and this led to the gradual growth of the world ocean. This growth is still continuing, and will continue in the future (Monin, 1977).

Not all the water was retained in the subsurface and surface hydrosphere. If 3.4×10^{18} tonnes of water have separated from the mantle since the Earth was formed, then according to J.L. Kulp (1951) 1.0×10^{18} tonnes have dissociated into hydrogen and oxygen, or have evaporated and vanished into space.

Taking the weight of the Earth's interior to be 47×10^{18} tonnes (Monin, 1977), and the quantity of water given out by the mantle to be 3.4×10^{18} tonnes, then the fraction of juvenile water in relation to the molten material is 7%. Almost the same quantity of water vapour came from volcanic eruptions. The figure of 7%, according to A.P. Vinogradov, quite accurately reflects the degree of permeation of the hypothetical primitive upper mantle material by pyrolite, which, according to A.E. Ringwood, participated in the formation of the crust and hydrosphere. Such a degree of permeation probably affected the characteristically softened and plastic rocks of the asthenosphere (the Gutenberg layer), and at the same time the rock at the base of the lithosphere, which, above the asthenosphere, is essentially dehydrated and contains even less water than the basalt layer of the crust i.e. about 0.2–1.0% (see Fig. 3.1).

In the opinion of a number of scientists (A.P. Vinogradov, N.M. Strakhov, A.B. Ronov, and others) dynamic equilibrium was established between the water on the one hand, and the issuing gases on the other hand about 2.5 billion years ago. It was then that, at the boundary of the Archaean and the Proterozoic, the most intensive granitization and the formation of the granite layer, the division of the Earth's shields and geosynclines took place, and also the inland seas were formed. All this laid down the beginning of an atmosphere and the regular circulation of water in nature. The hydrosphere is an early product of the differentiation of the material of the Earth in the evolution of which exogenic processes dominated (Sidorenko & Borshchevskii, 1979).

There is another point of view according to which the

hydrosphere was created gradually. In opposition to the supporters of the theory that practically all the water mass was formed in the Archaean, W.W. Rubey (1951) and Y. Miyaki (1969) have proposed that the process of the degassing of the mantle proceeded at more or less a constant rate throughout the whole of geological history. In the opinion of these scientists the renewal of the volume of the hydrosphere is proceeding at this same rate even today.

Independently of the creation of the hydrosphere, including the world ocean, there existed a well known constancy in the relationship between the initial water and the gases which were emitted during the degassing of the mantle, mainly acidic vapours. The juvenile solution, having first risen to the Earth's surface as a consequence of the neutralization of the acidic vapours by the bases in the consumed rocks, already contained salts. However, its composition was well defined.

As A.P. Vinogradov (1967) has stated, volatile mantle materials (HCl, CO, CO_2, CH_4, S, H_2S, NH_3, HF, HBr, HI) became the sources of the anions in the saline fraction of the ocean water (Cl^-, SO_4^{2-}, Br^-, etc.). All the important cations (Na^+, K^+, Ca^{2+}, Mg^{2+}) were formed by the destruction of the rocks and only partly introduced as a result of the processes of melting and degassing of the mantle. At an early stage there was almost no oxygen in the hydrosphere, hence easily oxidizable components could still be found in the water. The main component of the dissolved gases was CO_2, followed by the hydrogen compounds (NH_3, CH_4, H_2S).

The initial (juvenile) water, having arrived at the Earth's surface, tried to adjust to the new conditions and come into equilibrium with them. What was its composition? It is difficult to answer that question. Using the data on the volatile components in chondrite meteorites, which are characteristic of mantle material, M.G. Valyashko (1971) attempted to establish the composition of the initial hydrosphere. If it is accepted that the volatile components were completely dissolved in the initial water, it is possible to obtain an idea of its composition (Table 3.1). It is difficult to make a judgement on how true this interpretation is.

Figure 3.1. Changes in H_2O content of crustal rock and upper mantle.

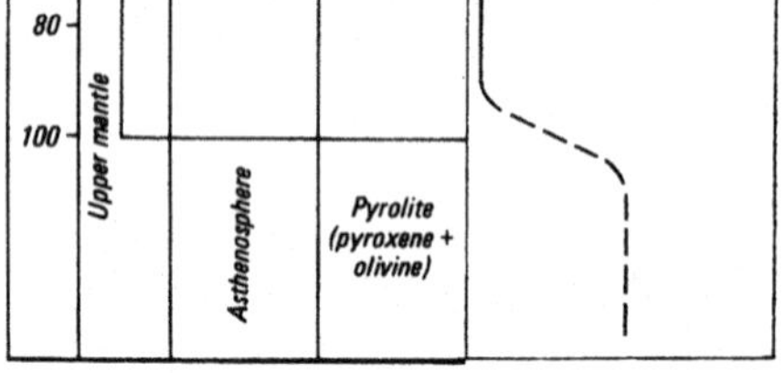

Changes in the composition and salinity of the ocean water. A comparison of the composition of contemporary ocean water with the hypothetical juvenile solution (see Table 3.1) shows that the chlorine and bromine contents have remained practically unchanged. The amount of the remaining components has reduced noticeably, and the carbon content most of all. Carbon loss is an important process which has proceeded in the world ocean throughout the whole of geological history, and is characteristic of it. The essence of the process of adjustment of the initial solution to the new conditions consisted of the retention of those components which formed easily soluble compounds, and the precipitation of those components of transitional compounds and those which were difficult to dissolve.

The composition of the oceanic water was formed after the stabilization of the oxygen atmosphere, when the easily oxidizable products had finally disappeared. The hydrogen of the atmosphere appeared, probably in the late middle Archaean, as a result of the photo-dissociation of water and other oxygen-containing compounds. In general the bulk of gases of the contemporary atmosphere are of biochemical origin, and are the result of the activity of plants. Gases which are now dominant in the atmosphere (N_2, O_2, and particularly CO_2) are manufactured. The present day features of the atmosphere were created at the end of the Silurian to the start of the Devonian period, when the land began to be colonized by land plants.

The separation of low solubility compounds to form deposits of precipitates, and oxidizing and biochemical processes led to the controlled evolution of the whole hydrosphere. The removal of carbon dioxide was caused by its transfer to sediments and its incorporation into the living material which was being formed. At the same time the nitrogen content fell (this was mainly connected with the development of life on Earth and the transfer of nitrogen to the atmosphere), and so did fluorine (which forms mineral compounds which do not easily dissolve), and iodine (which migrated with the organic material). The weathering of rocks played an important role; with the breaking down of basalt, more magnesium was released than in the case of the weathering of granites and the ancient sedimentary rocks. The last were the sources of the potassium, sodium, and calcium of the hydrosphere. In the Proterozoic era a surplus of oxygen brought about the disappearance of ammonia and the intensive

Table 3.1. *Comparison of compositions of hypothetical juvenile solutions and present day sea water (Valyashko, 1971)*

Component	Chondrite (%)	Initial solution (g/100 g)	Water of present-day ocean (g/100 g)
H_2O	0.5	–	–
C	4×10^{-2}	8	3.5×10^{-3}
F	$2-8 \times 10^{-5}$	6×10^{-3}	2.4×10^{-5}
Cl	7×10^{-3}	1.4	1.9
Br	5×10^{-5}	1×10^{-2}	7×10^{-3}
I	4×10^{-6}	8×10^{-4}	5×10^{-6}
B	4×10^{-5}	8×10^{-3}	4×10^{-4}
N	2×10^{-3}	4×10^{-1}	5×10^{-2}

formation of sulphates. With the appearance of oxygen in the water of the world ocean, life began to evolve.

The present day state of ocean water has been created gradually. However, the question of the constancy (at least over the last 500–600 million years) of the volume, salinity, and composition of the world ocean cannot be considered as solved and proved. V.I. Vernadskii in *The history of natural waters*, defending the concept of the constancy of the salinity and the composition of oceanic water, drew attention to the invariability of the average value only. This has now been further developed; some research workers (N.M. Strakhov, M.G. Valyashko, and others) do not perceive any changes in the composition of the water of the world ocean since the Cambrian, but others (A.L. Yanshin, Yu.P. Kazanskii, and others) on the other hand, propose that it has undergone variations and considerable changes. Those holding the second view are probably right: it is possible to put forward a multitude of facts which demonstrate the evolution of the world ocean during the Phanerozoic. We will confine ourselves to the fundamentals of the matter.

Beginning with the Cambrian, the volume of the oceanic water has increased by 2–3% (Kuenen, 1955), but the consequences of such a comparatively small increase have been enormous, because without it the level of the world ocean would be more than 100 m lower than it is today. Life in the ocean underwent a fundamental reconstruction; this alone bears witness to the changes in salinity and the composition of the ocean.

Opinions diverge concerning the original salinity of oceanic water. It is not clear whether it was greater or less than it is today. But there is no doubt that it has been continually changing, at times quite markedly. As A.L. Yanshin (1962) has proposed, halogenesis served as a regulating mechanism. In the Precambrian the salinity of the water was less than that of the present day ocean water, and then it rose, but after every great epoch of salt accumulation (in the Cambrian, Devonian, Permian, etc.) it fell, and again increased, until the advent of the next epoch of salt reduction. Apparently, the salinity of the water of the world ocean varied with time about some

average value, which can be provisionally taken as the magnitude of the present day degree of mineralization (Table 3.2).

Changes in the composition of the water of the world ocean (Table 3.3) had a controlled character. Along with the loss of carbon dioxide in the Phanerozoic a reduction in the calcium content took place, and this was compensated for by the increase of the concentration of sodium and magnesium. A.C. Lane (1945) pointed out the tendency of the Na^+/Cl^- in the water of the world ocean to increase. According to his calculations, it was 0.30 in the Early Palaeozoic, 0.45 in the Late Palaeozoic, 0.5 in the Mesozoic, and is 0.55 in present day ocean water. In the Precambrian ocean by comparison with the present day situation, it is proposed that there was a deficiency of sulphate and an excess of hydrogen carbonate. A characteristic feature of the Phanerozoic is the increase in concentration of sulphate. Because the environment was more acid than it is today, iron and other metals were present in great quantities.

The biogeochemical formation of compounds of calcium caused by the appearance of a fauna with skeletons, was accompanied by a loss of carbon dioxide. Calcium is being lost even today. In the opinion of N.N. Gorskii (1962), the store of calcium in the ocean will be run down to such an extent that many marine organisms will have to do without it or perish.

The change in the potassium concentration appears to be different. In the process of the evolution of the world ocean it initially increased continually (from the Archaean to the Late Precambrian), but in the Phanerozoic it fell gradually to the value obtaining in the Archaean. The cause of the changes remains to a great extent unexplained.

Thus the ion-salt composition of the ocean water even in the last 500–600 million years has not remained stable. Some ions have been replaced by others, and changes in the pH and Eh of the environment and other parameters have determined the varied character of the marine sediments in the history of the Earth. Yu.P. Kazanskii (1977) has established five hydrogeochemical types of ocean water, which have interchanged one with the other during the evolution of the hydrosphere from the Archaean to the Cenozoic, and as can be seen from

Table 3.2. *Ionic composition of present day ocean water, after S.V. Bruevich (1966), for salinity 35% or degree of mineralization 36g/l*

Anions				Cations			
	Content				Content		
Component	%	mg eq.	% eq.	Component	%	mg eq.	% eq.
Cl^-	19.353	545.82	90.0	Na^+	10.764	468.18	77.5
SO_4^{2-}	2.701	56.23	9.2	Mg^{2+}	1.297	106.67	17.6
HCO_3^-	0.143	2.34	0.4	Ca^{2+}	0.408	20.36	3.3
CO_3^{2-}	0.070	2.33	0.4	K^+	0.388	9.92	1.6
Br^-	0.066	0.83	–	Sr^{2+}	0.014	0.31	–
F^-	0.001	0.053	–	$M_{36} \dfrac{Cl90SO_49}{Na77Mg18}$ pH 7.5			
H_3BO_3	0.026	–	–				

Table 3.3, the present day (sulphato-) chloride (magnesium-) sodium composition appeared in the Permian.

With the transgressions and regressions and variations in the level of the world ocean and the associated epicontinental seas, a great deal of salt remained in the rocks. In turn the surface and underground streams carried no smaller quantity of salts to the world ocean. In this way salt exchange began: it existed in the past and continues today between the ocean and the subsurface hydrosphere.

The enormous water mass of the world ocean possesses a colossal amount of energy and reacts sluggishly to external influences. Its composition therefore reflects the conditions of bygone geological epochs. There is no doubt that this is the reason for the constancy of the isotopic ratios $^2H/^1H$ and $^{18}O/^{16}O$ of ocean water over the last 500–600 million years, which are used as a standard mean ocean water SMOW (Craig, 1961; Ferronskii, 1974).

3.2 Views on the formation of the subsurface hydrosphere

The data on the genesis of the hydrosphere is fragmentary and is in many respects contradictory. The problem itself is not only complex but is also the least studied in hydrogeology. In essence it has not yet got out of the stage of hypotheses and conjectures and this forces us to limit its exposition to the most general statements.

The role of juvenile fluids. Since Suess (1902) expressed the idea of juvenile water the arguments about this have never ceased. Nowadays, as we have seen, the mantle is considered to be the source of water, and the generation of water molecules is associated with (and in this the opinions of scientists are to some extent uniform) the most volatile products of the degassing of the material of the mantle. These are called *fluids* (see Section 2.2).

A fluid is not a definite concept. Both multi-component gas–liquid solutions and material in the supercritical state in which the differences and the features associated with the phases liquid and gas have been obliterated are called fluids. This phenomenon is associated with great depths and high

Table 3.3. *Changes in the ionic composition of sea water (after Kazanskii, 1977)*

Age	Ions	
	Dominant	Subordinate
Cenozoic–Permian	$\dfrac{Cl^-, SO_4^{2-}}{Na^+, Mg^{2+}}$	$\dfrac{HCO_3^-, CO_3^{2-}, F^-}{Ca^{2+}, K^+}$
Carboniferous–Silurian	$\dfrac{Cl^-, HCO_3^-}{Na^+, Mg^{2+}}$	$\dfrac{SO_4^{2-}, CO_3^{2-}, F^-}{Ca^{2+}, K^+}$
Ordovician–Cambrian	$\dfrac{Cl^-, HCO_3^-}{Mg^{2+}, Na^+, Ca^{2+}}$	$\dfrac{SO_4^{2-}, CO_3^{2-}, F^-}{K^+}$
Late–Middle Proterozoic	$\dfrac{HCO_3^-, Cl^-}{Ca^{2+}, Mg^{2+}, Na^+}$	$\dfrac{CO_3^{2-}, SO_4^{2-}, F^-}{K^+, NH_4^+}$
Early Proterozoic	$\dfrac{HCO_3^-, CO_3^{2-}, Cl^-}{Ca^{2+}, Mg^{2+}, NH_4^+}$	$\dfrac{SO_4^{2-}, F^-}{Na^+, K^+}$

pressures, and according to the apt remark of A.P. Vinogradov, is a kind of 'geological plasma'.

In the light of present day concepts of the fluid regime, the mantle at the dawn of geological history gave rise to chemically reducing fluids in which H_2 or hydrogen compounds of the CH_4, NH_3 types, etc. were dominant. The most probable mechanism of formation of water is the oxidation of hydrogen during its interaction with aluminosilicates, silicates, and oxidizing agents (Letnikov, 1977). The quantity of water liberated rose quickly, so that at the beginning of the Proterozoic this led to the change in the fluid regime and to the appearance of an oxidizing fluid.

During the Archean, water was absorbed mainly by the formation of the Earth's crust and was only partly ejected on to the Earth's surface and into the world ocean. The process of hydration of the rock-forming minerals must be considered as the fundamental absorbing mechanism. The granite layer in the continental crust was formed with the participation of water and serpentinization of the basic oceanic crust took place. If the action of the juvenile fluids on the granitization is set aside by the comparatively great thickness of the continental crust, then the direct action on the oceanic crust of the serpentinizing solutions which penetrated downwards becomes clearer.

$$4(Mg,Fe)SiO_4 + 4H_2O + 2CO_2 \rightarrow (Mg,Fe)_6Si_4O_{10}(OH)_8$$

olivine   serpentine

$$+ \quad 2(Mg,Fe)CO_3$$

magnesite, siderite

There have even been attempts to evaluate the quantity of water expended in serpentinization. According to the calculations of A.N. Pavlov (1977), it turns out to be 3.6×10^{16} tonnes, or about 3% of the total volume of the present day world ocean. Although this figure seems small, the geochemical consequences of serpentinization are quite great, because this process is responsible for the entry of considerable quantities of magnesium, iron, and other elements into the ocean water.

At the boundary between the Archaean and the Proterozoic fundamental changes took place. Apart from the juvenile water, *vadose* water began to acquire great significance in the upper sedimentary layer, i.e. 'near-surface' water, the appearance of which is responsible for processes of sediment accumulation in sedimentary basins and the regulated circulation of water in nature. This water may penetrate also into the deepest horizons or participate in the process of metamorphism. Simultaneously with metamorphism, in conjunction with the reduction of temperature and pressure, chemically bound water began to separate out from minerals of magmatic origin. The transport of fluids from the mantle, apparently, changed both qualitatively and quantitatively at this time.

F.A. Letnikov (1977) pointed out the unique ability of hydrogen to migrate through dense rocks. The interaction of H_2O with elements of variable valency and the appearance of water in the fluid led to the heterogeneity of rocks, the reduction of their strength and consequently the appearance of brittle and plastic deformations. In this case it is as though hydrogen prepared the road for a rising current of fluid.

It has already been noted that in post-Archaean time the quantity of water liberated from the mantle fell by a large factor. According to some evaluations (Kuenen, 1955; Sydykov, 1973), it did not exceed the present day value, i.e. $0.1-1.0$ km^3/year. If during the Archaean era fluids 'penetrated' through the Earth's crust, penetrating it more or less evenly, then in the following eras the ascending current became more and more localized to the zones of weakness. The body of magmatic intrusions and the deep fractures formed in the Proterozoic are similar conductors of fluids (Pospelov, 1963). Not for nothing is the magmatic body called a conduction path along which water, which has entered into the melt at great depths, is brought to the surface regions (Kadik *et al.*, 1971). As for the persistent deep fracture zones, they function even in plastic rocks.

Thus in the geological history of the Earth water has been continuously generated at the most varied levels (in the mantle and the crust), and in different currents (local and regional). Having separated from the magma and possessing great dissolving power, it took over the role of universal carrier of chemical elements both in taking them out from the magma and distributing them through the crust.*

The mechanism of the transport of fluids, in the first instance water, from the mantle to the upper horizons of the crust has not yet received unified treatment. Together with the recognition of the leading role of fluid conduction of the zones of weakness, certain research workers believe in the presence of compact restricted 'filtering' currents (Grigor'ev, 1971), or diffusing currents (Gavrilenko & Derpgol'ts, 1971). Opinions on the interaction of the rising juvenile solutions with the rocks also differ.

The hypothesis of a drainage shell or envelope. One of the original although contentious attempts to discover the character of the interaction between the water of the interior of the Earth and rock is S.M. Grigor'ev's hypothesis. This ascribes to water the role of the important factor in the evolution of the crust, the fundamental driving force in the formation of its materials and of the processes which take place within it.

Using the idea of V.I. Vernadskii concerning the influence of water on the course of geological processes and the present day views on the building of the Earth's crust, S.M. Grigor'ev expressed the opinion that a special drainage shell existed in the interior of the Earth, which in essence also determined the role of water in the evolution of the crust. The drainage shell is situated at the base of the continental crust in the region in which the temperature changes from 374 °C to 450 °C and is assigned to the basalt layer (Fig. 3.2). Above

---

*At the high temperatures and pressures which obtain at the boundary of the crust and the mantle, water possesses a unique dissolving ability. Thus for NaCl at a temperature of 600–700 °C and a pressure of 6–10 kbar, which obtains at the Moho, the dissolving power increases to more than three times that under 'normal' conditions. Water in this case enters as a component part into the fluid or melt.

---

this, because of the action of gravity and the gradual increase in density caused by dissolved components, outgoing currents of free (liquid) water are formed. On reaching the drainage shell, such solutions eject steam upwards, but at a temperature of 425–450 °C themselves change to the vapour state and free themselves from the chemical elements. The steam tries to expand and moves upwards.

Between the isotherms 374 and 450 °C the vertical currents of water are constant (see Fig. 3.2). S.M. Grigor'ev considers that there is also set up, in the drainage shell, a horizontal movement of fluid from under the continent towards the ocean and through the oceanic crust, connecting the formation of volcanoes with the action of the drainage envelope. He calls these volcanoes centres of water discharge.

Material carried away from the land by rivers enters the ocean. Consequently, as S.M. Grigor'ev proposes, the continents become lighter and rise. On the other hand the ocean bed sinks. It is here that the solutions from the drainage shell are conveyed, which transform the sedimentary rocks of the ocean into basalts and mantle material. Hence the drainage shell mainly discharges itself into the ocean. The essence of the changes of the continental crust is granitization: the outflowing solutions remove magnesium, calcium, and iron from the rocks which are located above the drainage shell, and the condensed vapours bring in silicic acid from the drainage shell. Hence the rising basalts change into granites.

According to S.M. Grigor'ev it is the drainage shell which formed the Conrad discontinuity, which separates the granite and basalt layers, and the Moho which is the boundary of the Earth's crust with the upper mantle (caused by the precipitation of compounds of heavy elements and the thickening of the rocks at this level). The depths of both boundaries are determined by the critical temperature of water (374 °C, for the Conrad discontinuity) and the critical temperature of aqueous solutions (450 °C, for the surface of the Moho). The drainage shell has moved throughout geological time: downwards under the continents and upwards under the oceans.

S.M. Grigor'ev's hypothesis is attractive because of its simplicity and its attempt to explain the mechanism of the circulation of material in the Earth's crust and the upper mantle. However, on the one hand the obvious desire of the author to explain complex processes too simply, and at times

Figure 3.2. Diagram of movement of water in the crust (after S.M. Grigor'ev, 1971).
1, original movement of water and aqueous solutions through the continent in the drainage shell; 2, horizontal displacement of water; 3, upward movement of vapours and aqueous solutions through the oceanic crust from the drainage shell; 4, the movement of descending aqueous solutions and rising vapours in the drainage shell.

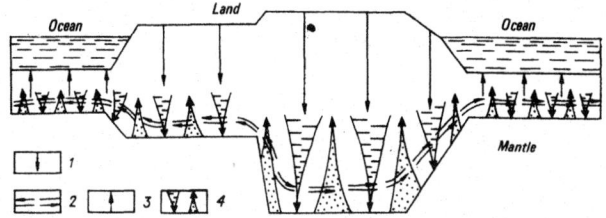

contrary to the facts, is obvious (Pavlov, 1977). For example, the Moho, the origin of which is so easily explained by the operation of the drainage shell, is not always associated with the 450 °C isotherm, but is found in the most varied temperature zones. On the other hand the passage of liquid water and steam may take place only under pressures of not less than 220 atmospheres, while at the depths of the basalt layer they are usually greater (Kafengauz, 1972). Therefore doubts about the possibility of the functioning of a drainage shell are very fundamental. Global tectonics explains the role of water in the evolution of the crust more convincingly.

The phenomenon of juvenile fluids in the light of global tectonics. From the position of modern concepts on the structure and the development of the Earth's crust, the rift zones may be regarded as deep drains. Rifts are deep V-shaped depressions which go like 'roots' into the mantle. They were formed as a result of horizontal stress forces in the Earth's crust. Mantle material is discharged along them. They are long 'slit-like wounds in the body' of the ocean or continental crusts. By comparison with the problematical drainage shell, the more real fluid conductors may be, in particular, the mid-ocean rifts and the Benioff–Zavaristkii zones around the margins of the ocean. In order to illustrate their role it is necessary to examine the basic ideas of the concept of 'spreading', i.e. the moving apart of the floor of the ocean (Hess, 1962), plate tectonics (Le Pichon *et al.*, 1973; Sorokhtin, 1974), and of the system of ideas which has exercised the minds of scientists over the last 10–15 years, and which have been called the new global tectonics (plate tectonics).

This new set of views was born out of the former much respected basis of A. Wegener's hypothesis of continental drift. The cause of the movement of 'neomobile' lithosphere plates is considered to be the convection current of material in the upper mantle. As a result of the stretching forces in the axes of the mid-oceanic ridges, the crust slides downwards and major fractures are formed, i.e. rift zones. Here the two processes of material rising from the asthenosphere and the formation of ocean crust take place. Together with the material from the mantle the condensed juvenile fluids are also discharged. From the rift zones, in both directions under the action of convection currents which act like conveyor belts, the oceanic plates move apart and at the margin of the ocean on the boundary with the continental crust, slide downwards to a depth of up to 120 km under the continental plates. The regions where this occurs have been named the Benioff–Zavaritskii subduction zones. These are sharply defined in the form of troughs or deeps which stretch across the land areas.

In this way one can explain the growth of the continental plates at the expense of the ocean crust which rises and joins them. New ocean crust is produced in the mid-oceanic rift zones. This process has been taking place at least from the Mesozoic and leads to even greater spreading of the ocean.

The new plate tectonics concept, in spite of the objections of its opponents (A. and G. Meinerhoff, G.D. Azhgirei, V.V. Beloussov, and others) provides a good explanation for the relationship between the centres of strong earthquakes and

volcanic activity to the Benioff–Zavaritskii zones. The presence of centres of hydrothermal discharge in submarine rift zones is also explainable by these ideas.

At the beginning of this section attention was drawn to the hydration of the ocean crust by the action of the water from the mantle in the formation of serpentine. With the sliding of the ocean crust under the continental plates and its melting, there also takes place the reverse process of dehydration. Although water is liberated by means of a multistage process (in which one can distinguish not only bound water but also vadose water from the sedimentary layer, and the absorption of oceanic water by the basalts), bound juvenile water also enters the volcanic mechanism from the Benioff–Zavaritskii zones, because the ocean crust in places consists of 70% serpentine which is formed by the aqueous fluid rising from the mantle (Monin, 1977).

The dehydration of serpentine and kaolinite from the second layer of the ocean crust is represented by the following reactions (Hess, 1962; Sorokhtin, 1974):

$$Mg_6Si_4O_{10}(OH)_8 \rightarrow 3Mg_2SiO_4 + Si(OH)_4 + 2H_2O$$
$$\text{serpentine} \qquad \text{forsterite}$$

$$Al_4Si_4O_{10}(OH)_8 \rightarrow Al_2O_3.3SiO_2 + SiO_2 + H_2O$$
$$\text{kaolinite} \qquad \text{sillimanite}$$

At present one can only guess at the subsequent fate of the water which is liberated in the Benioff–Zavaritskii zones. A significant part of it probably reaches the surface or enters the ocean via volcanoes during the emission of andesite lavas, which are formed at relatively shallow depths. According to the depths at which they are situated and the increase in temperature (over 500 °C), the molten material from the ocean crust is squeezed upwards and together with the water takes part in the formation of the granite layer of the continental crust. Finally, some part of the water remains in the upper mantle and is entrained in the subcrustal current.

The schematic diagram of the path by which water enters the crust (according to the new plate tectonics concept) is shown in Fig. 3.3. It is based, according to H.H. Hess (1962) and A.S. Monin (1977), on the formation of the serpentine layer of the ocean crust mainly as a result of rising juvenile fluids. This is in contrast with the ideas of O.G. Sorokhtin (1974) and the model of the geological circulation of water proposed by A.N. Pavlov (1977), in which ocean water is considered to be the agent of serpentinization. To recognize this as the fundamental process is impossible if only because the serpentine layer does not come into direct contact with the ocean water. Of course ocean water would have participated in the serpentinization, but in comparison with the water originating in the mantle was only a small fraction.

Another path by which juvenile fluid can enter is its direct discharge into the rift zones together with the molten mantle material (see Fig. 3.3).

Studies on the rift systems of the Earth point to the possibility of the entry of fluids from the mantle into both the ocean and the continental crusts. It is true that in the continental rifts, because of the reduced cover, the demarcation of the fluids is not so well defined, and even their composition is distinguished by greater oxidizing power – they contain significantly more H_2O and CO_2 (Letnikov *et al.*, 1977). In this case the enrichment of the water, more than anything, occurs during the rise of the cooling fluids in the crust, i.e. the water is of mixed origin. The fluids of the oceanic rifts come into contact with the cover material to a much smaller extent and therefore no doubt carry more juvenile water.

Juvenile water, bound in serpentine, is transferred from the ocean crust to the Benioff–Zavaritskii zones in what might be called a conserved form. On being liberated, it mixes here, as has been shown, with the vadose water. In moving to the surface its dilution increases. Hence the juvenile water content in the hydrothermal discharges in regions of volcanic activity, by the most optimistic calculations, which are obviously exceeded, hardly reach 5–10% and in rare cases 25% (Aver'ev, 1966; White, 1969).

An interesting part of the discharge, which also carries metal-bearing thermal solutions, is to be found on the floor of the Red Sea rift. This rift is transitional between the continental and the oceanic, and the hydrothermal discharge on its floor, in the opinion of many geologists and oceanologists, contains chemical elements which are brought from the mantle (White, 1969; Dzotsenidze, 1972). The water itself, judging by the isotope ratios, is probably vadose. 'Pure' juvenile water has so far not been found in any rift system, although such a possibility cannot be excluded and is very possible in the rift depressions of the mid-oceanic ranges. The problem is to find it in such places in which it has not yet become thoroughly mixed with ocean and other water.

The model of deep systems of drains and troughs. Information on the appearance and transport of water from the mantle is sufficiently unequivocal to point to the existence in the Earth's interior of a system of deep drains and troughs (Fig. 3.4). Drainage systems drain the upper mantle and conduct the water to the Earth's crust, where it either becomes bound into the rocks, or enters the upper horizons, or arrives on the surface of the Earth where it enters the hydrological cycle. The water penetrates into the depths of the Earth from the surface along a system of channels, travels into the interior and is subsequently discharged on to the surface, or into the world ocean.

What are these paths of deep drains and channels like? First and foremost the above-mentioned zones of weakness, magmatic bodies, and deep fracture zones, which not only drain the mantle but also facilitate the penetration of the vadose water into the interior of the Earth are among their number. There are furthermore the oceanic and continental rifts along which both outward and inward flow movements of water are possible, and magma and fluid channels also, which are formed in the Benioff–Zavaritskii zones. The models of the deep drains and channels is shown in Fig. 3.4. This shows the essential features clearly. Probably one ought to dwell only on the vadose water channel.

Vadose water enters the interior of the Earth in two ways: (1) in the process of the accumulation of deposits (sedimentation), and (2) as a result of the percolation (infiltration) under influence of the hydrostatic head.

Water is buried and stored in the sedimentary basins along with the marine, lagoonal, and other deposits being laid down. As the sediments become more dense it is partly squeezed back into the main body of water and partly preserved in the rocks in the form of 'relics' in the sedimentary basin. Some hydrogeologists, geochemists, and petrologists are inclined to believe that similar fossil water is to be found in the deep horizons of the sedimentary cover of the Earth, constituting the legacy of bygone geological epochs.

In this connection one should mention the hypothesis of 'chlorine-calcium' (calcium chloride) seas. This idea was put

Figure 3.3. Diagram of the rise of water from the mantle, from the viewpoint of the new global tectonics.
1, current of juvenile fluids which cause serpentinization of the base of the oceanic crust; 2, movement of juvenile fluids during degassing of the mantle material in the mid-oceanic rift zones; 3, currents of H_2O from the Benioff–Zavaritskii zone which arise during melting and dehydration of the oceanic crust; 4, convection currents of asthenosphere material; 5, direction of movement of oceanic plates.
I, oceanic plate; II, continental crust; III, lithosphere(substrate); IV, asthenosphere; V, Benioff–Zavaritskii zone; VI, deep water in trench; VII, mid-oceanic ridge rift zone; VIII, volcanoes; IX, basalt magma chamber; X, escape of andesite; XI, granite intrusion.

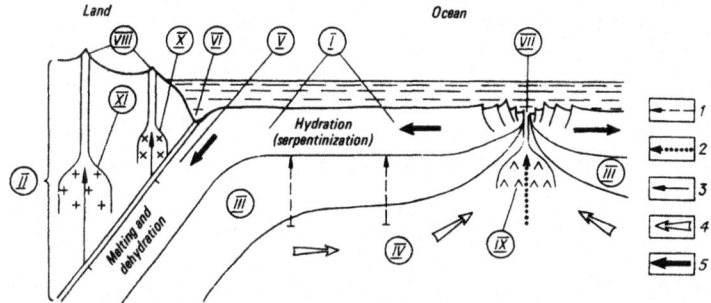

Figure 3.4. Model of the deep drainage and runoff systems.
1, movement of juvenile fluids from the mantle; 2, rising deep currents of water in the continental crust; 3, movement of vadose water; 4, displacement of chemically bound water from the oceanic crust; 5, fluid conductors (weakened zones) in the continental crust; 6, sedimentary layers; 7, granite layer; 8, basalt layer; 9, oceanic crust; 10, melting and dehydration of rocks in the Benioff–Zavaritskii zone; 11, upper mantle; 12, groundwater outlets; 13, volcanoes.

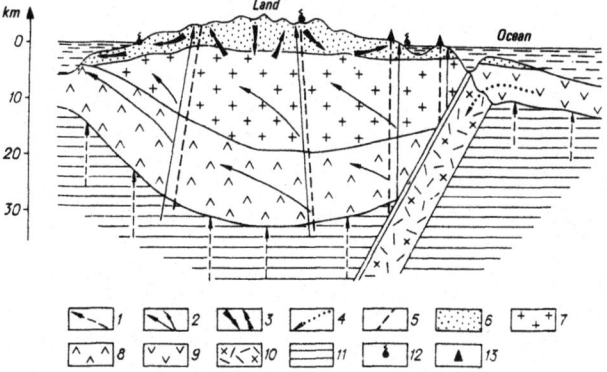

forward by the American hydrogeologists, supported by V.N. Sulin and A.N. Buneev, and has recently been developed by E.V. Posokhov (1962, 1975). The basis of this concept is as follows. Because it can be assumed that the world ocean contained much more calcium than it does at the present day, the water buried in the Palaeozoic and partly in the Mesozoic, must have contained calcium chloride. Supporters of this hypothesis assign groundwater of the deep horizons of sedimentary layers which contain one or other admixture of $CaCl_2$, to the inherited almost unchanged 'relics' of the ancient sedimentary basin. There is a direct link between the composition of the surface and subsurface hydrosphere, the unity of the waters of the world ocean and fossil water.

These models ignore the changes in composition of the deep groundwater and its dilution by young water entering the depths. The underestimation of these two factors makes the hypothesis of ancient 'calcium chloride' waters very vulnerable.

How exactly does vadose water penetrate into the deep layers? Infiltration is localized to areas where porous and fissured rocks outcrop on the surface. Exposed fault zones are good channels for the penetration of water into the depths of the Earth. The movement of water under the action of hydrostatic head proceeds from higher places to lower. In the course of geological time the infiltrating waters have penetrated comparatively deeply. The depths to which they have descended reaches 5–6 km. The driving force in this case is considered by White (1967) to be not so much hydrostatic head as the difference in the density of the infiltrating 'heavy' cold water in comparison with the 'lighter' water which is warmed in the depths: the former sinks under its own weight, and because of the difference in density forces the warm water upwards (Fig. 3.5).

Thus the sedimentary layer of the crust consists mainly of a drainage channel. The vadose water flows along the tectonically weakened zones and penetrates as far as the roof of the granite layer. It very rarely penetrates to a greater depth than this.

3.3 The genetic subdivisions of groundwater
The history of water in the crust, according to the ideas of V.I. Vernadskii, must be studied together with the geological history of the other mineral bodies.

Figure 3.5. Diagram of the formation and discharge of thermal waters (after White, 1967).
1, rocks with low permeability; 2, permeable rocks; 3, crystalline rocks; 4, direction of water movement; 5, faults.

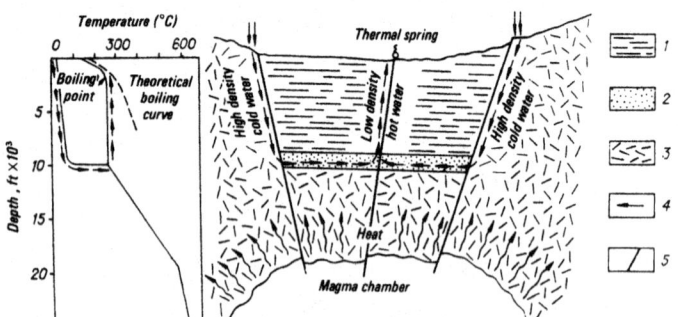

If groundwater is considered as a mineral, the concept of 'genetic cycle' proposed by A.E. Fersman is applicable. The 'genetic cycle' is the totality of the geological processes and their consequences. The application of this concept enabled G.N. Kamenskii (1947) to arrange the ideas about the origins of groundwater into an elegant system. He outlined the following genetic cycle of formation of groundwater:
(1) the infiltration (or continental) cycle which is determined by the infiltration of atmospheric precipitation and the influence of processes which take place in the upper parts of the crust; to this is assigned the condensation of water vapour in the atmosphere;
(2) the marine (or sedimentary) cycle associated with the penetration to the depths of marine (lagoonal, lake, etc.) waters during the course of sediment formation and the conversion of them during lithogenesis;
(3) the metamorphic and magmatic cycles embracing the formation of groundwater of the deep layers, which are produced by metamorphism and magmatism; the formation of deep hydrothermal waters which also include restored (i.e. liberated) water from the rocks under the influence of metamorphism and juvenile water come into this category.

The systematics of the genetic cycles were worked out by G.N. Kamenskii more than thirty years ago and retain their significance today. In essence the changes concern the division of the enumerated cycles into stages, which allow the basic genetic variety of groundwater to be distinguished.

Because the subsurface hydrosphere is situated in the sphere of action of surficial and deep factors, groundwater may be divided into *exogenic* waters, which penetrate the interior of the Earth from above, and *endogenic* waters, which enter from below, from the magma and the mantle (Kartsev *et al.*, 1969; Pinneker, 1977). We may recall that E. Suess (1902) called these two groups of water vadose (vadoses Wasser) and juvenile water (juveniles Wasser). The fraction of the latter in the total water balance is small. For comparison we point out that if the quantity of the yearly output of endogenic water from magma is evaluated roughly as being $0.6–0.7$ km^3 (Makarenko, 1973), then approximately $10\,000$ km^3 of water per year enters into the zone of intensive water exchange as the result of infiltration from the surface of the land.

The exogenic and endogenic waters comprise four genetic types (see Fig. 3.6), depending on the degree to which each takes part in the natural circulation of water. The features of the ways in which they enter the hydrogeological reservoir allow these types to be divided into genetic varieties.

The scheme of genetic classification of groundwater which is given in Fig. 3.6 does not by any means include all the possible varieties, but only includes the basic genetic types of groundwater. For example, some types of groundwater, which are generated in the interior of the Earth (let us say organic, etc.) are absent.

In accordance with the genetic cycles of G.N. Kamenskii, exogenic groundwater is divided into meteoric (of atmospheric origin) and thalassogenic (of marine origin). In this treatment the accent is given to the source of the groundwater: (1) fresh water on the land, closely linked with the atmosphere, (2) saline water of marine basins (Degens, 1965; Kartsev *et al.*, 1969). However, in the hydrogeological structures the classification more often has to resort to differentiation based on the ability of water to penetrate into rocks, distinguishing between *infiltrogenic water*, which enters from the surface into previously formed rock, and *sedimentogenic water*, which has been preserved in the rocks from the moment of sedimentation, or which appears as a product of lithogenesis.

As a consequence of the different principles used for the division, the above categories of exogenic groundwater do not coincide completely. These differences must be borne in mind when using the scheme given in Fig. 3.6, which shows both genetic varieties of exogenic groundwater, the paths by which they enter the interior of the Earth, and the source of the water concerned.

Infiltrogenic groundwater is formed from surface water of atmospheric origin (rain, snow, river, and lake water), the overwhelming mass of which, with the exclusion of water which percolated to the bottom of the seas and into the surrounding land, comes into the category of meteoric water. This genetic type of water is the basis of the underground branch of the hydrogeological water cycle (Chapter 4). The resupply of groundwater is performed by the percolation of water in the liquid state (infiltrated water), or in the state of condensed water vapour in the conditions obtaining in the near-surface layers (condensation water). Only a very small part of the meteoric water (water of lake basins, buried with sediments) comes into the category of sedimentogenic water. Infiltration water and water of atmospheric origin are practically the same. As a consequence of man's industrial activities these waters in a number of places have become simultaneously technogenic.

Sedimentogenic groundwater, which is sometimes called fossil water, buried, or relict water, has not been in contact with the atmosphere for a long time, in other words has been excluded from the water exchange with the surface of the Earth for a long period of geological time (White, 1957). This water has been captured by being drawn down into the Earth's

interior when sediments were being formed as the residues of vegetation, or squeezed out as the rocks became compacted. When fossil water is of the same age as the rocks containing it, it is called syngenetic (the English synonym is connate water, which means 'water formed at the same time'). There is also another form of sedimentogenic water. This is epigenetic water, i.e. water squeezed out during the process of lithogenesis from the overlying or underlying layers, and which has migrated from younger sedimentary rocks into more ancient, or conversely, from ancient rocks into younger. This water is younger or more ancient than the rocks containing it; its most important feature is that it appears after sedimentation.

Because the accumulation of sediments with the burial of fossil water occurs mainly in marine basins, almost all groundwater of sedimentary origin is assigned, because of its source, to the category of thalassogenic water. An exception, as we have already seen, is the meteorogenic groundwater incorporated with the sediments in fresh lake basins.

Magmatogenic water, more accurately called *mantinogenic* water enters periodically into the circulation. Molecules of such water are generated in the mantle or magma, from hydrogen and oxygen. It is necessary to distinguish to a first approximation (Pinneker, 1977) volcanic water which is formed as supercritical water or vapour from the magma as it rises and cools and then condenses in the upper layers, and cross-magmatic or transmagmatic liquid–gas solutions in the form of regionally rising currents from the centres of magmatism.* The latter, after D.S. Korzhinskii (1962), is the cause of the metamorphism and granitization of sedimentary rocks. Subsequently separating from the granite melt, such water enters the Earth's crust.

As was pointed out in a previous section, the water ejected during volcanic eruptions contains volcanic water proper (in the sense cited earlier, i.e. water which has been separated from the magma during the differentiation process), and also adjacent water of other genetic varieties mobilized from the contiguous rocks. It is in precisely this way that during the crystallization of the granite melt not only water of juvenile origin, but also vadose water captured by the melt is liberated from the rising transmagmatic solutions. The question of the degree to which one or the other takes part in postmagmatic processes must be decided separately in any particular case.

Metamorphogenic groundwater to some degree or other is associated with exogenic and endogenic water; it appears during metamorphism from bound water of the sedimentary or igneous rocks involved. The dehydration of rock-forming minerals usually occurs near magmatic centres at great depths. The reason for this is the increase in temperature and pressure. Such water once more (restored from sedimentary rock), or

Figure 3.6. Genetic classification of groundwater.

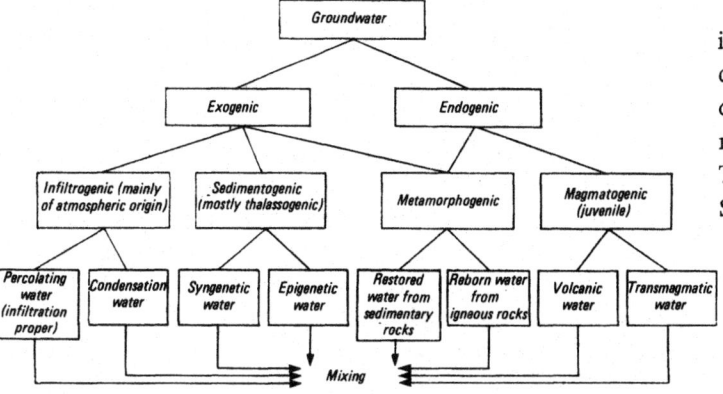

*Sometimes varieties of igneous rock are given other names or they are differentiated from completely different points of view (White, 1957; Kapchenko, 1966; Smirnov, 1969; Gavrilenko & Derpgol'ts, 1971), however, the two named varieties include all forms of water of juvenile origin.

for the first time (reborn from the igneous rock), enters the circulation. In particular the water of crystallization in minerals or the gas—liquid inclusions in them can be a source of metamorphogenic water. The dehydration processes (for example the transformation of gypsum to anhydrite) may have a regional character and continue over a long period, leading to the formation of huge accumulations of groundwater.

The diagnostic aspects of the genetic varieties we have quoted are extremely complex. The pure form is perhaps only found in water which has infiltrated (this takes place usually to a depth of 2 km), and syngenetic water (in 'young' artesian basins, which have recently emerged from below sea level). The remaining genetic varieties, as a rule, are found in mixtures with either type, mainly infiltrogenic water, because immediately after formation they become mixed and hence do not exist in the 'pure' form. The presence of magmatogenic water in general raises doubts among some scientists.

Up to now there has been an absence of reliable criteria for the differentiation of groundwater sources. Palaeohydrogeological reconstructions, coefficients of proportionality of chemical elements, and isotope data which are used with this in mind do not as yet give simple results, although success has already been obtained by using isotope ratios of hydrogen and oxygen.

The trend of the historical development of the natural stores of groundwater (the artesian basins and hydrogeological massifs), amounts to the displacement of sedimentary and magmatic water by infiltration water. Sometimes metamorphogenic water also takes part in this displacement. Therefore even in the deep-lying layers of the platform depressions, or centres of volcanic activity, a mixture of groundwater of different origins is found. In this category come hydrothermal solutions, which are hot metal-bearing solutions formed from the percolating magmatogenic, sedimentogenic, and metamorphogenic varieties of groundwater (Saukov, 1966; Perel'man, 1968; Smirnov, 1969; Ovchinnikov, 1970; Naboko, 1963). The deep-lying water of the sedimentary layers too are usually of mixed sedimento—infiltrogenic origin (Pinneker, 1977). This of course does not exclude the occurrence of hydrothermal water in one or other stage of the genesis of juvenile water, but there is sedimentogenic water within the deep layers of the sedimentary cover.

3.4 The nature of the regional (zonal) distribution of groundwater

Studies of the zonation (belt-like distribution) of natural phenomena have a long history. The German naturalist A. Humboldt in the first half of the nineteenth century noted the climatic zonation, which is caused by the action of the sun's radiation, and which leads to the ordered distribution of the vegetation and animals of the world. This idea was later applied to soils, and then the unconscious connection between natural phenomena led to the conclusion that there existed a natural law of zonation. This was formulated at the end of the nineteenth century by the soil scientist V.V. Dokuchaev. At the same time Suess noted the vertical zonation of the material of the globe.

The demarcation of natural zones or belts is a fundamental tenet of natural phenomena in general and the hydrosphere in particular. In the wide sense all natural waters will be zonal, and may be divided into atmospheric, surface, and subsurface. The zonation is clearly traceable in the subsurface hydrosphere. It is, in essence, one of the basic laws of distribution for subsurface water.

We will first summarize the data on the zonation of the subsurface hydrosphere. According to the phase-aggregate composition, as was shown in Chapter 2 (see Fig. 2.5), macrozones can be distinguished within it (for convenience these are usually called simply zones):

(1) The cryolithic zone, or zone of 'solid' water, which is bounded below by the ice—water phase transition and which has a thickness of up to 1 km or a little more.

(2) The saturation zone, or the 'liquid' zone, which is bounded by the isotherms of the phase transitions ice—water and water—steam, occupying about half of the Earth's crust. The limiting value of the temperature is 374—450 °C, and pressure of 20—25 kbar. The thickness of this zone reaches 20 km and more (in regions of Precambrian folding). On island arcs and in regions of recent volcanic activity the lower boundary is not deeper than 5 km.

(3) The zone in which the water is in the supercritical state, or, according to F.A. Makarenko et al. (1972), 'the zone of dense fluids'. It is located between the isotherms 450 and 700 °C and extends for tens of kilometres into the mantle where the pressure rises to 50 kbar. As a consequence of this the water becomes dense. Lower still the hydrogen bond is broken.

The zonation just examined concerns free water. The forms of existence of physically and chemically bound water are also limited by dynamic parameters. The quantity of bound water decreases, and in the regions of supercritical temperature water becomes practically free (Makarenko et al., 1972), with the exception of some part of the water of crystallization and constitution, which is retained up to 700—1000 °C.

Groundwater properly belongs to the zone of saturation, which is different on the continents and under the oceans. Even on continents this zone is very heterogeneous, being dependent on the physical-geographical, geological-structural, and thermodynamic conditions which act upon it.

The demarcation of natural zones according to latitude of the locality and the manner of deposition is applicable to groundwater. Therefore the following are distinguished: (1) latitudinal (climatic) zonation is applicable mainly to groundwater near the surface; (2) vertical (geological) zonation which is to be found in the deeper groundwater layers. The vertical zonation may be stretched out (massifs of crystalline rock), or deep (basins of sedimentary layers). In the case of the latter it is more correct to speak of vertically extended zones of groundwater (Pinneker, 1977).

The effect of geographical latitude operates most

strongly on groundwater down to a depth of 25–50 m, although there is some indirect effect to a much greater depth. Latitudinal groundwater zones, related to definite landscape features, succeed each other from the poles to the equator. As P.V. Ototskii showed as far back as 1906, shallow groundwater along this axis gradually 'deepens' and becomes mineralized. The work of V.S. Il'in, O.K. Lange, F.P. Savarenskii, and others is of great importance in the development of the concept of the latitudinal zonation of groundwater.

Considering the total state of the water, the humidity of the region, and the temperature changes in the northern hemisphere, the following macrozones of shallow groundwater succeed each other in a southward direction:

(1) Ice (frozen), in the limits of which groundwater of the upper layers is completely or partly frozen and exists in the liquid state only during the short summer; the recharge of these zones is difficult, and the average yearly temperature is below freezing.

(2) Humid, characterized as a rule by excess moisture because of the excess of precipitation over evaporation, which leads to intensive leaching of salts from the rocks and favours the replenishment of the resources of shallow groundwater. The average yearly temperature is above freezing.

(3) Arid (dry), in which evaporation exceeds precipitation, and as a consequence salination of the groundwater occurs in the upper layers (salts are precipitated on and near the surface), and the recharge of the groundwater resources is restricted. The temperature of the groundwater exceeds 20 °C.

The nature of the zonation of groundwater is explained quite simply – it is determined by the climate, or more accurately by the humidity and the average yearly temperature of the air. There are other exogenic factors which influence this (the degree to which the region is broken up, the types of relief, geological structure, etc.), but this influence is always subordinate to the climatic conditions.

The nature of the vertical zonation of groundwater is much more complex – there is no unanimity here. How does the vertical differentiation of groundwater arise, and upon what does it depend? We will examine this question by using the example of the strata of sedimentary basins.

In a vertical section of the sedimentary envelope of the Earth the changes in speed of movement, the degree of mineralization, the ion and dissolved gas content, and the temperature of the groundwater are easily determined. V.I. Vernadskii in 1931 at the first All-Union Hydrogeological Congress noted the trend of these changes: (1) the deeper it is the lower the mobility of the groundwater; (2) the degree of mineralization increases with depth, and at the same time the ion content changes; (3) the composition of the dissolved gas content changes completely with depth; (4) the temperature of stratal water rises from negative values (in the region of permafrost) to 100 °C and more (in deep layers). The systematization of data on the above laws led to a study of the vertical zonation of groundwater – hydrogeodynamic, hydrogeochemical, gaseous (hydrogasogeochemical), and hydrogeo-

thermal. Great credit for this is due to N.K. Ignatovich, B.L. Lichkov, and F.A. Makarenko for the founding of these studies in the 1930s and 1940s.

A general feature of the above-listed types of vertical zonation is the presence of three (or more) zones, which succeed each other from above downwards.

The vertical hydrogeodynamic zonation consists of the ordered vertical change of zones with different rates of water exchange for upper, middle, and lower zones.

The upper zone, the zone of *intensive* (active) water exchange, occupies strongly fissured rocks where the groundwater moves with high velocity. Here drainage takes place freely. Lower down there is the zone of reduced (impeded) water exchange. As a consequence of the smaller amount of fissuring of the rocks the velocity of groundwater movement is reduced and the drainage rate falls. Lower down still are the deepest parts of the sedimentary layers where drainage is very low or totally absent. The renewal of groundwater takes place on the scale of geological time. Therefore the lower zone is called the zone of *passive* (greatly impeded) water exchange.

The rate of renewal of groundwater in each zone differs markedly: (1) the zone of intensive exchange – hundreds and thousands of years (average 330 years); (3) zone of reduced exchange (tens and hundreds of thousands of years); (3) the zone of passive water exchange (millions of years).

Still more marked is the exchange of the dissolved ion content and the degree of mineralization of the groundwater, and this is seen in the vertical hydrogeochemical zonation. In the general case it is fixed according to the exchange of fresh hydrocarbonate (mainly calcium carbonate) water initially by saline water of mixed ion content, and then by solutions of sodium chloride or calcium chloride.

The zones have been named as follows (Zaitsev & Tolstikhin, 1972):

(1) upper – fresh water (degree of mineralization less than 1g/kg);

(2) middle – saline water (1–35 g/kg);

(3) lower – brines (more than 35 g/kg).

The hydrogeochemical zone (and the subzones into which it can be divided under detailed examination) occupies partly or completely the vertical section of the sedimentary envelope. The fixed combination or succession of zones and subzones which characterizes the hydrogeochemical section of the whole thickness of the sedimentary cover is called by N.I. Tolstikhin and I.K. Zaitsev the *hydrogeochemical belt* (see Fig. 3.7). The belt may be single-zoned, with a zone containing fresh water only; double-zoned, with fresh water and saline water zones; and triple-zoned, when all three zone types are present in the section. The single-zone belt usually surrounds the margins of sedimentary basins, the double-zoned belt is located in the interior of the field, and the central part contains all three zones.

Gaseous zonation of groundwater makes its appearance in the replacement with increasing depth of gases of atmospheric origin (mainly nitrogen and oxygen) by those of reducing, metamorphic, and magmatic conditions, which are characterized by the presence of CH_4, H_2S, CO_2, H_2, etc. In a similar way, the succession of the hydrogeothermal zones proceeds with increasing depth: above there is supercooled water (temperature below 0 °C), or cold water (0–20 °C), replaced at first by thermal water (20–100 °C), and then by superheated water (more than 100 °C).

Thus the trend in all the above-mentioned zonation types is the same: it is the replacement of surface conditions by the conditions of the Earth's interior. Therefore it is necessary also to look for the reasons which give rise to zonation in the processes which affect the subsurface hydrosphere. Among the causes of the vertical differentiation of groundwater (depending on the intensity of the exchange, dissolved ion and gas content, degree of mineralization or temperature) the following are most often cited: (1) the force of the Earth's gravitational field (Filatov, 1956; Samoilov & Sokolov, 1957; Valyashko, 1963); (2) geostatic pressure, which causes the compaction of the rocks and the removal of water by pressure (Mukhin, 1965; Zaitsev & Tolstikhin, 1972); (3) the thermal conditions in the Earth's interior (Al'tovskii, 1958). The points of view just enumerated are attractive mainly because they provide an explanation for hydrogeochemical zoning.

Without dwelling in detail upon these extremely contradictory concepts, we will point out that in the vertical differentiation of groundwater a complex of physical-geographical, geological-structural, and thermodynamic factors play a part, the influences of which may vary with time. In an attempt to explain, for example, hydrogeochemical zoning by one cause only (let us say, the gravitational stratification of water by specific gravity, or its being squeezed out of the compacting sedimentary layers), it is doubtful whether this has a good basis because it does not take into account all the driving forces involved in this phenomenon. Even in the forming of the zonation not only does the heat of the Earth's interior take part, but the influence of the temperature regime of the surface also plays a part.

It is impossible to explain every phenomenon of the

Figure 3.7. Concentric hydrogeochemical zoning of artesian basin, (a) cross-section, (b) plan.
Hydrogeochemical zones: I, single-zoned (fresh water); II, double-zoned (fresh and saline water); III, triple-zoned (fresh and saline waters, and brines); IV, hydrogeochemical anomaly. 1–3, hydrogeochemical zones (1, fresh water, 2, saline water, 3, brine); 4, discharge centre of saline water and brine; 5, brine emerging via fault; 6, crystalline basement.

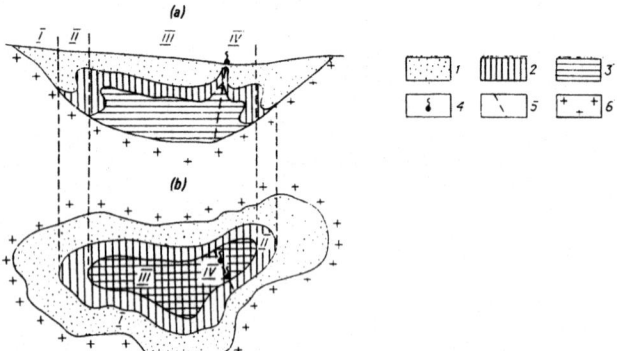

vertical zonation of groundwater by some single cause or other. Here it is necessary to consider all the natural factors.

V.A. Krotova (1962) conceived the vertical zonation of groundwater as the result of a prolonged struggle between two opposing controlling principles: (1) the surface factors which are a complex of physical-geographical and other exogenic factors and which are connected with the introduction of infiltration water into the Earth's interior, the renewal of chemical elements of the rocks, and with low pressures and temperatures; (2) depth factors, which are made up of the action of endogenic forces, and which are characterized by the good protective power of the interior, the compaction of rocks, the accumulation of chemical elements in groundwater, high temperatures and pressures.

If present day geological processes are dominant in the upper parts of the section, then in the deep layers the inheritance of former geological epochs is clearly felt. This inheritance makes its presence felt in the water which has been squeezed out and which is syngenetic with the rocks containing it, in the preservation of relics of magmatic activity, etc. In the course of geological history the struggle between exogenic and endogenic processes has acquired a complex character, which depends wholly upon the physical-geographical, geological-structural, and thermodynamic conditions.

Between the different forms of the vertical zonation of groundwater there is a cause and effect relationship; therefore they have much in common. Because there is in the basis of all of them the struggle between two forces — those of the surface and those of the depths, in each case three zones are formed: the upper, the intermediate, and the lower. Then it follows that the indicators of, let us say, hydrogeochemical, gas, or temperature zonation in some measure or other reflect the hydrogeodynamic zonation. The upper zones, which are formed under the influence of one exclusively surficial basis, i.e. the exogenic factor, correspond well so far as their thicknesses are concerned. For example, the zone of intensive water exchange in many artesian basins contains fresh and cold water with oxygen and nitrogen gases. However, there is no complete correspondence between the different forms of vertical zonation of artesian basins, especially when comparing the lower intermediate and the lower zones. The absence of similarity, for example, between the hydrogeodynamic and hydrogeochemical zones is explained by the fact that the water content of the former is determined by the intensity of the water exchange and the properties of the water collector, and that of the latter by the mixed composition of the rocks. Consequently it is necessary to bear in mind that the indicators of every form of vertical groundwater zonation are produced not only by general causes, but also particular ones, which determine each zonation.

References
Al'tovskii, M.E. (1958). The significance of natural conditions, of the physical, chemical, and biochemical processes in the formation of groundwater. *Trudy Laboratorii gidrogeol. problem.*, Moscow, 16, 34–7.

Aver'ev, V.V. (1966). The hydrothermal process in volcanic regions and its association with volcanic activity. In *Proceedings 2nd All-Union Conference on Volcanology*, pp. 118–28. Moscow.

Craig, H. (1961). Standards for reporting concentrations of deuterium and oxygen-18 in natural waters. *Science*, 133 (3467), 1833.

Degens, E.T. (1965). *Geochemistry of sediments*. Englewood Cliffs, New Jersey: Prentice-Hall. 342 pp.

Derpgol'ts, V.F. (1962). Major sources of the natural waters of the Earth. *Izv. AN SSSR Ser. geol.*, No. 11, 18–31.

Dzotsenidze, G.S. (1972). Hot springs of the Red Sea and the problems of volcano-sedimentary genesis of minerals. *Geol. rudn. mestorozh.*, No. 5, 3–21.

Ferronskii, V.I. (1974). The origin of the Earth's hydrosphere according to data on the isotope composition of water. *Vodnye resursi*, No. 4, 21–34.

Filatov, K.V. (1956). *The gravitational hypothesis of the formation of the chemical composition of groundwater in platform depressions*. Moscow: Izd-vo AN SSSR. 208 pp.

Gavrilenko, E.S. & V.F. Derpgol'ts (1971). *The deep hydrosphere of the Earth*. Kiev: Naukova dumka. 272 pp.

Gorskii, N.N. (1962). Water – nature's miracle. Moscow: Izd-vo AN SSSR. 224 pp.

Grigor'ev, S.M. (1971). *The role of water in the formation of the Earth's crust. The drainage shells of the Earth's crust*. Moscow: Nedra, 263 pp.

Hess, H.H. (1962). History of ocean basins. In *Petrologic studies. A volume in honor of A.E. Buddington*, ed. A.E.J. Engel *et al.*, pp. 599–620. New York: Geol. Soc. Am.

Kadik, A.A., E.B. Lebedev & N.I. Khitarov (1971). *Water in magmatic melts*. Moscow: Nauka. 267 pp.

Kafengauz, N.L. (1972). S.M. Grigor'ev's hypothesis on the drainage shells of the Earth's crust. *Izd. AN SSSR, Fizika zemli*, No. 2, 100–1.

Kamenskii, G.N. (1947). *Prospecting for groundwater*. Moscow, Leningrad: Gosgeolizdat. 313 pp.

Kapchenko, L.N. (1966). The present state of the problem of the age of deep-lying juvenile groundwater. *Litologiya i polezn. iskopaemye*, No. 4, 75–87.

Kartsev, A.A., S.B. Vagin & E.A. Baskov (1969). *Palaeohydrogeology*. Moscow: Nedra.

Kazanskii, Yu.P. (1977). Variations in the gas and salt composition of ocean water. *Geol. i geofiz.*, No. 8, 56–66.

Korzhinskii, D.S. (1962). The behaviour of water in magmatic and post-magmatic processes. *Geol. rudn. mestorozh.*, No. 5, 3–12.

Kropotkin, P.N. (1964). *Evaluation of the Earth: origin, structure, and geological history of the Earth*. Moscow: Znanie. 95 pp.

Kuenen, Ph.H. (1955). *The realms of water. Some aspects of its cycle in nature*. New York: Wiley. 327 pp.

Kulp, J.L. (1951). Origin of the hydrosphere. *Bull. Geol. Soc. Am.* 62 (3), 326–30.

Lane, A.C. (1945). The evolution of the hydrosphere. *Am. J. Sci.*, 245-A.

Le Pichon, X., J. Francheteau & J. Bonnin (1973). Plate tectonics, 300 pp. In *Developments in geotectonics*, No. 6. Amsterdam: Elsevier.

Letnikov, F.A. (1977). Features of the fluid regime in endogenetic processes in the crust and mantle. In *The fluid regime of the crust and upper mantle*, pp. 5–9. Irkutsk.

Letnikov, F.A., N.A. Logachev, E.M. Emel'yanov *et al.* (1977). Fluid regime of the rift zones. In *Major problems of the genesis of rifts*, pp. 51–60. Novosibirsk.

Makarenko, F.A. (1973). Some general laws governing the underground runoff. In *Proceedings of the Seventh Conference on the Groundwater of Siberia and the Far East*, pp. 6–9. Irkutsk, Novosibirsk.

Makarenko, F.A., V.A. Il'in, V.N. Kononov & B.G. Polyak (1972). Physical model of the subsurface hydrosphere. In *Hydrogeology and engineering geology*, pp. 15–25. Moscow.

Miller, A.R., C.D. Densmore, E.T. Degens *et al.* (1966). Hot brines and recent iron deposits in deeps of the Red Sea. *Geochim. et Cosmochim. Acta*, 30 (3), 341–60.

Miyaki, Y. (1969). *Fundamentals of geochemistry*. Moscow: Nedra. 327 pp. (Translated from the English.)

Monin, A.S. (1977). *The history of the Earth*. Leningrad: Nauka. 228 pp.

Mukhin, Yu.V. (1965). *The compaction processes of argillaceous sediments; applicable to the problems of geology of gas, petroleum, hydrogeology, and engineering geology*. Moscow: Nedra.

Naboko, S.I. (1963). *The hydrothermal metamorphism of rocks in volcanic regions*. Moscow: Izd-vo AN SSSR. 172 pp.

Ototskii, P.V. (1906). Groundwater, its origin, life, and distribution. *Trudy opytnykh lesnichestv*, No. 4, 1–300.

Ovchinnikov, A.M. (1970). *Hydrogeochemistry*. Moscow: Nedra. 200 pp.

Pavlov, A.N. (1977). *Geological water cycles on the Earth*. Leningrad: Nedra. 144 pp.

Perel'man, A.I. (1968). *Geochemistry of epigenetic processes. Zone of hypergenesis*. Moscow: Nedra. 331 pp.

Pinneker, E.V. (1977). *Problems of regional hydrogeology. Laws governing the occurrence and formation of groundwater*. Moscow: Nauka. 196 pp.

Posokhov, E.V. (1962). Theory of metamorphism of natural waters and the genesis of deep brines of the chloride-calcium type. *Trudy Novocherkassk. politekh. in-ta*, 128, 43–84.

Posokhov, E.V. (1975). *General hydrogeochemistry*. Leningrad: Nedra. 208 pp.

Pospelov, G.L. (1963). Geological prerequisites of the physics of ore-controlling fluid conductors. Communication 1. *Geol. i geofiz.*, No. 3, 18–38; Communication 2. *Geol. i geofiz.*, No. 4. 24–41.

Rubey, W.W. (1951). Geologic history of sea water. *Bull. Geol. Soc. Am.*, 62 (9), 1111–48.

Samoilov, O.Ya. & D.S. Sokolov (1957). On the possible causes of the vertical hydrogeochemical zonation of artesian waters. *Izv. AN SSSR Otd. khim. nauk*, No. 3.

Saukov, A.A. (1966). *Geochemistry*. Moscow: Nauka. 485 pp.

Sidorenko, A.V. & Yu.A. Borshchevskii (1979). The problem of the geochemical evolution of the Earth in the light of data from the isotope geology of the Precambrian. In *Problems of Precambrian sedimentary geology*, pp. 34–44. Moscow: Nauka.

Smirnov, V.I. (1969). *Geology of minerals*, 2nd edn. Moscow: Nedra. 687 pp.

Sorokhtin, O.G. (1974). *Global evolution of the Earth*. Moscow: Nauka. 184 pp.

Suess, E. (1902). Ueber heisse Quellen. *Verhandl. Gesell. deutsch. Naturforsch. und Aerzte (Leipzig)*, 71, 133–51.

Sydykov, Zh.S. (1973). Deep-lying waters – source of the hydrosphere. *Izv. AN Kaz. SSR, Ser. geol.*, No. 1, 1–12.

Valyashko, M.G. (1963). Genesis of the brines of the sedimentary shell. In *Chemistry of the Earth's crust*, Vol. 1, pp. 257–77. Moscow: Izd-vo AN SSSR.

Valyashko, M.G. (1971). The evolution of the chemical composition of the waters of the ocean. In *History of the world ocean*, pp. 97–104. Moscow.

Vinogradov, A.P. (1959). *Chemical evolution of the Earth*. Moscow: Izd-vo AN SSSR. 44 pp.

Vinogradov, A.P. (1962). The origin of the Earth's shells. *Izv. AN SSSR, Ser. geol.*, No. 11, 3–17.

Vinogradov, A.P. (1967). *Introduction to the chemistry of the ocean*. Moscow: Nauka. 215 pp.

White, D.E. (1957). Magmatic, connate, and metamorphic waters. *Bull. Geol. Soc. Am.*, No. 12, Part 1, 1659–82.

White, D.E. (1967). Some principles of geyser activity, mainly from Steamboat Springs, Nevada. *Am. J. Sci.*, 265, 641–84.

White, D.E. (1969). Thermal and mineral waters of the United States. Brief review of possible origins. *Internat. Geol. Congress, Report of twenty-third session*, Prague. Vol. 19, pp. 269–86.

Yanshin, A.L. (1962). Prospects of discovering deposits of potash salts in Siberia. *Geol. i geofiz.*, No. 10, 3–22.

Zaitsev, I.K. & N.I. Tolstikhin (1972). *The laws of the distribution and formation of mineral (industrial and medicinal) groundwater in the USSR*. Moscow: Nedra. 279 pp.

4

The circulation of water in the interior of the Earth

4.1 A short account of geology and the movement of water

The idea that the fundamental laws governing the movement of water in the Earth's interior constituted the theme of a special discipline — the dynamics of groundwater — which is concerned with the study of the general laws governing the motion of fluids of the interior (water, oil, gases) has long been firmly rooted in hydrogeology.

It is impossible to agree completely with such an approach to hydrogeodynamics (and hence with the consideration of the mobility of the components of the subsurface hydrosphere) because its laws of motion govern only 'liquid-droplet' water, one of the various forms.

D.I. Gordeev (1954) turned his attention to this view a quarter of a century ago, remarking that the classical hydrodynamics did not take into account the changes which took place in the water as it moved in the Earth's interior. According to the classical theory, the water moved but did not undergo any change. The error of this assumption when considering motion in the subsurface hydrosphere was later justifiably pointed out by G.Yu. Valukonis & A.E. Khod'kov (1973).

Historically it came about in this way: when observing the subsurface branch of water circulation, scientists by and large established the facts of the free circulation through the pores of the rocks, and obtained notable successes in their studies of this phenomenon. These can be seen in the discovery of the basic laws of infiltration, which became almost completely accepted as the fundamental laws of motion of groundwater, and of water in general.

The laws of classical hydrodynamics do not take account of changes of the geological environment with time, and consider it to be invariable. But in fact the water in the Earth does not simply move, but it interacts with the rocks continuously, influencing the dimensions and character of the pores, and also influencing the filtration properties of the rock itself; furthermore, it changes from one state to another. The bulk of the water in the Earth's crust moves in a bound state together with rock particles (inside the crystal lattice or on its surface). In certain geological conditions such water either migrates from the rock, or, conversely, is absorbed by it. The scale of these phenomena is vast. It is sufficient to say that all sedimentary water is formed not as the result of infiltration, but by being buried along with the sediments. Metamorphogenic water is also formed in precisely the same way. Hence (in addition to infiltration) the movement of water together with the rock, i.e. along with the geological environment in which this water is entrained, plays an important role in geological history.

Water moves in various ways in the interior of the Earth. It would be a mistake to miss seeing in the variety within this movement the unity, the interaction, and the interacting conditions of its forms. It is perfectly obvious, for example, that bound water becomes free water, and that the character of its movement changes in accordance with strict laws. The same thing occurs in the transition of the liquid phase to steam, and hygroscopic into pellicular water, etc. It is therefore proper to talk about the single *geological type of water movement* in the interior of the Earth as the most important geological form of the movement of materials.

According to B.M. Kedrov (1964); who proposed the necessity of distinguishing the geological forms of movement of material, this form is of a new and integral character. It is impossible to reduce it to a simple sum of the mechanical, physical, and chemical movements which geologists have traditionally raised to the rank of controlling or determining movements. It is precisely the geological form of water in the Earth's crust, not the mechanical sum of simpler forms; it is a qualitatively new form having its own laws of motion and development, although not as yet, it is true, sufficiently studied.

G.Yu. Valukonis & A.E. Khod'kov (1973) distinguish: (1) the movement of water as a physical body, (2) the movement of water in the broad sense, and (3) the movement of water in the geological context. The movement of huge masses of groundwater along permeable strata falls into the first category. The movement of water in the broad sense includes its transition from one phase to another (solid, liquid, gas/vapour), from one condition (free, bound) to another, from one sphere (atmosphere, lithosphere, biosphere) to another, etc. Finally, the movement of water in the geological context is none other than the history of natural water, including groundwater, the development with time of the water–rock–gas–living material system, including both the first two forms of migration and also the complex processes of metamorphism in actual geological environments. The movement of water in the geological context is the complex form of its migration (Valukonis & Khod'kov, 1973, pp. 10–11).

The proposed division of the movement of water into types with a strict approach gives rise to a methodological objection: mechanical, physical, and geological forms of movement enter on a practically equal footing, describing the individual facet of a more general phenomenon. In such a case the specific movement of water in the Earth as a geological process

is lost and the simpler forms (infiltration, osmosis, the transition from one state to another) come to the fore, the sum of which replaces the complexity of the phenomenon observed. Nevertheless, A.E. Khod'kov's and G.Yu. Valukonis's approach to the movement of water has a geological basis; it is based on the history of groundwater and the consideration of its interaction with the rocks.

The geological movement of groundwater includes not only simple forms (mechanical, physical, and chemical), but also the more complex ones (biological, technogenic, or noospheric). The study of each of these separate types of movement and its place among the others in the general scheme of things is undoubtedly necessary, but it is impossible to consider the whole movement as the sum of its simple parts. In order to better discuss this, let us take actual examples.

Sedimentogenic water is widely distributed in the Earth's crust, occupying the lower part of the section of basins containing stratal water. In what way and in what form did sea water reach a depth of 3–5 km and more? What is it – intrusion, infiltration, diffusion, convection? It is impossible to explain this fact by one of these kinds of movement alone. Neither does the bringing in of biological, physical, chemical, thermal, and other forms of movement clarify the problem. There remains only the geological path to the solution of this question.

The forming of sedimentary water is caused by the circulation of material in the interior of the Earth, in which water, together with the rocks containing it, sinks to considerable depths with the subsidence of certain blocks of the Earth's crust, compensated for by the accumulation of sedimentary rock of a corresponding thickness. In this way different forms of water, both free and bound (and not only physically but also chemically), become buried. With the sinking and burying of sediments the relationship between the different forms of water changes continuously: one type changes to another; some of the water molecules are chemically dissociated, some are mechanically squeezed out from the system, etc. The chief feature of the system examined is the process of transport of water to great depths as the result of geological movements of material. The movement of water itself is part of more general processes of sedimentation, diagenesis, katagenesis, and metamorphism, i.e. of lithogenesis in general. Hence this form of water movement should be called *lithogenic*.

In its turn lithogenic water movement as a type of geological form of movement may be subdivided into sedimentary, diagenetic, katagenic, and metamorphic, each of which includes in strict proportion simpler mechanical, physical, and chemical forms. Thus at the stage of forming of sediments the dominant process is that of the physical and chemical bonding of water by the rock particles and the complex redistribution of molecules corresponding to the porosity and organo–mineral type of the forming sediments. At the stage of diagenesis the dominant process is that of exclusion, i.e. the squeezing out of water, with its consequent infiltration into permeable deposits. At the stage of katagenesis of sediments the squeezing process is gradually replaced by ionic dissociation processes and the binding of water anew by the forming of clay and carbonate minerals with the simultaneous incorporation of free water into the forming minerals. During metamorphism bound water is reformed, as a consequence of the processes of dehydration, dehydroxylation, and molecule synthesis, forming zones with anomalously high water pressure, or water-bearing 'lodes' saturated with gas which migrate upwards. Consequently, during all stages of lithogenesis the movement of water is accompanied by a continuous interaction with the rocks, in which not only the rock, but also the water itself, is transformed (Shvartsev, 1975).

We will turn to the near-surface zone of the Earth's crust. It would seem that here the dynamics of the groundwater amount to the mechanical and physical forms of movement. However, that this is an illusion becomes obvious when the movement is considered as a geological process of the interaction of atmospheric water with the material of the Earth's crust.

In fact, simple forms of water movement may be distinguished in the upper parts of the Earth's crust: infiltration, diffusion, capillary rise, etc. However, neither these nor other forms determine the process of water movement. With infiltration or diffusion the water interacts with the rocks, gases, and organic material, dissolves or leaches out some rocks, cements others; i.e. not only water itself changes, but also the medium which contains it. In other words, there is evidently a complex variety of the geological form of water movement, in which not only are its simple forms present, but also the character of the movement appears qualitatively completely different.

What is this variety of geological form of movement to be called? It must be remembered that probably the most important feature of the dynamics of groundwater in the upper parts of the Earth's crust is the entry of atmospheric water. This process combines the action of exogenic factors and may be considered to be fundamental. It probably ought to be named after the variety of the geological form of movement being considered, i.e. it ought to be named *meteorogenic* or *atmospherogenic*.* Such a name simultaneously characterizes the nature of the movement too.

Up to this point far from every variety of the geological form of water movement has been studied. There are several of them. Apart from the varieties already considered, we will mention one more, which arises as a result of magmatic phenomena: the entry of magma into the Earth's crust, its crystallization and dehydration, the formation of postmagmatic deep solutions of juvenile or, in a number of instances, of regenerated origin. The mechanism of such a movement has not been studied in detail; however, it is hardly possible to doubt its great significance and its specific aspects. Such water is no doubt of wide occurrence in regions of volcanic activity, recent rift zones (including ocean regions), belts of Alpine folding, and active plate tectonic movements.

*The atmospherogenic form of water movement to an ever greater extent is changing in form because of the activities of man. In this connection, probably, one ought to talk also of the *technogenic* form of water movement (Editor-in-chief).

Its formation is accompanied by temperature diffusion, osmosis, molecular transport, and other processes. Such a variety of movement of deep water, because it is associated with magmatism, may be called *magmatogenic*.

Thus the geological form of the movement of water has a complex nature which is connected with the movement of water through rock. Together with material from the Earth's crust, it starts from the surface of the Earth and finishes in the deep zones of metamorphism and magmatism, i.e. it enters the geological circulation of material as a component part. An integral feature of the geological form of water movement in contrast with the mechanical or physical forms is the continuous interaction with the rocks, in which the water changes its physical and chemical composition as it moves. Thanks to geological forms of movement, water may turn out to be under conditions where the upward pressure exceeds the hydrostatic pressure.

The above enables us to outline, to a first approximation, three types of the geological form of water movement:

(1) The atmospherogenic, which is observed in the near-surface part of the Earth's interior. It is characterized by the prevalence of free infiltration water as a consequence of the variety of the hydrostatic pressures, but is accompanied also by other forms of movement which cause change both of the water and its environment. The pressure on the groundwater in this case does not exceed the hydrostatic. This type of movement takes place in the hypergene zone where the temperature and pressure are close to those of the surface (usually down to a depth of 0.5–1 km, and rarely to 3 km), but in favourable conditions it reaches a depth of 5–8 km.

(2) Lithogenic, when the transport proceeds in a continuous interaction with the processes of lithification of the sedimentary layers. The squeezing out of some of the water in conditions of the compaction of the sediments, the physical and chemical bonding of water by the rock and its subsequent renewal during the recrystallization of the latter come into this category. The abnormal pressures in this instance are, as a rule, higher than the hydrostatic pressures. Such a form of movement is to be found in the submarine regions of the Earth's crust and in the lower layers of the sedimentary envelope at depths of not less than 1–3 km.

(3) Magmatogenic, which is characteristic of the deep parts of the subsurface hydrosphere. The typical features of such a form of movement of water are connected with the influence of high temperatures and pressures, the separation of the water from the magmatic melt, or from the metamorphosing rock in conditions of high gas saturation, and the formation of long-living hydrothermal spring systems. Hydrothermal spring systems are, as a rule, in a state of strain; they contain water in the liquid form, a water–steam mixture, and hot vapours.

'It is easier to study the movement of the satellites of Jupiter than the flow of water', said the great naturalphilosopher G. Galileo several centuries ago (Biswas, 1975). He had in mind the movement of water in a stream, the behaviour of which has now been thoroughly studied. However, water exchange in the Earth's crust and the paths of water distribution in the Earth still contain much which is not clear. Therefore regarding the study of the subsurface hydrosphere, especially its deep regions, the words of Galileo still ring true. To become acquainted with the laws of motion of water in the Earth in all their variety is one of the most important tasks of hydrogeology. We will begin with the basic problem — the general circulation of water.

4.2 Modern ideas on the circulation of water

The famous hydrogeologist B.L. Lichkov (1962, p. 26) wrote: 'The circulation of natural water on the planet, although not as yet completely understood, must be recognized by all as critically significant for hydrogeology, and for the land areas; the primary significance must be given to their moisture content. Thus a new hydrogeology must be constructed on three principles — the unity of water, its circulation, and the moisture content of the land. In a word, the circulation of water is one of the cornerstones of modern hydrogeology.'

Studies of the circulation of water over geological time have been examined, and in a number of cases even now continue to be examined one-sidedly, from the hydrogeological point of view only, whose circulation mechanism amounts to the movement of water under the action of the thermal energy of the sun (evaporation and movement to the atmosphere) and the force of gravity (the movement of atmospheric precipitation, water in rivers, and groundwater). The circulation of water usually amounts to this: that the atmospheric precipitation falling on to the land partly forms the surface runoff and partly, percolating into the ground, forms groundwater; the surface water and groundwater under the influence of gravity flow downwards into depressions — river valleys — and form rivers. The water in rivers, descending into the terminal basins of runoff and evaporating, constitutes the source of new atmospheric precipitation (see Fig. 2.1).

M.I. L'vovich (1974) distinguishes the following fundamental links: atmospheric, oceanic, land (including lithogenic), soil, river, lake, glacier, biological, and agricultural. Each of these links plays its role in the chain of water circulation.

In spite of the extreme importance of moisture and the wide distribution of such a circulation, it far from exhausts all the varieties of water movement. It completely ignores the circulation of material, including water, within the Earth's crust. Therefore modern hydrogeology cannot be based only on the concepts of water exchange with the surface of the Earth, but must be based on a study of all the varieties of the cyclic movement of water in the Earth's crust.

One of the first deficiencies of a similar circulation system was recognized by V.I. Vernadskii, who urged that existence of thermodynamic membranes (or envelopes) be considered when investigating all geochemical (including aqueous) phenomena. He distinguished primary cyclic processes, which take place in the limits of several thermodynamic membranes,

and secondary ones which take place in a single membrane (Vernadskii, 1954).

Basing their arguments on his ideas, A.E. Khod'kov & G.Yu. Valukonis (1968) proposed the distinguishing of four basic types of water cycle: climatic, geological (sedimentary), metamorphogenic, and hydrogeological. The first type of cycle corresponds completely with the circulation system as understood by M.I. L'vovich and other hydrologists. The geological cycle is effected by processes which form sediments; together with the deposition of material in the epicontinental seas, lagoons or lakes there is buried syngenetic water. Subsequently, after the regression of the seas the buried water remains in the rocks of the land and takes part in the formation of water-bearing strata. As for the metamorphogenic cycle, this is considered to be the binding or freeing of groundwater by the rocks during the course of the geological development of the Earth's crust: the hydration of the rocks in the hypergene zone and their dehydration in the zone of metamorphism. Finally, the hydrogeological cycle is characteristic of the interaction of infiltrogenic, sedimentogenic, and metamorphogenic water.

From consideration of the cycle types proposed by A.E. Khod'kov and G.Yu. Valukonis, as described above, it is not difficult to see that neither the metamorphogenic nor the hydrogeological types proceed from the major circulation system to the full cyclic processes, and therefore they are not of independent significance, and must be regarded as part of a more general circulation system. In fact, of the four cycle types proposed by these authors only two meet the requirements demanded by cyclic processes.

The problem of the circulation of water in the Earth's shells has been discussed in more detail by A.N. Pavlov (1977). He justifiably regards it as amounting to water exchange between the ocean and the land. The first, the climatic cycle, proceeds, according to A.N. Pavlov (1977), under the influence of meteoric and hydrologic factors; the second, the lithogenic cycle, is the result of geological processes such as sediment accumulation, tectonic movements, volcanic activity, metamorphism, and granitization. In each type of cycle several minor cycles can be distinguished, the sum of which determines the individuality of each of them. By the word cycle is understood the totality of the processes which cause the continuous exchange of water between the ocean and the land, and which proceed with more or less uninterrupted sequence and rate in the climatic circulation of the atmospheric water via surface and subsurface runoff; in the lithogenic circulation there are hydrogeological and geological (*sensu stricto*) cycles. Of the cycles just enumerated we are interested in three: the subsurface infiltration, the hydrogeological, and the geological (*s.s.*). The cycles of the subsurface infiltration include groundwater mainly in the zone of active water exchange, which takes part in the climatic circulation. The hydrogeological cycle is an interval of time between two marine transgressions across a basin of deposition, separated by a phase of uplift, folding, and erosion. During this time the sea water, contained in the sediments, is pressed out, and then becomes entrained in the climatic circulation. A.N. Pavlov restricts the geological cycle (*s.s.*) to the movement of the plates of the lithosphere, the bonding of water in serpentine and its dehydration (see Section 3.2).

A.N. Pavlov's scheme of the lithogenic circulation of water with all its novelty and progressiveness is the result of the consideration of the interdependence of the water of the mantle and of the Earth's crust, and also the transport of water in a conserved form, but it has nevertheless fundamental flaws and cannot be accepted completely in the form proposed by the author.

The basic defect of this scheme is that in a single circulation the combined cycles are not absolutely connected with one another and have different characters. Thus, the hydrogeological cycle depends upon the accumulation of and burial of sedimentary rocks in sedimentary basins, and their subsequent dewatering is sedimentary in character. Properly the geological cycle depends upon the entry of material from the mantle into the Earth's crust and its interaction with the rocks and the water of the oceans. Water is involved in the lengthy process of ocean crust formation and, possibly, subsequently in the continental crust. These two cycles are branches of water circulation systems which are different both in the nature, duration, and scale of the phenomena involved, and in their geological role.

Another defect is that many important links in the circulation of water in the geosynclinal stage of Earth processes are missing from A.N. Pavlov's viewpoint. In fact the buried deposits after sedimentation are far from always found in the zone of influence of the climatic circulation without extensive metamorphism; this is characteristic only of a part of platform formations and structures. But in geosynclinal conditions sediments become buried to such a depth that the processes of metamorphism and local melting of material are possible, i.e. the formation of magmas. But what happens in this case to the water which is entrapped in the sediments? Pavlov does not answer this question, because his scheme does not take account of processes of this sort. If these are not considered the circulation of water cannot be considered complete, even in the most general way.

Studying the existing schemes, it is possible to propose a more fundamental model of the circulation of water in the Earth. In this scheme there are two circulations, the *hydrologic* (climatic), and the *geological*. Both of these are connected in the closest possible way with the water of the oceans: the first depends upon the processes of transport of moisture from the ocean via the atmosphere to the continent; the second is the result of the burial of marine (or other) waters during sedimentation and their taking part in the geological processes right up to the final stages of metamorphism and the formation of regenerated water. The other abovementioned cycles are only a part of these two circulation systems.

4.3 The hydrologic (climatic) circulation of water

One of the most important properties of water, the one to which we have become accustomed, is its ability to change its state in the different thermodynamic conditions which

occur on Earth. The constant influx of energy onto the surface of the Earth results in the evaporation of vast amounts of water, which leads to the supply of atmospheric moisture. This evaporation, naturally, proceeds more intensively in the hot equatorial regions, and reduces towards the poles. It is increased in windy conditions, and reduces with increasing salinity. Plants play a major role in the evaporation of water: trees of the equatorial rain forests are very rightly compared with gigantic pumps, pumping moisture from the soil to the atmosphere. Even the birch in middle latitudes evaporates up to 20 kg of water per day. But of course the main supplier of moisture to the atmosphere is the ocean.

Having entered the atmosphere, water, together with air, is entrained in a complex system of air currents. In certain conditions the water vapour begins to condense and gather into drops, which fall on to the Earth in the form of rain or snow, and in cold regions as so-called 'horizontal precipitation' (hoar frost, dew). The amount and type of precipitation which falls depends upon the actual geographical conditions: distance from the sea, height above sea level, the aspect of high ground relative to the air currents bringing moisture, and many other factors.

The bulk of the atmospheric precipitation which falls on the land because of gravity can be grouped into streams, rivulets, and rivers, flowing into the ocean, completing the hydrologic cycle (see Fig. 4.1). The intensity and scale of this circulation depends entirely on the amount of solar energy reaching the Earth, which, via a complex mechanism of movement of water masses, determines the climate. Hence it is possible to call this climatic too.

As a result of the hydrologic circulation the quantity of groundwater is constantly being renewed. The atmospheric precipitation (X) is divided on the surface of the Earth: one part of the precipitation (V_1) evaporates on the spot back to the atmosphere; a second part, flowing over the surface of the Earth to the world ocean, constitutes the surface runoff (V_2), and finally a third part percolates through the soil into the rocks, forming the subsurface flow (V_3). Consequently, the

Figure 4.1. The hydrologic cycle. The values given are the corresponding amounts of the world water balance: without partheneses – in km³, inside parentheses – mm (after M.I. L'vovich).
1, precipitation; 2, surface runoff; 3, underground runoff; 4, evaporation.

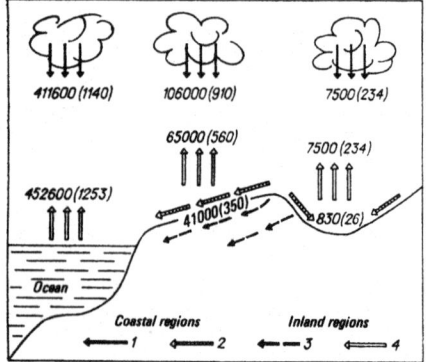

following relationship holds:

$$X = V_1 + V_2 + V_3 \qquad (4.1)$$

The values of these three components vary and depend upon the actual conditions: the relief, the type of rocks, their porosity and the degree to which they are fissured, the air temperature, the amount of precipitation, evaporation, the type of vegetation, etc. Thus for the European part of the USSR, G.V. Bogomolov (1975), quotes the figures for the infiltration of atmospheric precipitation (in percentage of annual precipitation): loess-like rocks – 15–20%, clay and loam – 10–12%, sandy rocks – 22–28%, fissured rocks – 35–45%, karst rocks – 50–60%. As a whole for the territory of the USSR the average value apparently does not exceed 30%. In Holland infiltration into the dune sands, which are denuded of vegetation, reaches 83% of the precipitation (700 mm/year), but in those areas which have a vegetation cover 48–52%. G. Druen quotes for the centre of France the value of infiltration for Albian age sands as 20% of the precipitation, and for the regions in the mountainous areas of Algeria for exactly the same rocks the figure is 10%. According to the observations of S.L. Shvartsev the magnitude of the infiltration in the humid tropical countries (Guinea) reaches 65% in porous laterites.

The rate of movement of water in rocks is considerably less than in open catchment areas. Therefore surface runoff water takes part in the circulation significantly more often than the water in the subsurface flow. Obviously, the deeper the water sinks, the slower it moves in the rocks. Sooner or later this water appears again on the surface and takes part in the hydrologic circulation. The discharge of the subsurface flow, i.e. its reappearance on the surface, may occur both above and below the level of the world ocean.

An enormous quantity of water is involved in the hydrologic (climatic) circulation. Altogether in the atmosphere, according to the figures given by M.I. L'vovich (1974), there is about 14 000 km^3 of water, about 11.6 times more than that contained in the rivers: and this amount is completely renewed roughly every ten days – 36 times a year. The volume of all the river water on Earth – about 1200 km^3 – is completely changed every 11–12 days, or on average 32 times a year. The water in lakes and marshes is renewed over a considerably longer period. The exchange of water in glaciers takes place at a very slow rate, but the circulation of groundwater proceeds still more slowly, changing over periods of 330 years (in the zone of active exchange) to 10 000 years, with an average time of 5000 years (see Table 2.1). Nevertheless, all these are links in a single interrelated circulation, which consists of three basic cycles: atmospheric, surface runoff, and subsurface flow.

The process of the hydrologic circulation of water and its individual parts may be expressed quantitatively by the water balance. The water balance of any territory, i.e. the accumulation and disposal of water within its boundaries and any intervals of time, depends on climatic factors and the character of the Earth's surface. The relationship of the elements of the water balance, precipitation, evaporation,

surface runoff, and subsurface flow, in given physical-geographical conditions, is, on average, practically constant over a period of years and determines the average flow of rivers and the water resources of the given region.

The laws governing the changes in the reserves of water are usually expressed in the water balance equation. In the general case this equation for any territory over any interval of time has the following form:

$$X + K + Y_1 - Y_2 - Z \mp W_1 \mp W_2 + U_1 - U_2 = 0 \qquad (4.2)$$

in which X = the amount of precipitation; K = the condensation moisture; Y_1 = the inflow of river water from other regions; Y_2 = the runoff and drainage of rivers beyond the boundaries of the given region (including water supply); Z = evaporation; W_1 = change in groundwater reserves; W_2 = change in moisture reserves on the surface of the catchment area; U_1 = inflow of groundwater from neighbouring regions; and U_2 = runoff and drainage of groundwater into neighbouring regions at a lower level than that of the drainage of their river beds.

For practical purposes some of the terms of the above equation are combined or made equal to zero. For example, because of the practical difficulties in determining the condensation, this component of the water balance is usually calculated together with the precipitation or the evaporation. The influx of river water in calculating the water balance of a river catchment area from its source to any section is omitted. The amount of accumulated moisture W_1 and W_2, and for sufficiently large basins the subsurface water exchange U_1 and U_2 as well, almost equalize over a period of years. Therefore, in practice, when calculating the average balance of any territory over a large number of years, the equation:

$$X + Y_1 - Y_2 - Z + U_1 - U_2 = 0 \qquad (4.3)$$

is used, and for the river catchment area or for the whole of an inland sea or marine basin an equation of the form:

$$X - Y_2 - Z + U_1 - U_2 = 0 \qquad (4.4)$$

is used. On the basis of the above the water balance of the USSR was calculated from the data accumulated over many years by the Hydrometservice of the USSR. The figures are given in Table 4.1, and those for the whole world in Table 4.2.

The figures in Table 4.2 show that more water evaporates from the surface of the ocean than falls as precipitation, and this difference is quite discernible and reaches the magnitude of the river runoff. As a whole for the Earth evaporation is equal to precipitation. It is important to emphasize that every year the rivers carry 41 000 km^3 of water to the ocean, of which about 12 000 km^3 (29.3%) consists of subsurface flow. This is the magnitude of the subsurface component of the hydrologic circulation.

The relationship between the subsurface flow and the surface runoff is easily found from Table 4.3. It is not difficult to see that the total runoff is 23% (Africa) to 43% (Europe) of the total precipitation. Of this quantity 48 mm of the African contribution is due to subsurface flow, and up to 210 mm from South America, or 24–36% of the total river flow.

South America is the richest continent in terms of water resources (per unit area): the total river and subsurface runoff of this land mass is almost twice that of Europe, which takes second place. Then comes Asia, North America, and Africa. Australia is the poorest in terms of river runoff. Asia is the richest in terms of volume of flow over the whole area of the continent, for which the figures exceed those of South America, Europe, and Australia.

However, it is necessary to bear in mind that all the figures quoted above have been obtained by analysing the hydrographic statistics, and therefore do not take into consideration the runoff, which, avoiding the rivers, moves directly to the epicontinental seas and the ocean. The latter, in round figures is, according to R.L. Neis, 7000 m^3/day, or 224 km^3/year for the whole planet (Dzhamalov et al., 1977). The determination of the subsurface flow to the ocean is one of the most important tasks for hydrogeology in the near future.

Thus the hydrogeological circulation of water, which embraces the upper part of the Earth's crust, leads to the formation of huge masses of fresh (i.e. non-saline) groundwater, which are distributed at relatively shallow depths. At the same time part of the water of this circulation penetrates to great depths in zones of deep fracturing and very permeable rocks; here it is heated and gives rise to the formation of thermal waters of various compositions which are formed extensively in fold mountain areas, and more rarely to the formation of solutions in the deep layers of platforms and deep narrow intermontane valleys. Ancient and recent groundwater of infiltration origin is formed as a result. This circulation is brought about by the infiltration of water through the rocks from regions of high pressure to those of low pressure, as a rule, from mountainous areas to the foothills, from continents to the seas and oceans.

4.4   The geological circulation of water

The geological circulation of water, in contrast to the hydrologic, is the result of continuous movement of the Earth's crust in the vertical and horizontal directions depending on the general geological processes of the Earth. The initial stages in structural development are basins of sedimentary accumulation at the stage of formation of geosynclinal depression, where the accumulation of huge thicknesses of sedimentary rocks takes place, mostly of marine origin.

The freshly formed sediment on the bottom of such basins is usually 'a fluid mass, which is very wet, rich in micro-organisms and consisting of very heterogeneous chemico-mineralogical material, partly solid, partly liquid and gaseous' (Strakhov, 1962). For this deposit the presence of a great quantity of water, which in a number of instances reaches 100% and more, is characteristic. Thus according to N.V. Tageeva & M.M. Tikhomirova (1962) the average natural moisture content and the quantity of free water in recent bottom deposits of the northern Caspian Sea are 71.4% and 66.7% respectively, but the maximum value reaches 140% and 122%. O.V. Shishkina (1972) cites a multitude of data on the moisture content of recent sediments of marginal depressions, marginal regions of open and inland seas, and the open ocean, an analysis of which shows that in all these cases there is a moisture content of 50–70%, and more rarely 30–40%.

As the depth to which the zones of sedimentary accumulation are buried increases, the water content of the sediment begins to decrease because of the pressure of the overlying layers, and the compacting of the sediments leads to their being changed into rocks. They undergo compaction and from the muds are formed clays and eventually shales. At the same time the porosity decreases and water is squeezed out. This is particularly characteristic of clay sediments.

By the time they have sunk to a depth of a few hundred

Table 4.1. *Water balance of the USSR by sea basins (Protas'ev, 1967)*

Sea basins	Area × 10^3 km^3	Elements of the water balance						Runoff coefficient
		Vol. (km^3)			Depth (mm)			
		Precipitation	Runoff	Evaporation	Precipitation	Runoff	Evaporation	
White Sea and Barents Sea	1 192	876	408	438	710	341	369	0.48
Baltic Sea	661	506	171	335	765	259	506	0.34
Black Sea and Azov Sea	1 347	889	159	730	660	118	542	0.18
Caspian Sea	2 927	1 440	300	1 140	491	102	389	0.12
Karsk Sea	6 579	3 640	1 324	2 316	553	201	382	0.36
Laptev Sea, East Siberian Sea, Sea of Chukotsk	5 048	2 135	1 038	1 097	423	206	217	0.49
Bering Sea, Sea of Okhotsk, Sea of Japan	3 269	2 126	890	1 236	652	273	379	0.42
Regions of little runoff, Kazakhstan and Central Asia	2 420	723	125	598	299	52	247	0.17
Territory of all the basins within the USSR	22 013	11 694	4 208	7 486	531	191	340	0.36
Total for USSR	22 013[a]	11 694	4 358	7 336	531	198	333	0.37

[a]Not including the islands of the Arctic Ocean.

Table 4.2. *Annual water balance of the Earth (L'vovich, 1974)*

Element of the water balance	Volume (km³)	Depth (mm)	% of total world precipitation
Coastal regions (116 800 × 1000 km²)			
Precipitation	106 000	910	20.2
River runoff	41 000	350	7.8
Evaporation	65 000	560	12.4
Inland regions (32 100 × 1000 km²)			
Precipitation	7 500	238	1.4
Evaporation	7 500	238	1.4
World ocean (361 100 × 1000 km²)			
Precipitation	411 600	1 140	78.5
Inflow of river water	41 000	350	7.8
Evaporation	452 600	1 254	86.3
Total surface of the Earth (510 000 × 1000 km²)			
Precipitation	525 100	1 030	100
Evaporation	525 100	1 030	100

metres the porosity of the clay sediments has already greatly decreased, and they have lost a considerable amount of free water. As they sink further, the rate of compaction of the clay and the emission of water from the mass decreases (Mukhin, 1965).

It is important to emphasize that the graphs of porosity against depth (Fig. 4.2) for one and the same rock (clay) differ sharply even in a single region, but with depth the porosity always diminishes. Thus according to N.B. Vassoevich (1960), at a depth of 400–500 m it amounts to 35–40%, at a depth of 2000 m it is already 20%, and when the sediments are buried to a depth of 3000 m and more it is less than 10%.

The porosity of sandy and carbonate rocks decreases with depth more slowly than does that of clays (Fig. 4.3). With limestones at a pressure of 900 bar the least porosity is found in the biogenic varieties, and the greatest in the fine-grained types. The porosity of karst-like sandstones depends not only upon the depths of the deposits, but also on the age of the rocks, and on the geothermal conditions. Under stable unvarying conditions in the sections that are characterized by

Table 4.3. *Relationship of the total surface and subsurface runoff (L'vovich, 1974)*

Element	Europe	Asia	Africa	North America[a]	South America[a]	Australia[b]	Total land mass[c]	USSR
Area × 10⁶ km²	9.8	45.0	30.3	20.7	17.8	8.7	132.3	22.4
				in millimetres				
Precipitation	734	726	686	670	1 648	736	834	500
River runoff								
total	319	293	139	287	583	226	294	198
subsurface	109	76	48	84	210	54	90	46
surface	210	217	91	203	378	172	204	152
Total moisture	524	509	595	467	1 275	564	630	348
Evaporation	415	433	547	383	1 065	510	540	300
				in km³				
Precipitation	7 165	32 690	20 780	13 910	29 355	6 405	110 305	10 960
River runoff								
total	3 110	13 190	4 225	5 960	10 380	1 965	38 830	4 350[d]
subsurface	1 065	3 410	1 465	1 740	3 740	465	11 885	1 020
surface	2 045	9 780	2 760	4 220	6 640	1 500	26 945	3 330
Total moisture	5 120	22 910	18 020	9 690	22 715	4 905	83 360	7 630
Evaporation	4 055	19 500	16 555	7 950	18 975	4 440	71 475	6 610
				Relative magnitudes				
Subsurface runoff % of total	34	26	35	32	36	24	31	25
Coefficient of recharge of rivers by groundwater	0,21	0.15	0.08	0.18	0.16	0.10	0.14	0.13
Runoff coefficient	0.43	0.40	0.23	0.31	0.35	0.31	0.35	0.40

[a] Excluding the Canadian Archipelago and including Central America.
[b] Including Tasmania, New Guinea, and New Zealand.
[c] Excluding the Antarctic, Greenland, and the Canadian Archipelago.
[d] Excluding 300 km³ of transitory runoff.

a high geothermal degree the porosity decreases more slowly than in sections confined to the thermally active regions.

G.I. Teodorovich & A.A. Chernov (1968) established three stages of compaction of lime muds, silts, and clays of the Apsheron oil-bearing region: (1) at depths from 0 to 10 m the porosity of clays is 66–40%, sandy material 56–40%; (2) at depths between 10 and 1400 m the porosity of clays and sandy sediments decreases almost in parallel to 21% and their moisture content approaches the hygroscopic; (3) from 1400 to 6000 m the porosity of sandy sediments decreases to 16% and that of clays to 8%. Analogous behaviour has been noted in the works of other authors.

Thus, as the depths to which sediments sink into sedimentary basins increases, their porosity steadily decreases. According to some research workers (Weller, 1961; Vassoevich, 1960; Mukhin, 1965), the porosity varies with depth either logarithmically:

$$m = m_0 - a \ln bH \tag{4.5}$$

or exponentially

$$m = m_0{}^{-cH} \tag{4.6}$$

in which m = the porosity of the rock at a depth H; m_0 = the initial porosity; and a, b, and c are numerical coefficients which are determined by statistical analysis of actual data.

However, as A.E. Gurevich *et al.* (1972) rightly emphasized, attempts to establish the relationship between the porosity of the rock and the depth of the horizon meet with insurmountable difficulties of a mathematical nature (it is impossible to give the parameters with the necessary precision and in sufficient detail). Hence various simplified and limited models have to be considered.

The general laws governing the reduction of porosity with depth break down because of the action of various factors: diagenetic changes, mechanical stresses within the rock, the scale of the destruction of the rock by water, age, the intensity of the hot currents, etc. Hence even at great depths (of the order of 8–10 km) the porosity, in a number of cases, remains quite high; this can be seen from the data on super-deep wells, which show at these depths considerable

quantities of water, oil, and gas. Thus in 13 of the main oil-producing regions of the USA in 1965–67 1024 wells of depth greater than 4.5 km were sunk: of these gas was found in 212, and oil in 60. Some authors, G.D. Afanas'ev (1966) for instance, admit that the material of the Earth's crust has micropores and microfissures right down to the Moho, i.e. to depths of 40–50 km on land.

Pore water, squeezed out of the clay sediments as they sink, returns initially to the sedimentary basin. However, as the sediment sinks to a great depth in conditions of inhomogeneity of the percolation properties, movement along the stratification gradually begins to dominate (Mukhin, 1965). Such a lateral movement leads to the transport of liquid to the marginal regions of the sedimentary basin, where, because of changes in the sedimentary conditions, shallow and coastal water sediments are usually coarser and consequently more permeable. Therefore it is through the coastal facies that the hydraulic connection between the water squeezed out of the silty sediments together with the water of the sedimentary basin, or the groundwater in the permeable sediments is formed.

There is sedimentogenic water in sandy areas from the very beginning of their formation, and this is of the same age as the sediments. But it gradually gives way to epigenetic water squeezed out of the clays because the geostatic pressure in the compacting layers of clay is twice as large, or more, than the hydrostatic pressure which is characteristic of poorly compressible aquifers. The geostatic pressure in clays is transmitted to the water which they contain and is rather higher than hydrostatic pressure in layered aquifers. This circumstance also determines the movement of water from the clays to the sands.

The bulk of the free water is squeezed out in the first few hundred metres as the sediment sinks. But the squeezing out of the water by no means ceases: subsequently both physically and chemically bound water are included in the process. The attention of research workers is drawn primarily to the inter-layer water of montmorillonite, which, as is well known, contains more than 20% of water in a bound condition.

Interesting aspects of the fluid-release mechanism in Gulf Coast (USA) mudrocks have been noted by M.C. Powers (1967), who was one of the first to draw attention to the

Figure 4.2. Graph of compaction of clays of different ages (after G.V. Bogomolov and Yu.V. Mukhin).

1, Lower Cambrian; 2, Lower Jurassic; 3, Upper Jurassic; 4, Maikopski; 5, Ashperonski; 6, Quaternary.

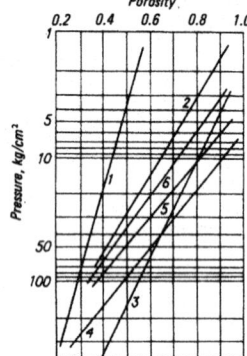

Figure 4.3. Graph of the compaction of sands and clays (after G. Mulde).

1, clays and clay-shales; 2, sands and sandstones.

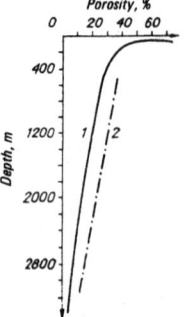

possibility that water is released when montmorillonite alters to illite. This process was examined in more detail by J.F. Burst (1969), who distinguished three separate stages in the dehydration sequence in clay sediments: (1) initial expulsion of pore water and partly bound (excess, more than two water interlayers) water down to a depth of 0.7–1.0 km, because of overburden pressure, this initial compaction reduces the water content from 80% to 30% by volume (stage I in Fig. 4.4); (2) at depths greater than 1.0 km the transformation of montmorillonite to illite begins with the displacement of considerable quantities of water, 10–15% of the compacted bulk volume; (3) very slow elimination of the last interlayer of water over tens to hundreds of millions of years.

Later E.A. Perry & J. Hower (1972) subdivided the stage of montmorillonite dehydration further: (1) an initial stage in which up to 65% of the montmorillonite is transformed into illite with a reduction in volume of the water in the montmorillonite of up to 10%; this stage is complete at depths of 2.0–2.7 km at a temperature of 90–110 °C; (2) an intermediate stage in which up to 80% of the montmorillonite is transformed into illite; (3) the moisture of the rock is now reduced to 5% by volume. A fourth stage is added to the above three stages (see Fig. 4.4) which, it is assumed, takes place at depths greater than 3.5 km and consists of the final crystallization of montmorillonite into illite with the loss of the remaining 5% of the interlayer water. Perry and Hower have calculated that the transformation of montmorillonite into illite is accompanied by the squeezing out of 6.6% by weight or 15.5% by volume of water into the reservoir rocks.

Consequently, the lithogenic changes of clays, which take place during the stages of diagenesis and katagenesis are accompanied by reconstruction of the crystal structure. It is because of this that the volume of the mineral portion of the rock reduces and water passes to the free condition. Such a process proceeds at various depths, but especially intensively between 2 and 3 km. It is also impossible not to consider the circumstance that at all stages of the sinking of the sediment chemical interaction takes place between it and the pore water, which leads on the one hand to the formation of secondary (authigenic) clays and carbonates which cement the sedimentary rock, and on the other hand to ionic dissociation of the molecules of the water that takes part in these reactions (Shvartsev, 1975). The scale of chemical dissociation of water in some cases can be significant and cannot be neglected.

Figure 4.4. Compaction curves for sediment dehydration.
1, according to J.F. Burst (1969); 2, according to E.A. Perry & J. Hower (1972).
I, II, III, and IV, stages of water expulsion from the sediment (see text).

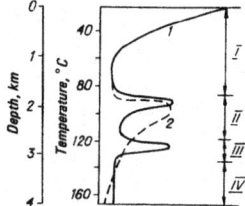

It must be pointed out that water, coming into close contact with clay particles, is characterized by peculiar structural properties; it has a reduced dielectric constant and increased density (up to 1.4 g/cm^3). Therefore during the transition to the free condition, in which the density is 1 g/cm^3, an increase in volume of up to 40% is observed, which must cause an increased pressure in the clay, forming channels in it along which the excess water is squeezed out into reservoir rocks.

Consequently, the formation of deposits in sedimentary basins is accompanied by the trapping and burial of huge volumes of water, and the sinking and compacting of these deposits by the expulsion of the water and by its ionic dissociation. The expelled water can return to the upper part of the basin in which the accumulation of sediments is proceeding, but subsequently it moves into the reservoir rocks, mainly sandstones which are laid down between the compacting layers of clay. In the sandy layers from the very start of their formation there is also some sedimentogenic water. This, however, gives way to water expelled from the clay because the geostatic pressure in the compacted clay layers is twice the hydrostatic pressure, which is the prevalent pressure in practically incompressible sandstones. The geostatic pressure in clays, being transmitted to the water they contain, creates an abnormal pressure which is rather higher than the hydrostatic pressure in the reservoir rocks. Subsequently the movement of water in water-bearing horizons proceeds in response to the hydraulic gradient, which is directed from zones of the greatest downwarping and dehydration to places which are tectonically relatively quiet.

The expulsion of sedimentogenic water from clay rocks also takes place at depths greater than 3000 m; however here the rate is considerably reduced because the porosity of clay at these depths is insignificantly small (2–5% and less). Nevertheless, a considerable quantity of sedimentogenic water is squeezed out from the clay rocks at great depths.

Because the first stage of the geological cycle is associated with the burial of sediments and corresponds to the dispersal of detritus stage, it may be termed *sedimentogenic*. In platform conditions it leads to the formation of a great quantity of sedimentogenic water which, after the regression of the sea, is found to be on the land area and becomes generally entrained in the hydrologic cycle. However, because of low rock permeability these processes (especially at considerable depth) proceed extremely slowly. Naturally, in stable, uniform conditions, the older the basin the greater the probability that it is permeated with infiltration water. However, in actual natural conditions the principle becomes complicated by the different structural features of the basins: their dimensions, the ratio of recharge to discharge, the length of the infiltration and sedimentation cycles, neotectonic features, etc.

In geosynclinal conditions the geological water cycle does not end with the sedimentary stage, because the downwarping of the region leads to further sinking of the sedimentary pile and the water which it contains, which is already being liberated during the process of metamorphism.

After the compaction and complete lithification of the sediments the physically bound water remaining in them constitutes about 5% by volume. When the sedimentary mass enters the zone of progressive metamorphism, which involves recrystallization within the rock, this water and also the water of crystallization and the hydroxyl content of minerals enter the composition of the clay minerals, and are distinguished as free water. Thus what occurs is not just a simple dehydration of the rocks, but their dehydroxylation too, i.e. the separation of hydroxyl groups (OH^-), and also of oxygen and hydrogen ions which, combining, synthesize water molecules (Shvartsev, 1975). Therefore the zone of metamorphism is not only a zone of rock dehydration but is also a zone of water synthesis. It is the synthesis of the water which makes the zone of metamorphism a qualitatively new stage in the geological circulation of water, for which the name *metamorphogenic* has been proposed.

The amount of water which separates during the metamorphic stage may reach 15–25% by weight. It is sufficient to note that the hydromica and montmorillonite contain only 10–12% hydroxyl, but kaolinite and chlorite contain 25–28%. Mica inevitably adds pore water, which is always present in one form or another. The separation of water during metamorphism takes place slowly according to the degree of recrystallization of the minerals, including the zeolite, greenschist, epidote–albite, amphibolite, and granulite facies (Sudovikov, 1964). Thus the water initially included in the sedimentary deposits gradually, during the metamorphic process, becomes completely free and occupies the fissures and intergranular spaces of the rocks; it then begins to move upwards. As a result the free water proceeds to the surface via the system of interconnecting pores. As S.N. Ivanov remarked (1970, p. 25), 'The rising current of water in the form of an independent fluid phase (strictly a liquid, supercritical fluid, in rare instances gaseous), which compensates for the sinking of the porous mass and the bound water of the sinking rock material, is extremely characteristic of and important for the Earth's crust.'

The processes of metamorphism are in the majority of cases caused by the sharp changes of the tectonic regime, when downwarp alternates with uplift and the regression of the sea. At the same time the fold mountain system is created. The process of mountain formation is accompanied by the breaking up of the region into pieces by the start of magmatic processes, with volcanic phenomena. All this leads to the buried sedimentogenic water entering into active interaction with the deep solutions and gases of magmatic origin. The magma in the zone of metamorphism either emerges from the subcrustal sources or is formed *in situ* as the result of the melting of the sedimentary and metamorphic rocks. In the latter case part of the water liberated is taken up by the magma with the isolation of the solutions during the solidification and crystallization of the melt and has in this instance juvenile aspects. However, such solutions are crustal by affinity; they are connected not with the mantle, but with the *magmatogenic* stage of the geological cycle.

Open fractures in the Earth's crust assist the active migration of buried water among the various rocks during the process of magmatic activity and its entrainment in the sphere of activity of the hydrogeological cycle. The rising currents of water are exposed to the influence of atmospheric factors which also favour the incorporation of the water into the general circulation.

Subsequently, according to N.K. Ignatovich (1947), a differential development of the fold structures takes place, which includes: (1) progressive unroofing of their central areas; (2) more intensive downwarping of the depressions of the foothills; (3) continual subsequent consolidation of the rock mass and the 'ageing' of the whole fold system. As a result the fractures and fissures, which go down to considerable depths, close up. It is true to say that some parts of the region develop differently; some 'age', others become 'rejuvenated' during tectono–magmatic activity.

The geological circulation of water, in contrast to the hydrologic, takes place in various thermodynamic shells of the Earth's crust; hence V.I. Vernadskii identified it with the initial cyclic processes of the crust. In the theory here developed, the complete geological cycle consists of three stages (sedimentation, metamorphic, and sometimes magmatic), each of which is, in a certain sense, independent. At the same time they are all part of the general circulation of material, which plays such an important role in the crust (Fig. 4.5).

The geological circulation, although it takes place within the crust, is by no means isolated from the other sources of water, exogenic (vadose), and endogenic (juvenile). At the

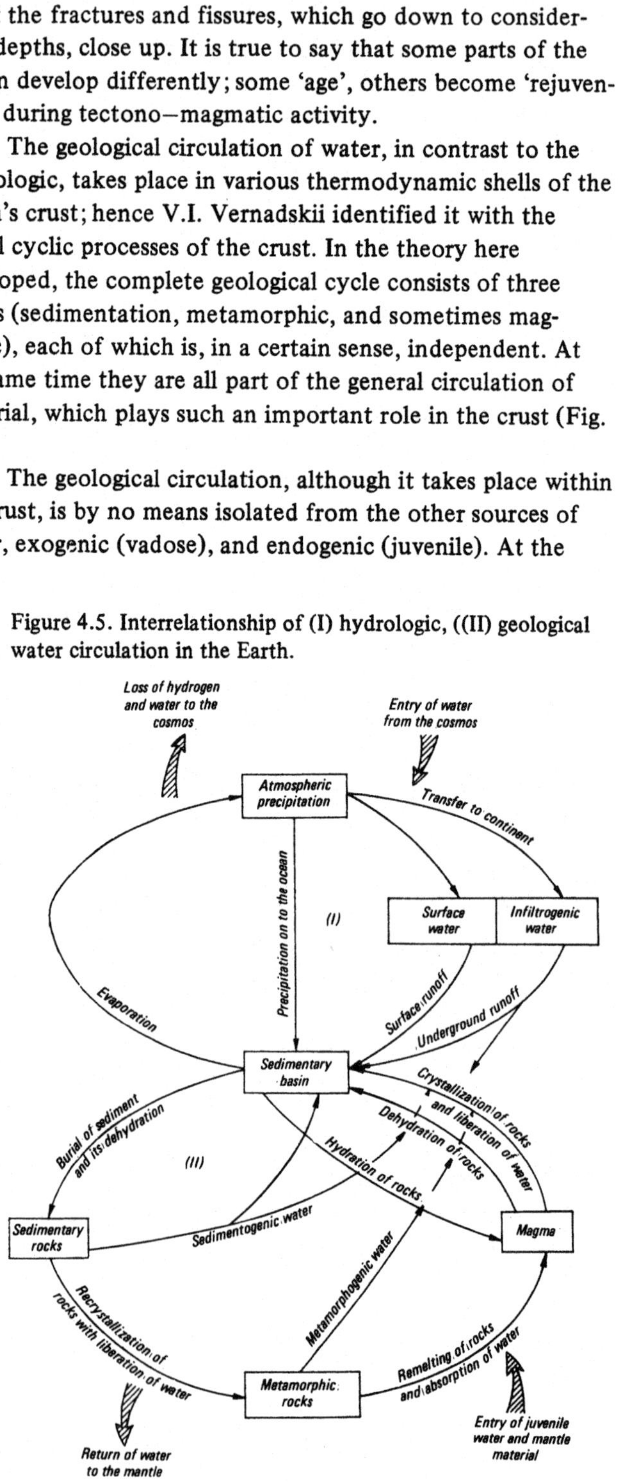

Figure 4.5. Interrelationship of (I) hydrologic, ((II) geological water circulation in the Earth.

same time some of the water is doubtless lost, returning to the mantle together with the rocks of the crust. Hence between the mantle and the crust there arises an uncompensated volume of water, the magnitude of which has not so far been determined. It may be that there is a circulation process existing between the mantle and the crust. But the solution to this problem is possible only by studying the mechanism of the formation of the crust as a whole, and in particular the interactions between the continental and the oceanic sectors. Therefore at the present time even the most general features of this process cannot be ascertained. Above all, the character and the intensity in the course of geological history have changed greatly.

On the other hand, the Earth exchanges material and energy with the cosmos. Water is one of the components of this exchange, it falls onto the Earth as part of stony meteorites and obviously in ice meteorites, and is lost in the form of oxygen and hydrogen ions which are formed when the water is dissociated in the ionosphere. And, although the scale of these phenomena remains uncertain, it is not possible to neglect them because, in the opinion of many scientists, the bulk of the hydrogen is dispersed and eventually leaves our planet. As is well known, V.I. Vernadskii attached great significance to the exchange of water between the Earth and the cosmos.

The hydrologic and the geological cycles take place mainly within the crust, but are connected with the water of the lower and the upper shells of the crust. Although water exchange takes place at different rates and on different scales, both cycles are closely interconnected (see Fig. 4.5) and in this interaction there lies the secret of the most important and characteristic features of the development of the Earth as a planet. The study of the interconnection of cycles is the key to the understanding of many geological processes, including the formation of groundwater.

References

Afanas'ev, G.D. (1966). New data on the relationship between the Earth's crust and the upper mantle. *Izv. AN SSSR, Ser. geol.* No. 11, 9–36.

Biswas, A.K. (1975). Man and water (from the *History of hydrogeology*). Leningrad: Gidrometeoizdat. 288 pp.

Bogomolov, G.V. (1975). *Hydrogeology and the fundamentals of engineering geology*, 3rd edn. Moscow: Vyshaya shkola. 320 pp.

Burst, J.F. (1969). Diagenesis of Gulf Coast clayey sediments and its possible relation to petroleum migration. *Bull. Am. Assoc. Petrol. Geol.*, 53 (1), 73–93.

Dzhamalov, R.G., I.S. Zektser & A.V. Meskheteli (1977). *Underground runoff into the seas and the world ocean.* Moscow: Nauka. 93 pp.

Gordeev, D.I. (1954). Main stages in the history of Russian hydrogeology. *Trudy Laboratorii gidrogeol. problem. Moscow, Izd-vo AN SSSR*, Vol. 7, 383 pp.

Gurevich, A.E., L.N. Kapchenko & N.M. Kruglikov (1972). *Theoretical basis of petroleum hydrogeology.* Leningrad: Nedra. 271 pp.

Ignatovich, N.K. (1947). Hydrogeological structures – the basis of the hydrogeological zonation of the USSR. *Sov. geologiya*, No. 19, 24–33.

Ivanov, S.N. (1970). The reasons for the formation of hydrothermal mineral deposits. In *Laws of distribution of minerals.* Vol. 9, pp. 20–47. Moscow: Nauka.

Kedrov, B.M. (1964). On the geological forms of movement in relation to other forms of movement. In *Interaction of sciences in the study of the Earth*, pp. 129–51. Moscow: Nauka.

Khod'kov, A.E. & G.Yu. Valukonis (1968). *The formation and geological role of groundwater.* Leningrad: Izd-vo Leningr. in-ta. 216 pp.

Lichkov, B.L. (1962). The value of a theory of the Earth and the necessity to create one. In *Geographical collection*, Vol. 15, Astrogeology, pp. 7–28. Leningrad: Izd-vo AN SSSR.

L'vovich, M.I. (1974). *World water resources and their future.* Moscow: Mysl'. 447 pp.

Mukhin, Yu.V. (1965). *The compaction processes of argillaceous sediments.* Moscow: Nedra. 200 pp.

Pavlov, A.N. (1977). *Geological water cycles on the Earth.* Leningrad: Nedra. 144 pp.

Perry, E.A. & J. Hower (1972). Late-stage dehydration in deeply buried pelitic sediments. *Bull. Am. Assoc. Petrol. Geol.*, 56 (10), 2013–21.

Powers, M.C. (1967). Fluid-release mechanisms in compacting marine mudrocks and their importance in oil exploration. *Bull. Am. Assoc. Petrol. Geol.*, 51 (7), 1240–54.

Protas'ev, M.S. (ed.) (1967). *Water resources and the water budget of the Soviet Union.* Leningrad: Gidrometeoizdat. 199 pp.

Shishkina, O.V. (1972). *The geochemistry of marine and oceanic silty water.* Moscow: Nauka. 228 pp.

Shvartsev, S.L. (1975). Dissociation and synthesis of water in the processes of lithogenesis. *Geol. i geofiz.*, No. 5, 60–9.

Strakhov, N.M. (1962). *Fundamentals of lithogenesis.* Vol. 1. Moscow: Izd-vo AN SSSR. 212 pp.

Sudovikov, N.G. (1964). *Regional metamorphism and some problems of petrology.* Leningrad: Izd-vo Leningr. in-ta. 550 pp.

Tageeva, N.V. & M.N. Tikhomirova (1962). *The geochemistry of porewater during the diagenesis of marine deposits.* Moscow: Izd-vo AN SSSR. 246 pp.

Teodorovich, G.I. & A.A. Chernov (1968). The character of changes with depth of the deposits of the productive level of the Apsheron oil-bearing region. *Sov. geologiya.*, No. 4, 83–93.

Valukonis, G.Yu. & A.E. Khod'kov (1973). *The geological laws governing the movement of groundwater, oil, and gas.* Leningrad: Izd-vo LGU. 304 pp.

Vassoevich, N.B. (1960). An attempt to construct a model curve for the gravitational compaction of argillaceous sediments. *Novosti neft. tekh., ser. 'Geologiya'*, No. 4, 11–15.

Vernadskii, V.I. (1954). *Collected works*, Vol. 1. Moscow: Izd-vo AN SSSR. 696 pp.

Weller, G.M. (1961). Compaction of sediments. In *Problems of petroleum geology in papers by foreign scientists*, pp. 84–137. Moscow: Gostoptekhizdat.

5

Subsurface water-bearing systems

The natural reservoirs of groundwater (subsurface water-bearing systems) have been described quite recently (Pinneker, 1977). This description with certain additions forms the basis of the present chapter.

5.1 The structural—hydrogeological subdivisions

In Section 1.3 attention was drawn to the vagueness of the term 'hydrogeological structure'. Like the term 'geological (tectonic) structure' it has several semantic meanings. However, if by the term geological structure is meant the totality of the tectonic forms which determine the features of a geological structure (Anon., 1955; Anon., 1970), and this term is not used as a synonym for the separate tectonic forms (Kosygin, 1974), then the concept 'hydrogeological structure' may also acquire uniqueness.

The hydrogeological structure must reflect the laws of distribution of groundwater. According to the semantic meaning it is not at all 'a natural reservoir filled with water', as certain hydrogeologists (Zaitsev & Tolstikhin, 1963, p. 16) consider, and neither is it a synonym for the term 'hydrogeological region' (Anon., 1978). The hydrogeological structure determines the spatial distribution of groundwater and its interrelationship with the rocks containing it. Subsequently this term was used only in such a sense. As for giving a name to a geological body which contains water, it must be said that another nomenclature is necessary.

The structural—hydrogeological subdivision is in many respects a vexed question and therefore not well worked out. Two approaches to the definition of geological structures can be discerned. If then attention is directed to the component parts of the hydrogeological structure (for example the hydrogeodynamic and hydrogeochemical structures), then there will be more such approaches.

We will now discuss the structural—hydrogeological classifications. One reflects the internal structure, i.e. the distribution of groundwater in the geological body containing it (Stepanovič, 1962; Pinneker, 1977), the others reflect the internal structure which demonstrates the interrelationship of the groundwater with the geological body (Ignatovich, 1945).

The classification shown in Fig. 5.1 is based upon the fundamental ideas of a geological structure as a concept of structural geologists (G.D. Azhgirei, Yu.A. Kosygin, and others) and is based on a distribution of groundwater which depends upon the water-collecting properties of the rocks containing it.

In this classification the geological space which is occupied by groundwater, and separated by various boundaries, is divided into simple and complex geological bodies. A simple body is a structural element of the hydrogeological structure. Properly a hydrogeological structure is distinctive of a complex body which is a combination of similar or different structural elements.

If we accept as a structural element the filling by water of an elementary geological space, a pore or fissure, we will obtain two simple forms of hydrogeological structure: *porous*, formed by water-bearing pores, and *fissured*, consisting of permeated fissures. When these structural elements are distributed in an orderly fashion throughout the groundwater reservoir, it is possible to speak of a complex hydrogeological structure. This or some other combination of them in the groundwater reservoir forms more complex varieties which are provisionally combined into two types of hydrogeological structure – *stratal*, and *vein-fissure* (see Fig. 5.1).

Just as geological structure enables us to call rocks porous, fissured, layered, etc., different kinds of groundwater – pore, fissure, stratal, vein-fissure, poro-stratal, etc. – are distinguished according to the geological structure.

The naming of the complex hydrogeological structure in Fig. 5.1 is made on the principle of 'shade and colour', i.e. the determining name is placed at the end of the complex qualifier.

The nomenclature of the types and classes of accumulations of groundwater (Zaitsev, 1961a) is very close to the above treatment of hydrogeological structure. Phonetically, it is true, they reflect the capacity of the medium, but in fact they describe the distribution of groundwater (Table 5.1).

Figure 5.1. Structural–hydrogeological subdivisions.

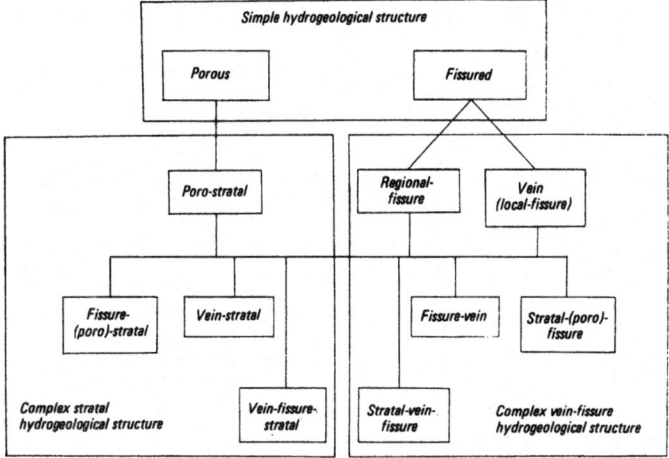

The Yugoslav scientist B. Stepanovič (1962) approaches the problem from rather different, but on the whole similar positions. He distinguishes three varieties of hydrogeological structure, *compact*, *non-compact*, and *combined*. A compact hydrogeological structure is characterized by water collectors with pore and microfissure permeability, when water forms a *mélange* with the rock which contains it. In the case of the non-compact hydrogeological structure the water in the reservoir rock is distributed unevenly (in veins, macrofissures, and voids) and cannot be connected in all directions into a unified 'mass': it is divided and 'broken up' by monolithic blocks. The combination of a compact structure of one aquifer with a non-compact one in a neighbouring aquifer is typical of a combined hydrogeological structure.

A more detailed systematization is given in Table 5.1; the poro-stratal and regional-fissure water is a compact hydrogeological structure, the local-fissure water is the non-compact, and the combined is a hydrogeological structure which combines the local-fissure water and the poro-stratal or regional-fissure water.

The hydrogeological classification of the structural–geological elements worked out by N.K. Ignatovich (1945), which F.P. Savarenskii (1947, p. 21) considered to be a 'living dynamic system of the accumulation and movement of groundwater in the Earth's crust', gives an idea of the structural peculiarities of the capacity of the environment. In other words, here we are dealing with a completely different approach to the understanding of the essence of the structural–hydrogeological subdivisions, with an approach which describes the spatial relationships of the water to the geological body containing it.

The following differences form the basis of N.K.

Table 5.1. *Classification of accumulations of groundwater (after I.K. Zaitsev)*

Type	Class	Division and subdivision
Stratal water	Poro-stratal Fissure-stratal Karsto-stratal	Soil Suspended water Groundwater Interstratal non-pressure water Interstratal pressure water (artesian): (*a*) exposed (*b*) closed
Fissure-vein water	Fissure water of the zone of weathering (regional-fissure water)	Water of the exposed weathering zone (fissure subsurface water): (*a*) stratal-fissure water (*b*) fissure-massif water (*c*) poro-fissure water, etc. Water of the closed weathering zone
	Local-fissure water	Water of tectonic faults Water of intrusive contacts Water of vein structures
	Fissure-karst water	Water of shallow karst (*a*) zone of vertical movement (*b*) zone of horizontal movement (karsto-subsurface water) Water of deep karst (*a*) zone of vertical movement (*b*) zone of horizontal movement (karsto-subsurface water) (*c*) zone of deep movement

Ignatovich's classification of structural-geological form:
(1) according to the degree to which they are closed: (*a*) open, (*b*) partly open, and (*c*) closed;
(2) according to the degree of ease with which they flow: (*a*) flowing, (*b*) partly flowing, and (*c*) non-flowing;
(3) according to the degree of flushing through: (*a*) flushed, (*b*) partly flushed (semi-flushed), and (*c*) unflushed (weakly flushed).

Properly a hydrogeological structure (an interrelationship of water with the geological body) is expressed by the degree of closure, although, strictly speaking, the structural element here is very provisional. Open structures (shields, organic massifs, shallow raised basins) are characterized by their considerable degree of erosion, and the open and hypsometrically high positions of the hydrogeological reservoirs. Upper and peripheral parts of deep valleys or depressions, and the slopes of uplifts have a partly open structure. A closed structure is characteristic of the deep parts of platform and intermontane depressions, which are sunk below the sea level and are isolated by confining beds.

5.2 The natural reservoirs of groundwater

The drawing of boundaries between geological bodies which are permeated with water to a large extent depends upon a clear idea of their systematics and their coordination with one another. The division of natural stores of groundwater and their spatial mapping is one of the priority tasks of hydrogeology.

Natural stores of groundwater have been given different names. As has already been noted, the name 'hydrogeological structure' is not appropriate. Many call it a 'pressure-water system' (A.M. Ovchinnikov, V.N. Kortsenshtein, and others), and sometimes a 'geohydrodynamic system' (P.F. Shvetsov and others). Having a special name, these terms only characterize the dynamics and do not touch upon the groundwater as a whole. The collective term (for the water-bearing layer, the groundwater basin, etc.) must be more embracing. In the foreign literature (USA, Czechoslovakia, West Germany) the term 'groundwater body' is used (Jetel, 1973; Richter & Lillich, 1975). But groundwater does not form an independent body; it permeates the geological body or, more correctly, is contained within it.

What then should a geological body that contains water be called? Perhaps the most suitable name would be the term 'water-bearing system'. However, it is too broad and general. Probably preference must be given to the term '*groundwater reservoir* (hydrogeological reservoir)'. This is a subsurface water-bearing system. The concept of natural reservoirs has been successfully used in oil and gas geology. In recent years this concept has also penetrated into hydrogeology (Pinneker, 1977). The word 'reservoir' with the addition of 'groundwater' or the definition 'hydrogeological' reflects the collective concept of the capacity of the medium of the subsurface hydrosphere sufficiently.*

---

*Nevertheless these names are not generally accepted. Therefore in the future the terms 'hydrogeological reservoir', 'groundwater capacity', and 'subsurface water-bearing system' will be used.

A hydrogeological reservoir, i.e. a subsurface water-bearing system, is not only a collector or store of groundwater. When the reservoir possesses a complex structure it combines both collectors and confining beds (or more precisely, isolators). A groundwater reservoir can be either an accumulator or a conductor of groundwater according to its position in space (see Fig. 2.7), whence it follows that this concept characterizes both static and dynamic groundwater.

Thus, a subsurface water-bearing system is a separate and water-containing geological body which is characterized by a veritable congeries of spatial distribution, transfer, and formation of groundwater.

5.2.1 The basic types of natural groundwater reservoirs

The differentiation of natural groundwater reservoirs takes into account the layered structure of the upper part of the crust. Within platform and folded regions two stages can be distinguished (Kamenskii *et al.*, 1959; Zaitsev & Tolstikhin, 1963, 1971):
(1) the fundament or basement is the lower stage, which is made up of mainly crystalline rocks, often crumpled into folds and intensively dislocated;
(2) the cover is the upper layer which is represented mainly by sedimentary rocks which are laid down in undisturbed conditions and which are dislocated only to a very small extent.

Young volcanic formations, dyke facies and fracture zones occupy a position outside the layer structure. The cover of unconsolidated Quaternary deposits is laid down from above (on the basement, the cover, the volcanic formations, etc.)

The basement where it emerges at the surface has an open hydrogeological structure, but the cover in the regions in which the basement sinks below the surface is distinguished by a definite degree of hydrogeological closure. In the basement there are concentrated fissures and vein-fissure water. At great depths it is permeated locally. Pore and various forms of stratal water are characteristic of the cover.

Depending on the relation between the structural–hydrogeological layers and the dominant type of groundwater, N.I. Tolstikhin, G.N. Kamenskii, and I.K. Zaitsev have established two basic types of natural water reservoirs:
(1) artesian basins are sunken, filled mainly with layered sedimentary rocks and consist of cover and the basement underlying it;
(2) hydrogeological massifs are basement outcrops, usually stripped of cover; fissured crystalline rocks have the greatest significance in this group.

In platform regions the predominance of huge artesian basins and the subordination to them of hydrogeological massifs is typical. The wide development of hydrogeological massifs and limited artesian basins is peculiar to folded regions.

Some hydrogeologists (Ovchinnikov, 1961; Kartsev *et al.*, 1971) avoid the term 'hydrogeological massif', calling it a *basin of fissure water*. However, a basin and a massif differ in the form of the geological body, the distribution, recharging mechanism, subsurface flow, and discharge of the groundwater. The water resources they contain are also different and in the final analysis so are the laws and the history of the formation of the groundwater they contain. Hence there is no basis for rejecting one of their fundamental subdivisions.

Artesian basins have negative tectonic forms – bowls, depressions, synclinal and enclosed sloping strata. They contain water-bearing layers. Poro-stratal groundwater is characteristic of the upper horizons; in the fracture zones, intrusive bodies, and rising basement, vein-fissure water is found, but the predominant form is rising stratal water. The artesian basin forms a reservoir of various forms of *stratal* water, according to the principal water-collecting properties of the rocks.

Hydrogeological massifs come into the category of positive tectonic forms. They are uprising folded structures in which the bedding has essentially lost its stratigraphical significance. The permeability of the rocks is determined by the degree to which they are fissured and the amount of fragmentation. The groundwater here consists mainly of fissure and ascending vein-fissure water, although in the overburden porostratal water is found. Consequently the geological massif is a reservoir of *fissure* and *vein-fissure* water according to the manner in which the groundwater is distributed within it.

Water in basins and massifs moves from the regions of recharge to regions of subsurface discharge, and drains the reservoir or emerges on to the surface in the discharge region. However, the recharge, the subsurface flow, and the discharge each has its own distinctive features (Fig. 5.2).

A centripetal subsurface flow is characteristic of artesian basins as a whole. It is true that in the early stages of development when the pore water is being squeezed out from the compacting sedimentary mass a reverse movement of groundwater arises. Usually the pressure is transmitted over vast distances and to considerable depths from the recharge regions. Artesian basins are accumulators of groundwater. Drainage, with the exception of the uppermost layers, is hindered.

The subsurface flow in the hydrogeological massifs is directed from the centre to the periphery and proceeds practically over the whole area. The depth to which groundwater penetrates, measured by the thickness of intensively fissured rocks, is comparatively small. Fragmentation is an exception. As the result of the extreme degree of dissection of the relief

hydrogeological massifs are deeply drained. The discharge region is situated at the base of the massif.

Artesian basins and hydrogeological massifs are found singly or in the form of complex systems. When they exist in complex combinations they are united into groups (Zaitsev & Tolstikhin, 1963), thus:
(1) an artesian region is a very widespread sinking of the sedimentary rocks of the cover, which combines a group of interconnected artesian basins;
(2) a hydrogeological folded region is a combination of hydrogeological massifs (outcrops of crystalline basement) and any intermontane artesian basins dividing them.

An artesian basin is a system of stratal basin formations, but a hydrogeological folded region is a system of massifs and groundwater basins and sometimes a system of fissure water massifs.

There are no strict criteria for the division of basins and massifs into ranks. Neither is there unanimity as to how the boundaries between basins and massifs, neighbouring basins or massifs, and even artesian and hydrogeological folded regions should be drawn.

The drawing of boundaries between natural reservoirs of groundwater to a large extent depends upon their volume and degree of coordination. Apart from the basic types of reservoirs there also exist other subdivisions, both smaller and larger. Hence we will examine the detailed systematics of hydrogeological reservoirs which enable us to obtain the most complete picture of subsurface water-bearing systems.

5.2.2 Systematics of subsurface water-bearing systems

Reservoirs of groundwater of different taxonomic rank can be characterized by type according to size and volume, construction (form and shape), composition, and properties. Naturally one feature alone will not give a clear picture of the geological body containing water. Even defining hydrogeological reservoirs according to several principal features cannot account for all the factors affecting the distribution and formation of groundwater.

Following the structural–material principle of differentiation which is known as the structural–hydrogeological principle in hydrogeology, we obtain the fullest possible collection of features. The following features at least must be considered applicable to the systematics of subsurface water-bearing systems: (1) the size and structure of the geological body; (2) the composition of the rocks, which determines the manner in which the groundwater is distributed; (3) the features of the recharge, subsurface flow, and discharge of the groundwater.

To a first approximation the several orders or ranks from the smallest to the largest reservoirs which meet the tectonic subdivisions of geological bodies which we have already defined are shown in Fig. 5.3. This systematic arrangement of subsurface water-bearing systems takes into account the size of the geological body (from top to bottom) and the way in which the groundwater is distributed (to the left reservoirs with stratal water, to the right those with vein-fissure water).

Figure 5.2. Diagram of the movement of groundwater (a) in a massif and (b) in a basin.
1, zone of intensive fissuring of crystalline rocks; 2, sedimentary aquifers; 3, saturated fault zone; 4, impermeable crystalline rocks; 5, impermeable sedimentary rocks; 6, direction of water movement; 7, groundwater outlet.

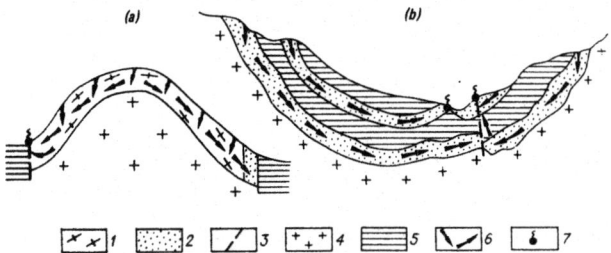

The finest subdivisions of reservoirs (the totality of the water-containing pores or fissures) form collectors of simple or complex form with more or less homogeneous distribution of groundwater.

In sedimentary rocks the combination of permeated pores and sometimes fissures will give a *water-bearing horizon*, or *complex*. In crystalline rocks which contain regional-fissure water a *fissured water-bearing zone* is a reservoir of the same rank. Both in the cover and especially in the basement there are extensive channels, cavities, and voids which are filled with water and which form a reservoir of the vein type, which is called a *water-bearing fracture zone* (see Chapter 2).

Collectors of mainly pore and stratal structure are found in the cover. The basement, by contrast, is distinguished by fissure and vein-fissure structures.

The hydrogeological reservoirs of higher rank in Fig. 5.3 are called collectively a *basin of stratal water and a massif of fissure water*.* In these the collectors alternate with confining beds. An alternation of the sedimentary rocks of water-bearing layers and complexes along the section is a property of a basin. The massif is an outcrop at the surface or a mountain uplift, more often than not of crystalline rocks, which consists of water-bearing zones of fissures and veins.

These reservoirs of groundwater contrast well with each other, which cannot be said when using the distributive terms 'artesian basin' and 'hydrogeological massif' of Zaitsev & Tolstikhin (1963). In fact an artesian basin by the definition given by K. Keilhack (1917) is nothing more than a system of rising water, which also includes the external recharge region, i.e. the slopes of the massif. According to N.I. Tolstikhin (1962), a hydrogeological massif is a massif of fissure water. Contrasting the system of pressured rising water (artesian basin) to the massif of fissure water, it seems that reservoirs of groundwater are compared by various features – pressure and collecting properties of the rocks. If however a name is to be given by comparing and contrasting features one with another, then it is more logical to use the term 'basin of stratal water' and 'massif of fissure water'. Then, incidentally, the problem of drawing up the borders of the basin and the massif is much simpler to solve.

A basin and a massif of considerable size and complex structure, in contrast to a *simple basin* and a *simple massif*, may be considered as a *complex basin* and a *complex massif*.

A complex basin of stratal water forms a vast platform or intermontane depression which has the form of syneclises, perikratonic sinking, and orogenic downwarps – intermontane or marginal (in the foothills). Characteristic features of its structure are a combination of several basins which are simple in form, coordinated or superimposed, divided in places by uplifts of the basement and separated along the upper horizons and united along the lower horizons of the sedimentary cover, or vice versa.

A combination of massifs which are more or less simple in form (mountain ranges, folded uplifts, or isolated intrusions)

*It is more correct to call it a 'massif of fissure and vein-fissure water'; the only advantage of the term 'massif of fissure water' is brevity.

which join territorially to an anticline of folded structures, a group of intrusive bodies, or an outcrop of the platform basement, falls into the category of a massif of fissure water. Simple massifs, in which the groundwater is similarly distributed, in such a hydrogeological reservoir, are indirectly coordinated with one another, and are sometimes divided by intermontane valleys.

Subsurface water-bearing systems of a higher rank include huge geological bodies: on platforms, plates, shields, and in geosynclinal (folded) regions, geosynclinal systems (branches of foldings) and central massifs (stable central massifs). In this, groups of basins – *a system of basins of stratal water*; groups of massifs – *a system of massifs of fissure water* or a complex combination of massifs and basins which are called *a system of massifs and groundwater basins* are distinguished (see Fig. 5.3). These subdivisions of groundwater reservoirs differ from complex basins and massifs not only in terms of size but also by the degree of complexity of their structural form. Other hydrogeological features which arise in the common character of the recharge, movement, and the formation of groundwater of each system, are also peculiar to them.

In the taxonomic sense the system of basins of stratal water is equivalent to the artesian region of I.K. Zaitsev & N.I. Tolstikhin (1963, 1971), it does not have the spatial concept of a 'region', but rather the capacitative volume concept of a 'reservoir of groundwater'. In precisely the same way the system of massifs of fissure water and the system of massifs and basins of groundwater have much in common with the term hydrogeological folded regions as used by the same authors. Here also an areal nomenclature has been replaced by a volumetric one.

The system of basins which corresponds to a platform is a sunken part of a platform, which contains several complex basins of stratal water, separated by uplifts or outcrops of crystalline basement. Sometimes a combination of sheets and submarginal troughs belongs to the system of basins. The cover

Figure 5.3. Ranking order of natural groundwater reservoirs, sizes and corresponding gradations of hydrogeological regionalization.

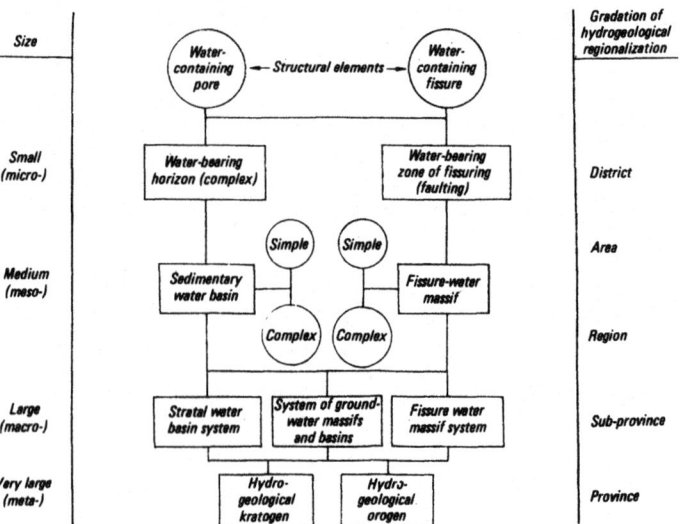

of the great central masses, the 'remnants of platforms' within folded regions, can sometimes be assigned to the stratal water basin system. The raised basements of the ancient platforms — crystalline plates — belong to the system of massifs. Plainly some geosynclinal systems within whose boundaries there are no artesian basins of stratal water, do not belong here.

However, frequently the huge fold mountain structure, called a geosynclinal system, as a hydrogeological reservoir, is a system of massifs and groundwater basins. This system includes massifs of fissure water and groundwater basins; the latter, as a rule, being of secondary significance. Sometimes plates form also a system of massifs and groundwater basins if superimposed depressions enter into their structure.

Platforms occupying stable territories, and geosynclines which are mobile regions of folding within the boundaries of mountain structures are the largest geological bodies which are distinguished by their structural features and the mineral composition of their rocks. As a reservoir of groundwater a platform is a hydrogeological kratogen, and a geosyncline (a folded structure) a hydrogeological orogen (Pinneker, 1977). The terms 'kratogen' as a concept of the stability (from the Greek $\kappa\rho\dot{\alpha}\tau o\varsigma$ meaning strength or stronghold) and 'orogen', a synonym for degree of folding and mountain-building ($o\rho o\varsigma$ meaning a mountain), describe the largest subsurface water-bearing systems (see Fig. 5.3), and are well differentiated according to the history of groundwater.

A hydrogeological kratogen is a combination of systems of basins and systems of massifs (sometimes with superimposed basins). Ancient platforms serve as similar groundwater reservoirs. But this concept is probably inapplicable to young platforms, outcrops of basement which emerge in geosynclinal (folded) regions. After all, young platforms are themselves geosynclines covered by platform cover. It is better to restrict the division of systems of groundwater basins.

A groundwater reservoir which includes part of a mobile planetary belt, i.e. the totality of geosynclinal forms, which are connected by the common character of their structural plan, and the age of the folds which it contains, falls into the category of a hydrogeological orogen. Therefore by the term hydrogeological orogen must be understood a combination of systems of massifs and groundwater basins which form united, in the structural–geological sense, and territorially distinct, geosynclinal (folded) regions.

Reservoirs of groundwater, from the smallest to the largest are coordinated in an orderly fashion (see Fig. 5.3). A complex of reservoirs of lower rank will give a reservoir of higher rank. However, coordination is not always observed. Sometimes independent reservoirs which are not of the preceding rank, but of a lower rank, enter a reservoir of a high rank as independent taxonomic units. Therefore the proposed systematics cannot be considered as a 'bed of Procrustes'. In every individual case the degree of coordination must take into account the hydrogeological features of the territory under consideration.

The differentiation of subsurface water-bearing systems on the basis of structural–material principles sometimes demands the consideration of other factors (climate, relief,

hydrogeography, etc.). The reservoirs in a similar geological structure may be situated in different climatic zones. The geomorphological conditions and the character of the hydrogeographical net which determine the position of the water divides of the subsurface flow play an important role. Only a comprehensive analysis enables the boundaries of a groundwater reservoir to be drawn.

The subsurface water-bearing systems just enumerated belong to the continental part of the crust. In the marine domain their gradation will be different. As we have seen (see Fig. 2.5), the structures of the continents and the ocean floor are fundamentally different. On a global scale *continental*, *marine*, and possibly at the first stage of study, *transitional* groundwater reservoirs can be distinguished. It is especially desirable to distinguish transitional reservoirs, part of which are on land, with the other part covered by sea or ocean. In such subsurface water-bearing systems the infiltration water is only beginning to enter, displacing water of marine origin. The study of transitional reservoirs has already begun in a number of places (the southern Caspian Basin, the North Sea shelf, etc.). The elucidation of the structure of groundwater reservoirs is a task for the near future (see Section 6.9).

5.2.3 The different types of basins and massifs

A basin of stratal water (artesian basin) and a massif of fissure water, considered as basic types of reservoirs are of the same rank. It is possible to speak of their varieties but not of the division of the different basic types of similar taxonomic rank and significance.

The varieties of basins and massifs are distinguished by the most varied features: the manner in which the groundwater is distributed, its position in space, the degree to which the rocks are frozen through, the pressure conditions, or even the rising conditions of the groundwater, etc.

The manner in which the groundwater is distributed in the geological body enables basins of *poro-stratal*, *fissure-stratal*, or *vein-fissure-stratal* water to be distinguished. In precisely the same way massifs of *regional-fissure* water, *fissure-vein* water, *stratal-(poro)-fissure* water, or *stratal-vein-fissure* water are observed.

Basins in which the hydrogeological significance of the degree of layering and degree of fissuring are equivalent, i.e. when they are characterized by both stratal and vein-fissure water, are called adartesian (*ad* from the Latin, meaning close, near). Into this category come basins filled with fissured and dislocated sedimentary deposits. It is proposed that massifs consisting of weakly metamorphosed and dislocated sedimentary rocks with vein-fissure and stratal-fissure water, be called hydrogeological admassifs (Zaitsev & Tolstikhin, 1971).

Thus a basin of vein-fissure-stratal water (an adartesian basin) and a massif of stratal-vein-fissure water (hydrogeological admassif) are virtually identical as to the manner in which the groundwater is distributed, but differ in respect of geological structure, forming transitional varieties from typical basins of stratal water to identical massifs of fissure water.

According to K.P. Karavanov (1977) these two subdivisions, called respectively an adbasin and an admassif, together with a basin and a massif, belong to the basic types of subsurface water-bearing systems (Fig. 5.4). It is scarcely possible to agree with this, bearing in mind their intermediate positions.

If the shape of the geological body is taken as a criterion for classification, then it is possible to talk of *bowl-, cup-, canyon-, graben-shaped*, and *monoclinal* (artesian slopes, according to A.N. Ovchinnikov) basins of stratal water. Among the massifs of fissure water, *dome-shaped* (quaquaversal), *anticlinal, horst-shaped*, and *block-shaped* varieties can be distinguished.

Depending on the genesis of the geological body, G.N. Kamenskii (1955) distinguished four basic types of structural—geological body, which may be called groundwater reservoirs: (1) artesian basins of platforms; (2) artesian basins of intermontane valleys; (3) platform-type uplifts and (4) folded regions.

In the light of modern data on the origins of geological bodies it is possible to distinguish:

(1) Basins of stratal water: *platform* — large depressions peculiar to the platform cover; *intermontane* — small and medium depressions within geosynclinal folded systems; *central* — which are part of the sedimentary cover of stable central masses of geosynclinal regions; *marginal* — large and deep downwarps in the articulation zone between a platform and a folded region.

(2) Massifs of fissure water: *basement* — outcrops of the basement of ancient platforms; *orogenic* — folded mountain structures of geosynclinal systems; *intrusive* — igneous bodies (dykes, sills, laccoliths, etc.).

The origin of groundwater reservoirs is very complex. During the course of geological history tectonic movements, magmatism, and various lithogenic and metamorphic processes have all influenced the formation of the geological body.

Figure 5.4. Classification of subsurface water-bearing systems (after Karavanov, 1977).
Types of subsurface water-bearing systems (SWS): B, basin; AB, adbasin; AM, admassif; M, massif.
Subtypes of SWS (the following letters signify): SS, subsurface water; S, subartesian water; A, artesian; SP, semipressure; P, pressure water.
Varieties of SWS (according to degree of mineralization): 1, fresh (pure); 2, brackish; 3, saline; 4, brine; 5, presence of SWS is very doubtful.

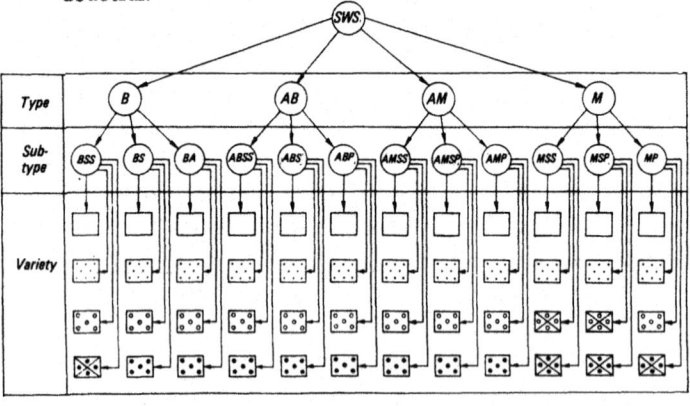

Present-day hydrogeological features of the reservoirs have, to a considerable extent, been predetermined by the influence of climate, neotectonism, recent volcanic activity, formations caused by glacial action, the emergence of thermal springs and carbonated mineral springs, etc.

The age of basins of stratal water is evaluated according to the epoch of folding, the creation of the basement, or according to the age of the rocks constituting the cover. The most ancient basins are of the same age as the cover of the Russian, Siberian, North American, and other platforms. There exist multi-stratal basins, covers of ancient platforms consisting of Palaeozoic, Mesozoic, and partly Cenozoic rocks. In their basements are found Precambrian deposits. The basins of young platforms (for example the Western Siberian Platform) are filled with Mesozoic and Cenozoic deposits. The Palaeozoic—Mesozoic intermontane basins of central Asia and the Sayan—Altai region have two layers. However, intermontane basins are often single-layered structures. The Mesozoic basins of the Trans-Baikal and the Cenozoic basins of the Pre-Baikal are of this type.

It is usual to give the age of massifs of fissure water according to the epoch of folding which created them. On the tectonic map of Eurasia drawn up in 1966 under the chief editorship of A.L. Yanshin, there can be seen massifs of the following folding epochs: (1) Pre-Riphean — on the Baltic, Aldan, and other shields; (2) Baikal (Riphean) — on the Timan, the Yenisei mountain range, and the Vitimo—Patomsk upland; (3) Caledonian (Early Palaeozoic) — in Central Kazakhstan, Western Sayan, Norway; (4) Hercynian (Late Palaeozoic) — on the Ural, Taimir and Central European massifs, etc; (5) Mesozoic — on the Sikhote—Alin, North-east SSSR and Alyask; (6) Alpine (end of the Mesozoic to beginning of the Cenozoic) — in the Carpathians, the Crimea, and the Caucasus; (7) Kamchatka (Cenozoic) — on Kamchatka, the Kuriles, and in Japan. Some massifs have undergone several epochs of folding and are called renewed or rejuvenated massifs. Renewal is associated with the last folding epoch.

Among basins and especially among massifs one can distinguish orographically *plains*, *low hills*, *medium hills*, and *high hills*. The height determines the recharge conditions, the rate of water exchange, the type of the discharge, and the chemical composition of the groundwater. Basins and ancient (Pre-Riphean and Palaeozoic) massifs, possessing smoothed relief are usually put into the category of plains and low hills. On the other hand, a dissected relief of low hills and high hills is a feature of young massifs. These are the most open and well-flowing groundwater reservoirs.

The position in the geological series enables us to talk about superimposed groundwater reservoirs, which can be sedimentogenic, and volcanogenic (see Fig. 5.5). *Superimposed sedimentary basins* have an upper structural stratum made up of eroded or tectonic depressions in complex basins, and forming basins of lower rank than the substratum. The basins of the southern parts of the Siberian Platform or the Kuzbass/Kuznetsk Basin, which are superimposed on Palaeozoic rocks, are of this type. They are also to be found on massifs.

The young cover of igneous, i.e. volcanic, origin also

shows superimposed groundwater reservoirs. They are widely developed in the Trans-Baikal, on the Siberian Platform, and in Kamchatka. In these it is difficult to distinguish the basins from the massifs because of their complex configuration, and the alternation of tuffs and lavas. Generally depending on the form of the geological body and the character of the distribution of the groundwater, there are *superimposed volcanogenic basins* of stratal water (sunken, and consisting of tuff material) and *superimposed volcanogenic massifs* of fissure water (uplifts consisting of lava streams, etc.).

On the basis of the nature of the boundaries, basins are classified as *open*, if the boundary with neighbouring basins is not sharply defined, and *closed*, when it is isolated on all sides by massifs (Tolstikhin, 1947). Obviously, approaching massifs from the same point of view, *associated* and *isolated* massifs can be differentiated.

The groundwater in permafrost regions is frozen to great depths. Water in the frozen layer has gone from the liquid-droplet state to the solid state. The basins and massifs are considered to be *deeply frozen* (cryogeologic, according to O.N. Tolstikhin), or *partly frozen*, according to the degree to which they are frozen. The Anabar Shield is an example of a frozen system of massifs of fissure water. The basins of western and eastern Siberia are partly frozen, and the basins of stratal water outside the permafrost regions are classified as *thawed*.

The use of the concept 'basin of stratal water' does not mean that the deeply rooted term 'artesian basin' has been replaced. The definition 'artesian' is applicable to basins and water, and has become firmly embedded in the hydrogeological literature and is used by specialists of different branches of knowledge. It is true that it is traditional, but not exhaustive, because it explains the pressure head of groundwater only in terms of hydrostatic pressure, which is transmitted from the recharge regions in the elevated rock formations. The artesian basin scheme neglects the pressure that arises in the early stages of the development of artesian basins because of the geostatic pressure and the squeezing out of water from the compacting sedimentary mass. Artesian basins arise only as a result of infiltration water.

A reservoir of artesian water is a product of geological history. Evolution, leading to the appearance of artesian water, has a directed character. It is determined by the entry of infiltration water and its continual displacement of water of other origins. Subsurface water-bearing systems with pressured rising water become, as a result of geological development, both basins and massifs. Therefore it is better to use the terms

Figure 5.5. Superimposed groundwater basins and massifs.
1, crystalline basement; 2, ancient sedimentary rocks; 3, young sedimentary rocks; 4, tuffaceous deposits; 5, lavas.
I, sedimentogenic basin of stratal water; II, volcanogenic groundwater basin; III, volcanogenic fissure water massif.

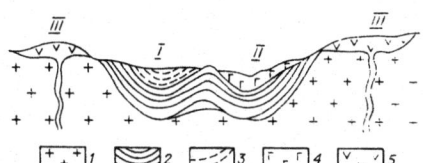

'artesian' and also 'ground' in addition to the terms 'basin of stratal water' and 'massif of fissure water' in order to show the hydraulic mechanism of the reservoir.

In the broad approach based on the features of the pressure of the groundwater, hydrogeological reservoirs are divided into pressureless (ground and interstratal), pressured (artesian and deep), and pressure–non-pressure (subartesian). K.P. Karavanov (Fig. 5.4) proposes a more complex grading of subsurface water-bearing systems, in which the subtypes are separated according to the principle of the degree of pressure acting on the groundwater. One of these definitions shows the hydraulic mechanism, illustrating the name of a basin of stratal water or a massif of fissure water.

A.A. Kartsev *et al.* (1971) propose that the accumulations of groundwater in which the pressure arises from similar conditions and in which the movement of water is similar, be grouped into *geohydrodynamic systems*. Geohydrodynamic systems include all free (gravitational) water in the crust. The following systems can be distinguished: (1) groundwater systems, i.e. groundwater with a free surface; and (2) pressure-water systems.

The concept of a pressure-water system was used by K. Keilhack, and has been developed and put on a firm footing by A.M. Ovchinnikov (1960, 1961). He treated it very broadly. Pressure-water systems – of groundwater basins – are divisible on the basis of the laws which determine the formation and distribution of groundwater. For A.M. Ovchinnikov the concept of a 'pressure-water system' is used not only for artesian basins, but also for basins of fissure water in the crystalline rocks of fold structures, and also in basins and currents of groundwater of the subartesian type (see Table 5.2). The pressure-water systems of the crust are interconnected.

It must be borne in mind that the concepts of a 'groundwater reservoir' and a 'pressure-water system' are separated on the basis of various features and hence they do not always correspond. A pressure-water system may combine several artesian basins of stratal water or artesian basins and the adjacent slopes of the neighbouring massifs. Furthermore, a large and deep stratal water basin consists of two or three pressure-water systems. This is classified as a complex massif of fissure water.

A 'water-exchange system' differs semantically from a groundwater reservoir. According to P.F. Shvetsov and his co-authors (1973, p. 50), 'this is the totality of the layers, thicknesses, series, and massifs of rock, soil, reservoirs and currents of water, which is peculiar to a certain geostructural form and litho-petrographic formation of the lithosphere'. The simplest forms of water-exchange systems are sandy soil–ground complex to lithosphere and sandy ground to reservoir. It is not difficult to see that a 'water-exchange system' includes not only the subsurface but also the surface hydrosphere.

5.3 The charge and discharge of groundwater systems

In any subsurface water-bearing system the following must be differentiated: (1) a recharge region (in subsurface water-bearing systems of the artesian type it is the region in which the pressure head is created; (2) a runoff region (distribution of the pressure); (3) a region of discharge (drainage). For some reservoirs these regions are established quite clearly (Fig. 5.6), in others they are blurred or superimposed one upon the other.

The recharge of subsurface water-bearing systems usually takes place in areas where water-bearing rocks emerge at the surface, if they occupy a relatively elevated position. Here the atmospheric precipitation and the surface water can penetrate into the depths of the Earth. Such regions of recharge are called *infiltration* in contrast to *elision*.* The latter are characteristic of stratal water reservoirs within which, under the influence of the pressure, sedimentogenic water is squeezed out from the compacting clay-containing bed. The elision areas of recharge are situated usually in the most deeply sunken parts of the depression. Finally, one talks sometimes about so-called *endogenic* 'regions' of recharge – the places in which the generation of juvenile fluids and metamorphogenic water takes place, which then enter the reservoirs along weakened zones as stratal or fissure water, are of this type (see Fig. 5.7). Groundwater of the elision and endogenic recharge regions is usually a mixture of water of different origins. In order to avoid vagueness in evaluating their genetic type the general name *deep waters* is used.

Present day replenishment of subsurface water-bearing systems is secured mainly by infiltration water. When referring to one of the basic types of reservoirs – the basin of stratal water (the artesian basin) – it is common practice to distinguish *external* and *internal* regions of infiltration recharge. The external recharge regions are found on the borders of the basin in the contiguous massifs of fissure water. The outcrops of water-bearing rock in the borders of the basin belong to the external recharge regions: they may be *basic* (the bulk along the periphery) and *local* (local uplifts within the basin). In contrast to a basin, a massif has only an internal recharge region, in which it coincides, as a rule, with the runoff region of the massif.

The groundwater of the basins and the contiguous massifs are hydraulically closely connected with one another. The massifs play a role with respect to the basins of external recharge regions. The groundwater of the massif flows over into the basin and accumulates there. In order not to isolate the external and internal regions of recharge from the runoff

region, the contiguous slopes of the massif are sometimes included in the artesian basin. Such a treatment of the concept 'artesian basin' has been developed by N.A. Marinov (1971), but, in spite of what would seem to be fundamental arguments, it did not obtain wide recognition.

It is common practice to draw the boundary between the basin and the massif along the line of contact of the cover and the basement at the surface. In essence such a boundary divides the reservoirs of stratal and fissure water. The border between the adjacent basins in areas in which the uplift of the basement does not emerge at the surface is well defined. Massifs can be comparatively easily separated one from the

Table 5.2. *Types of pressure-water systems of the Earth's crust (Ovchinnikov, 1960)*

Nomenclature	Area classification

Large artesian basins of platform regions:
   (*a*) Palaeozoic formations
   (*b*) Mesozoic formations
   (*c*) Cenozoic formations
   (*d*) Undefined formations
	more than 100 000 km²

Average artesian basins of local down-warps, and of large intermontane depressions	10 000–100 000 km²
Small artesian basins, usually 'super-imposed' on large and average basins	less than 10 000 km²

Fissure pressure-water systems in crystalline rocks (ancient shields and massifs):
   (*a*) without young rock deformation
   (*b*) associated with young movements and faults
	varied

Interconnected groundwater basins, mainly fissure water of mountain structures (folded blocks, older faults, etc.)
   (*a*) Palaeozoic folding
   (*b*) Mesozoic folding
   (*c*) Alpine folding
      without the appearance of young magmatism
      with young magmatism
      with active volcanism
	varied

Large basins and currents of subsurface water of the subartesian type	usually less than 1 000 km²

Figure 5.6. Idealized structure of an artesian basin (after A.M. Ovchinnikov).
a, recharge area; b, head; c, discharge point; P_1, pressure level higher than the surface of the ground which produces gushing water from boreholes; P_2, pressure level below the surface.
1, aquifer; 2, impermeable rocks; 3, groundwater level; 4, discharge point; 5, groundwater recharge; 6, direction of artesian water movement.

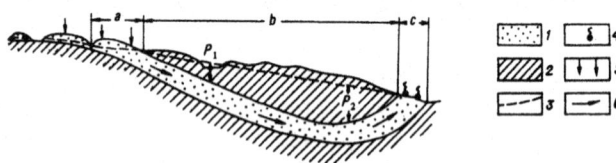

Figure 5.7. Diagram of the recharge of a 'young' (i.e. recently elevated from below sea level) basin of stratal water.
1, infiltration recharge resulting from percolation of atmospheric water; 2, elision recharge resulting from expulsion of sedimento-genic water from compacting clays; 3, endogenic recharge by juvenile fluids and metamorphogenic water; 4, front of entering infiltrogenic water; 5, water-bearing sandstone bed; 6, clay confining beds; 7, crystalline basement.

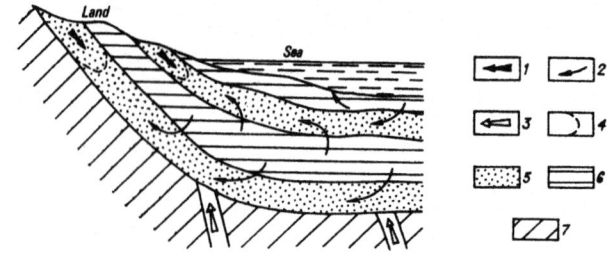

*From the Latin '*elisio*', meaning 'I squeeze out', 'squeezing out'.

other if there are intervening depressions. However, in the majority of cases the delineation of basins and massifs gives rise to some difficulties. Here careful analysis and consideration of factors is necessary, first and foremost the whole of the geological–structural features and the direction of the sub-surface flow.

If the infiltration recharge region occupies an elevated position, then the discharge region occupies a lower one. Depressions, river valleys, and the sea floor are natural drains. Within reservoirs tectonic fractures or fault zones constitute the drainage.

The recharge region consists of groundwater *discharging centres*, by which are understood places at which water emerges. A.M. Ovchinnikov divides discharge centres into ancient and recent; the latter may be natural or artificial. A natural groundwater exit (a place where groundwater flows out on to the ground surface) is called a *spring* or *source*. Excavations (wells, boreholes, etc.) from which groundwater is drawn, come into the category of artificial discharge centres.

There are a large number of classifications of springs. One of the earliest is the detailed classification of K. Keilhack (Table 5.3). Different classifications were subsequently put forward by J. Stiny (1933), A.M. Ovchinnikov (1955), M.E. Al'tovskii (1961), A. Thurner (1967), and others.

The points at which groundwater (descending) and artesian water (rising) emerge are easily distinguished. Stratal, fissure-vein, and karst water are distinguished according to the nature of the exit and the composition of the rocks containing it. They are dispersed or concentrated according to the conditions under which they appear at the surface. Among the groundwater exits which are confined to places whose relief has been lowered by erosion, or to contact with water-bearing rocks which have confining beds, there are along with typical descending (outflowing) sources, overflowing sources, which because of the presence of a barrage (an overflow instead of an exit), are characterized by an upward movement which is reminiscent of artesian springs. The popular karst spring of Vaucluse in the south of France is of this type (Fig. 5.8). Discharging centres of rising water are more varied (Table 5.4

and Fig. 5.9). They form erosion barrier and structural–tectonic springs (Ovchinnikov, 1968).

In natural discharge centres the groundwater exits on to the surface in the form of springs (freshwater, carbonated, saline, thermal, etc.), which pour out at the bottom of rivers and seas, or overflow into water-bearing layers of a higher elevation. This enables them to be classified according to the manner in which they drain on to the surface and via hidden discharge centres. Among the latter are external (sub-fluvial and submarine) and internal (subsurface) reservoirs (See Table 5.4).

The conditions of groundwater discharge in permafrost regions and those in regions of contemporary volcanic activity are regionally specific. As a result of freezing of the water-bearing layers in the active zone above the permafrost in winter, many springs function only seasonally in such regions; only discharging centres of unfrozen water below the permafrost are in action all the year round, and these form icings round the discharge point. Fig. 5.10 shows the scheme of recharge and discharge of groundwater in the cryolithosphere.

Table 5.3. *Classification of springs (after Keilhack, 1912)*

Descending (non-pressure water)	(1) formed as a result of the squeezing of the water-bearing stratum (2) in places of natural wedging-out of the water-bearing stratum (3) stratal in places where the water-bearing stratum has been exposed by erosion (4) overflowing (5) dam or sill (6) fissure (7) waste water effluent		
Ascending (water under head of pressure)	(1) because of the occurrence of hydrostatic pressure	(a) stratal (b) waste water effluent	
	(2) water moves under the force of gas pressure	(a) water vapour (b) carbon dioxide (c) hydrocarbon	

Figure 5.9. Types of exposed artesian water outlets (after Klimentov & Bogdanov, 1977).
(A) erosion in (a, a river valley, b, in an erosional–tectonic depression of relief); (B) groundwater barrier (c, dam of igneous or other rock in the path of the groundwater, d, anticlinal structure of impermeable rock in the path of the groundwater); (C) structural–tectonic (e, exposure of an anticlinal aquifer, f, discharge of emerging groundwater).
1, aquifer; 2, impermeable rocks; 3, relatively impermeable rocks; 4, dense igneous rocks; 5, piezometric surface; 6, direction of groundwater movement; 7, artesian water discharge point.

Figure 5.8. Periodically flowing springs.
(a) spring from an overflowing groundwater reservoir, (b) siphon spring of the Vaucluse type.
P_1, level of the groundwater before recharge; P_2, groundwater level after overfilling.
1, dry unconsolidated layer; 2, limestone; 3, unconsolidated layer (saturated); 4, basement rocks; 5, water table; 6, groundwater flow direction.

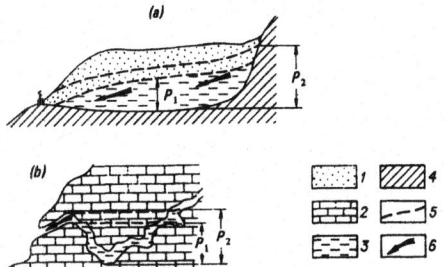

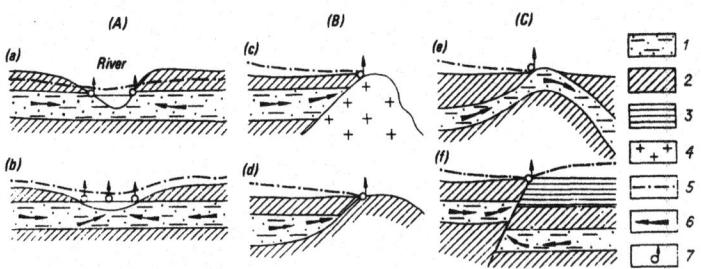

Table 5.4. *The types of pressure water discharge centres (after A.M. Ovchinnikov)*

Age	Type		Characteristics
Ancient	Quiescent or extinct		Oil and gas deposits in places with impeded discharge, in places with traps and internal discharge points
			Sulphur deposits at discharge points of hydrogen sulphide water
			Hydrothermal ore deposits, calcite veins, etc.
			Skarn zones at the contact of intrusions and carbonate rocks
			Tufa deposits (travertine) and aragonite at exit points of carbonated water
Active	Exposed	Erosional	Localized discharge points of stratal and fissure water in large river valleys; in some desert regions – depressions with no runoff (solonchaks)
		Barrier	Discharge points of fresh and mineralized pressure water near zones of uprise and barriers
		Structural–tectonic	Groups of springs in zones of tectonic faulting and cores of anticlines on platforms; linear discharge zones in folded regions
	Concealed	External	Subfluvial: river beds below Quaternary alluvial deposits; submarine (close to the coast); hidden finely dispersed – percolation of water through almost impermeable layers
		Internal	Groundwater: in places of unconformity between the rock groups; in places of 'facies windows'; in zones of faulting in axial parts of anticlines, cupolas; uplifted areas

In regions of contemporary volcanic activity the discharge centres expel gravitational water and a vapour–liquid mixture. Here geysers, which produce periodic hot fountains are a unique form of discharge centre (see Section 6.8).

5.4 The general mapping of subsurface water-bearing systems

Subsurface water-bearing systems are represented on hydrogeological maps of various types. The content of a hydrogeological map is determined by its scale. M.R. Nikitin (1969) differentiates the following types of hydrogeological map based on the scale of the topographic and geological features: (1) synoptic (summary) (less than 1 : 1 000 000); (2) small scale (1 : 1 000 000–1 : 500 000); (3) medium scale (1 : 200 000–1 : 100 000); (4) large scale (1 : 50 000–1 : 25 000); and (5) detailed (1 : 25 000 and greater). Strictly hydrogeological cartography can be general, or it can be specific.

Water-bearing layers and confining beds, the permeability and the abundance of water in the rock, the position of the bed and the direction of movement of the groundwater, the degree of mineralization and composition, the water discharge points – springs, boreholes etc. (Nikitin, 1969), included on such a map give a complex qualitative and quantitative picture of subsurface water-bearing systems.

On the other hand, particular maps show only certain hydrogeological parameters. M.R. Nikitin provisionally combines these varieties of map into three groups: (1) maps showing separate elements of the subsurface water-bearing system; (2) maps showing the different aspects of the hydrogeological conditions; and (3) maps showing the different elements of a hydrogeological section (water-bearing horizons, complexes, formations, zones).

When a particular hydrogeological map has a completely defined purpose it becomes a specialized map. For example

Figure 5.10. Diagram of formation of high permanently frozen hydrogeological belts (after Tolstikhin, 1974).
I, hydrogeological massif; II, artesian basin.
High frozen geological belts: belts of a, hydrothermal accumulation; b, infiltration and inflow; c, transit and accumulation; d, groundwater discharge points.
1, river bed and terrace of gravels; 2, (a) sand, (b) sandy loam; 3, deformed rocks and fault zones; 4, intensively permeated rocks and the border of the saturated zone; 5, permafrost rocks and their boundary; 6, icings; 7, spring; 8, discharge and infiltration (inflow) of groundwater (the length of arrow shows the dominance of one or the other process); 9, direction of groundwater movement.

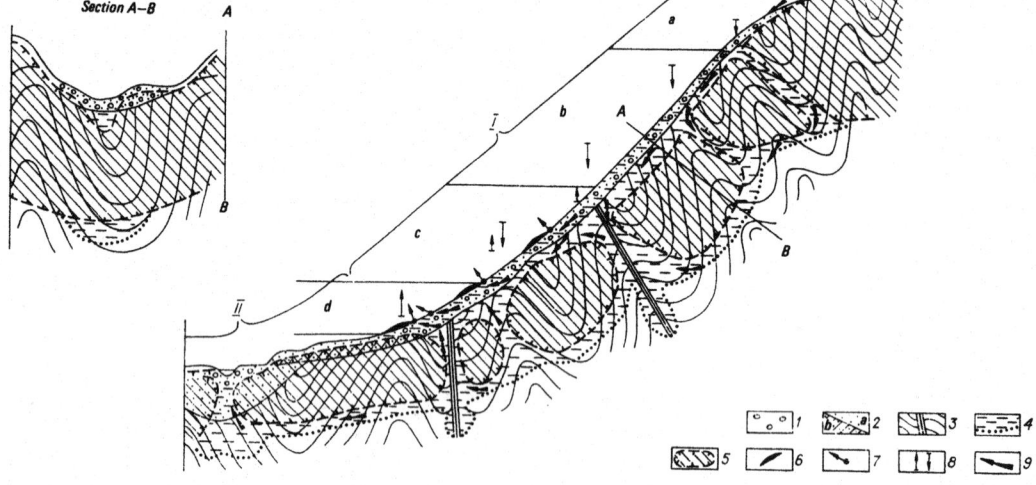

maps of the basic horizons of groundwater, suitable for the supply of potable water, or the subsurface disposal of harmful industrial effluents are of this type. They will often be maps made for the solution of engineering problems (the evaluation of water flows in excavations, mining, irrigation conditions, etc.).

In hydrogeological research, maps are made to the most varied scales, and are intended for a great variety of purposes. However, their basis is always the general synoptic or review survey maps of the general type. Practically all the gradations of subsurface water-bearing systems are shown on these, and so we will confine ourselves to an examination of this type of hydrogeological map.

General synoptic hydrogeological maps show the distribution of the quality and quantity of the groundwater over quite a large area. The content of such maps must reflect, in a concentrated form, the groundwater data, inseparably linked with the geological environment. Various hydrogeological indices can serve as an important element of mapping in coloured maps. There is no unified system for captions and legends on general synoptic hydrogeological maps, but different systems are under consideration in the USSR and abroad (Zaitsev, 1961b; Rogovskaya & Kunin, 1969; Nikitin, 1969).

Soviet hydrogeologists have accumulated a very great amount of experience in synoptic hydrogeological mapping. Two variants of hydrogeological maps of the territory of the USSR have been made to the scale of 1 : 2 500 000.

(1) On the hydrogeological map of the USSR which was prepared under the editorship of I.K. Zaitsev (1959), the most important elements are the types and classes of 'accumulation' of groundwater (see Table 5.2). The colouring is used in the same way in the hydrogeological maps of the USSR on scales 1 : 5 000 000 and 1 : 7 500 000, which were also prepared under the editorship of I.K. Zaitsev. The age of the water-bearing rocks which are combined into water-bearing complexes is shown by a geological index for each type of groundwater. The experimental pattern of the hydrogeological map of Europe is closest to these in the way in which the basic hydrogeological index is represented. On this the subject of the map is the form of the 'accumulation' of groundwater in the geological formations (unconsolidated, fissured, etc.).

(2) A different method of demonstrating the important elements, which can be called stratigrapho–hydrogeological, has been used for the hydrogeological map of the USSR produced under the editorship of N.A. Marinov (1964). The areas of distribution of the reservoirs of groundwater nearest the surface (water-bearing horizons, complexes, fault zones, etc.) and impermeable layers are shown in colour. The laws of distribution of groundwater are shown most fully by such a method.

Both variants of hydrogeological map have defects, and advantages. Without dwelling on this method in detail, we will point out that the second ('geologized') variant is to be preferred. It is easy to read, and in contrast to the first it shows the basic controls of distribution of groundwater less subjectively.

The stratigrapho–hydrogeological principle of synoptic hydrogeological mapping has been used extensively in recent years in the making of maps for the monograph *The hydrogeology of the USSR*. On these maps, with the object of giving them universality, apart from the water-bearing layers and confining beds which lie closest to the surface, the following are also shown:

(1) the lower-lying water-bearing horizons and confining beds (in the form of a section – a 'window', or by contours);

(2) the natural resources of the groundwater or the abundance of water in the rocks;

(3) the degree of mineralization, composition, and temperature of the groundwater both for the water-bearing layer that lies closest to the surface, and for the lower-lying layer;

(4) natural and artificial outlets of water (springs, exploration wells, etc.);

(5) the parameters of the geological conditions (the distribution of the permafrost, the extent of the fresh water zone, the presence of karst, etc.);

(6) hydrogeological zonation.

The hydrogeological map is accompanied by hydrogeological sections and inset maps; on the latter is shown any hydrogeological index (the distribution of brines, hydrogeological regions, etc.), which because of lack of space or for some other reason cannot be shown on the basic map.

There is a vast literature devoted to the methods of comparing hydrogeological maps (M.E. Al'tovskii, I.K. Zaitsev, N.A. Marinov, M.R. Nikitin, N.V. Rogovskaya, and others). We refer to the detailed works of I.K. Zaitsev (1961b) and M.R. Nikitin (1969) who were interested in this matter, in which both general and specialized methods of hydrogeological mapping are examined. A review of the hydrogeological maps of foreign countries was made by N.V. Rogovskaya (1970).

References

Al'tovskii, M.E. (1961). Classification of springs. *Voprosy gidrogeologii i inzhenernoi geologii, Moscow*, No. 19, pp. 49–53.

Anon. (1955). *Geological dictionary*, 2 vols. Moscow: Gosgeoltekhizdat. Vol. 1, 403 pp., Vol. 2, 445 pp.

Anon. (1970). *Reference book on tectonic terminology*. Moscow: Nedra. 582 pp.

Anon. (1978). *Geological dictionary*, 2nd edn, Vol. 1, 486 pp., Vol. 2, 456 pp.

Ignatovich, N.K. (1945). The hydrogeological classification of geostructural elements. *Dokl. AN SSSR*, 49 (4), 292–5.

Jetel, J. (1973). Logicky system pojmu-zakladni padminka formalizace a matematizace hydrogeologii. *Geol. pruzk.*, No. 1, 13–17.

Kamenskii, G.N. (1955). The principles of hydrogeological zonation in the USSR. In *Problems in groundwater studies and engineering geology procedures*, pp. 3–13. Moscow: Gosgeoltekhizdat.

Kamenskii, G.N., M.M. Tolstikhina & N.I. Tolstikhin (1959). *Hydrogeology of the USSR*. Moscow: Gosgeoltekhizdat. 366 pp.

Karavanov, K.P. (1977). *Groundwater basins in the fold mountain regions of Eastern Asia*. Moscow: Nauka. 142 pp.

Kartsev, A.A., V.V. Kolodii, V.A. Kudriakov & R.G. Semashov (1971). The types and evolution of geohydrodynamic systems. *Izv. AN SSSR, Ser. geol.*, No. 6, 122–7.

Keilhack, K. (1912). *Lehrbuch der Grundwasser- und Quellenkunde*. 1st edn. Berlin.

Keilhack, K. (1917). *Lehrbuch der Grundwasser- und Quellenkunde*. 2nd edn.

Klimentov, P.P. & G.Ya. Bogdanov (1977). *General hydrogeology.* Moscow: Nedra. 357 pp.

Kosygin, Yu.A. (1974). *Fundamentals of tectonics.* Moscow: Nedra. 215 pp.

Marinov, N.A. (1961). Hydrogeological formations. *Razvedka i okh. nedr*, No. 2, 40–3.

Marinov, N.A. (ed.) (1964). *Hydrogeological map of the USSR. Scale 1 : 2 500 000.* Moscow: Nedra.

Marinov, N.A. (1971). Principles of and a scheme of hydrogeological zonation of Asia. In *Groundwater of Siberia and the Far East*, pp. 33–43. Moscow: Nauka.

Nikitin, M.R. (1969). Fundamental problems of hydrogeological cartography. In *Problems of regional hydrogeology and methods of hydrogeological cartography*, pp. 187–215. Moscow.

Ovchinnikov, A.M. (1955). *General hydrogeology.* Moscow: Gosgeoltekhizdat. 380 pp.

Ovchinnikov, A.M. (1960). Basic principles of hydrogeological zonation. In *Problems of hydrogeology*, pp. 106–9, Moscow: Gosgeoltekhizdat.

Ovchinnikov, A.M. (1961). Pressure water systems of the Earth's crust. *Izv. vuzov. Geologiya i razvedka*, No. 8, 85–90.

Ovchinnikov, A.M. (1968). Types of discharge points of pressure water. *Sov. geologiya*, No. 7, 136–42.

Pinneker, E.V. (1977). *Problems of regional hydrogeology. Laws governing the occurrence and formation of groundwater.* Moscow: Nauka. 196 pp.

Richter, W. & W. Lillich (1975). *Abriss der Hydrogeologie.* Stuttgart: E. Schweizerbartsche Verl. 281 pp.

Rogovskaya, N.V. (1970). Review and analysis of the UNESCO map collection as at 1 Nov. 1968. *Sov. geologiya*, No. 1, 148–52.

Rogovskaya, N.V. & V.N. Kunin (1969). An international hydrogeological map. *Sov. geologiya*, No. 11, 140–3.

Savarenskii, F.P. (1947). The principles of hydrogeological zonation. *Sov. geologiya*, No. 19, 19–23.

Shvetsov, P.F., P.A. Konoplyantsev & V.M. Shvets (1973). The present content, basis, branches, and organizational forms of the development of hydrogeology in the USSR. *Izv. AN SSSR, Ser. geol.*, No. 2, 56–66.

Stepanović, B. (1962). *Principi opšte hidrogeologije.* Beograd: Rad. 144 pp.

Stiny, J. (1933). *Die Quellen. Die geologischen Grundlage der Quellenkunde für Ingenieure aller Fachrichtungen sowie für Studierende der Naturwissenschaften.* Vienna: Springer-Verlag. 255 pp.

Thurner, A. (1967). *Hydrogeologie.* Vienna, New York: Springer-Verlag. 350 pp.

Tolstikhin, N.I. (1947). Artesian water in the permafrost zone of the USSR. *Merzlotovedenie*, 2 (1), 31–5.

Tolstikhin, N.I. (1962). The principles of the structural–hydrogeological zonation of Siberia. In *Regional hydrogeology of Siberia and the Far East*. pp. 3–9.

Tolstikhin, N.I. (1974). *Icings and groundwater of the north-east USSR.* Novosibirsk: Nauka. 164 pp.

Zaitsev, I.K. (ed.) (1959). *Hydrogeological map of the USSR. Scale 1 : 2 500 000.* Moscow: Gosgeoltekhizdat.

Zaitsev, I.K. (1961a). Some problems of terminology and classification of groundwater. In *Material on regional and prospecting hydrogeology*, pp. 111–60. Leningrad.

Zaitsev, I.K. (1961b). Methods of making hydrogeological survey maps. In *Material on regional and prospecting hydrogeology*, pp. 7–48. Leningrad.

Zaitsev, I.K. & N.I. Tolstikhin (1963). The fundamentals of structural-hydrogeological zonation of the USSR. In *Material on regional and prospecting hydrogeology*, pp. 5–35.

Zaitsev, I.K. & N.I. Tolstikhin (1971). The classification of groundwater and hard rocks – the basis of hydrogeological mapping and zonation. In *Problems of hydrogeological mapping and zonation*, pp. 4–15. Leningrad.

6

The varieties of groundwater on a basis of depositional origin

6.1 Schemes of groundwater classification

The classification of groundwater according to the way in which it is deposited* is widely accepted. This is connected with the fact that the physical-geographical, structural—geological, and thermodynamic conditions determine, in the final analysis, the features of the formation, movement, and distribution of groundwater, and the character of the bedding most fully combines all of these natural conditions, and features of the life of water in the interior of the Earth.

The difficulty of drawing up such a classification has been pointed out by K. Keilhack (1935), A.M. Ovchinnikov (1955), O.K. Lange (1969), and other researchers. Its drawing up is complicated by the fact that groundwater is found in the crust almost everywhere, and in the most varied natural conditions, varying in quantity and quality both temporally and spatially. The basic requirement of such a classification is universality, but this is precisely what cannot always be attained. The situation is complicated by the absence of a generally accepted hydrogeological terminology and by the different approaches of researchers to the principles by which groundwater should be classified.

The proposed schemes for the classification are extremely varied. Many of them take into account, along with the geological conditions under which they are laid down, the pressure of the water, the thermodynamic features of the environment, etc. Until now a single classification of groundwater according to the character of the rock bedding, based on the way in which it is deposited, which would satisfy the interests of the existing scientific groups and be generally accepted does not exist, although proposals along these lines have been put forward many times by both Soviet geologists and by those of other countries.

*To talk of 'deposition' of groundwater is not strictly speaking true because it is not only deposited but also moves. However, regarding the classification of groundwater, this term is firmly entrenched and authors are compelled to use it.

We will examine the role of individual researchers in establishing the basic types of groundwater according to the way in which they are deposited and the classifications most widely used.

From about 1860 the French term 'artesian water' quickly became established in the Russian literature in connection with the widely practised use of gushing groundwater. At the same time, in contrast to the term 'artesian' water, which described pressure water, the term 'soil' water, corresponding to the concept of 'non-pressure' water arose. Subsequently, after V.V. Dokuchaev established a new concept of 'soil', together with the term 'soil' water, the term 'subsoil' water was used, but during the 1880s it was gradually replaced by the term 'subsurface' water (Gordeev, 1954).

Credit for the final consolidation of the terms 'subsurface' water and 'artesian' water in the modern sense in the literature, the precise definition of these concepts, and the comprehensive description of the two basic types of groundwater goes to the great scientist–hydrogeologist S.N. Nikitin (1900).

He called subsurface water that water which was formed as a result of the absorbed atmospheric precipitation in the water-bearing layer nearest the surface, which is situated in the subsoil or in the deeper bedrock, on the impermeable layer nearest the surface – water which 'remains free after the last moisture content of the rocks has been satisfied'.

S.N. Nikitin's definition is wider than the similar concepts of one of the earliest authors – Haas (1895) whose *Quellenkunde* limited groundwater only to accumulation in unconsolidated deposits (Haas considered water in dense and fissured rocks to be of a different variety).

Nikitin regarded as artesian not only water which rose to the surface through boreholes, but also that whose piezometric level was higher than the roof of the water-bearing horizon, even though it was below the surface.

The principle on which the French geologist Daubrée (1887) distinguished groundwater was similar to that of Nikitin's with the sole difference that he called subsurface water *phreatic* water (from the Greek φρεατος – of a well).

The groundwater classification of the German scientist Steuer (1907) deserves mention. Calling liquid water situated below the surface 'subsoil' water, he divided it into subsurface water, percolating water, stratal water, and fissure water.

Steuer's definition of subsurface water is close to that of Haas. By percolating water he understood infiltration water, which is in process of percolating and which has not yet attained the state of groundwater below the water table. Groundwater in the pores and voids of sedimentary rocks from the time of its formation is called stratal water. Water which is concentrated in the spaces of fissured rocks of secondary origin he classified as fissure water.

Consequently Steuer considered the way in which groundwater moves, and the water-collecting properties of the rocks to be the basis of groundwater classification. Only the use of the term 'soil' water for groundwater was unsuccessful.

From about 1910 to 1940 a number of groundwater classification schemes based on the geological factors of the

way in which it is deposited were proposed by the Germans (Keilhack, 1912; Keller, 1931; Kühne, 1932) and the American Meinzer (1923). A very original classification which has stood the test of time is that of Meinzer (see Table 6.1). It is widely used in the USA to the present day with some minor changes and additions (Davies & De Wiest, 1967). The majority of geologists in non-Russian Europe use the term 'groundwater' as a general concept, but in a different sense from the use of this term in the USSR. Meinzer proposed that all water below the surface be called groundwater. This concept embraces any variety of water in the interior of the Earth in contrast to surface water; it also includes 'gravitational subsurface water'. The concept of the 'zone of aeration' was introduced by Meinzer and rapidly became widely used in America, Europe, and the USSR.

Groundwater in the broad sense[*] is classified by Meinzer (1923) according to the conditions in which it is found in the crust, as follows.

(I) Water in rock voids

    (*A*) Suspended groundwater (vadose water)
(1) Soil water
    (*a*) water available to plants
    (*b*) water unavailable to plants; water which can be separated by evaporation (hygroscopic water)
(2) Water of the intermediate zone
(3) Water of the capillary zone

    (*B*) Subsurface (phreatic) water
(1) Gravitational subsurface water
(2) Subsurface water that is independent of gravity

    (*C*) Subsurface ice (ice in the rock voids)
(II) Water in minerals, which enters into the composition of rock (water in a solid solution, water of crystallization, etc.).
(III) Deep-lying water (and magmatic water higher than the zones of plastic rock).

With this approach Meinzer calls the water in the zone of aeration suspended groundwater, or vadose water (as was mentioned in Section 3.3, Soviet scientists have a completely different concept of this term). Meinzer classifies all water in the zone of saturation as subsurface water, dividing it into free water and pressure water. Meinzer applies the term 'phreatic water' to all the water in the zone of saturation (which, in the view of Soviet hydrogeologists is also incorrect). Water in the form of a vapour is absent from the classification: this variety of water comes into the so-called subsurface water, which is independent of gravity, and into deep-lying water.

The division of groundwater into subsurface water, infiltrating between the particles of disintegrated rock, and underground water currents (which move in the fissures and voids of hard rocks) was put forward in 1919 by E. Prinz.

In 1928 in the USSR it was proposed by A.M. Zhirmunskii and A.A. Kozirev that an attempt be made to draw up a comprehensive classification of groundwater on the basis of its

---

[*]In this sense the equivalent of Meinzer's 'groundwater' is the 'subsurface hydrosphere', as the authors of the present monograph treat it (see Chapters 1 and 2).

genesis and the conditions in which it is deposited. In this the groundwater was subdivided according to its origin into vadose, juvenile, fossil, and mixed, and also according to the hydrogeological and stratigraphical features, i.e. free (upper (suspended water) and lower), and pressure water (subartesian and artesian). In each group of groundwater, stratal water, associated with sedimentary rocks, and underground water currents, were distinguished.

General classifications of groundwater according to the geological conditions under which it was deposited have been proposed by many Soviet scientists (B.L. Lichkov, V.I. Vernadskii, N.N. Slavyanov, F.P. Savarenskii, O.K. Lange, G.N. Kamenskii, N.N. Bindemann, I.K. Zaitsev, A.M. Ovchinnikov, A.N. Semikhatov, N.I. Tolstikhin, M.E. Al'tovskii, P.P. Klimentov). Several of the classifications used in the USSR are shown in Tables 6.2–6.4.

The classifications of F.P. Savarenskii and A.M. Ovchinnikov are based upon the same principles of the separ-

ation of the determining types (subtypes) of groundwater, in the basis of which lies a complex approach.

F.P. Savarenskii distinguishes five types of groundwater (Table 6.2). A brief description of these types is given, based on the great number of common indices (the position of the recharge regions and the distribution of the given type of water, its hydraulic character, the way in which it moves, origin, the geological conditions under which it was formed, the climatic zonation, temperature, geochemical zones, and the saline character of the water).

In developing Savarenskii's ideas, A.M. Ovchinnikov reduces the number of basic types distinguished according to the way in which they are deposited to three. These are — suspended water, subsurface water, and artesian water. These basic types are subdivided, depending upon the nature of the water-bearing rocks, into subtypes — pore water and fissure water. The need to separate the basic types arose from the specific features of groundwater in permafrost regions and

Table 6.1. *Groundwater classification scheme (after Meinzer, 1923)*

Zone of fissured rocks	Zone of aeration	Soil water zone	Soil water	Water in voids in rocks (pore, fissure, cavern water)
		Intermediate zone	Intermediate zone water	
		Capillary zone	Capillary water	
	Zone of saturation		Groundwater (phreatic water)	
Zone of rock flowage			Deep water	

Table 6.2. *Classification of groundwater (after F.P. Savarenskii, 1939)*

Water type	Relationship between recharge and discharge areas	Pressure characteristic	Characteristics of the movement of the current	Origin	Hydrogeological conditions of formation	Climatic zonation	Temperature	Geochemical zones	Chemical characteristics
Soil, marsh, suspended water	Three types coincide (water near the surface)	Descending, non-pressure	Laminar	Vadose	Surface formations	Intrazonal	Subject to seasonal variations	Zone of weathering and in places, saturation	Fresh, in places saline
Subsurface	Usually coincide (not far below surface)	Descending, non-pressure, sometimes with local pressure	Mainly laminar	Vadose	Surface deposits and upper layers of the weathered crust	Zonal	Subject to seasonal variations	Zone of weathering and in places, saturation	Fresh, in places saline
Karst	Close but separate (water is mainly not deep-lying)	Usually descending, non-pressure	Mainly turbulent	Vadose	Sandstones, dolomites, and other rocks undergoing weathering	Azonal	Usually unstable	Zone of weathering	Fresh, usually aggressive
Artesian	Do not coincide (waters mainly deep)	Ascending, pressure; pressure is hydrostatic	Laminar in unconsolidated rocks, and may be turbulent in fissured rocks	Vadose	Structures of sedimentary rocks (basins)	Azonal	Increasing with depth	Zone of weathering and cementation	Fresh, sometimes mineralized
Vein (fissure)	Do not coincide (waters mainly deep)	Ascending, pressure; pressure is hydrostatic or gas	Mainly turbulent	Vadose and juvenile	Mainly zones of tectonic fissuring	Azonal	Increasing with depth	Zone of cementation	Fresh and mineralized

regions of recent volcanic activity (see Table 6.3). The advantages of Ovchinnikov's scheme are simplicity and clarity.

In N.I. Tolstikhin's classification (in Al'tovskii, 1962) Meinzer's scheme is developed in which groundwater is formed in three zones of the crust and the underlying plastic zone (see Table 6.4). N.I. Tolstikhin adheres to the detailed division of groundwater separately for territory where there is no permafrost, and for permafrost regions. In this scheme hot vapours from the deep zones of the crust and the mantle are distinguished, although to call these 'groundwater' is hardly correct. The scheme is rather unwieldy.

The classification of O.K. Lange is very convenient and is often used by hydrogeologists. In its simplest form (Lange, 1969) three basic groups of groundwater are recognized according to the way in which they are deposited: soil water, subsurface water, and interstratal water.* O.K. Lange considers soil water to be suspended water, subsurface and interstratal non-pressure water to be descending, and interstratal pressure water to be ascending.

A unique classification of groundwater was proposed by I.K. Zaitsev (1961). He recommended that the basic types of groundwater be distinguished on the basis of the collecting properties of the rocks (see Table 5.1). Classes can be distinguished according to the morphology of the accumulations of groundwater. The degree of isolation of groundwater from the ordinary surface, the nature of its regime and features of formation are accounted for by dividing into sections and subsections. Furthermore, I.K. Zaitsev recommends that different kinds of locality at which groundwater is found be recognized. This classification did not achieve wide recognition but is used

Table 6.3. *Groundwater classification (after A.M. Ovchinnikov, 1955)*

Basic type of water	Subtypes		Special types	
	In porous rocks (pore water)	In fissured rocks (fissure water)	Permafrost regions	Regions of young volcanic activity
Suspended water	Soil Marsh Water perched on lenses of impermeable rock Clay-surfaced desert (takir) and hills of sand (in deserts) Sand accumulations and dunes on the coast	Weathered crust of fissured rock Upper (drained) horizon of karst massifs Impermeable roof layer of lava flows and tuff breccia	Active layer	Derived water from hot springs Intermittently active fumaroles in wet periods
Subsurface water	Alluvial deposits Diluvial, proluvial, and lacustrine deposits Ancient alluvial deposits Fluvio-glacial deposits Supra-, inter- and submoraine sandy pebbly accumulations Basal deposits	Fissure groundwater in covering layer of erupted basalt rock and in the base of lava flows Stratal-fissure and fissure-stratal sedimentary deposits Karst massifs of carbonate rocks (and gypsum and salt-bearing rocks)	Suprapermafrost Interpermafrost	Water at elevated temperatures, enriched with gases Small fumaroles and geysers
Artesian	Artesian basins (in sandy strata) Artesian slopes (in monoclinally deposited and wedged-out sandy pebbly suites in foothills)	Waters of artesian basins (in strata, massifs, and stocks of fissured hard rocks) Waters of artesian slopes (in limestones and tuffaceous beds and massifs of intrusive rock)	Subpermafrost	Emitting gases, mineral waters (sometimes hot), which ascend along tectonic faults and contacts of different rock types Artesian systems of complex intrusions of igneous rock, which are sometimes rich in specific (sometimes rare) elements

*The term 'interstratal water' is not used by all researchers. In the opinion of A.M. Ovchinnikov (1955), it cannot be contrasted with the term 'subsurface water', because the latter may also, in certain circumstances, be interstratal. Essentially, interstratal non-pressure water is water from the recharge region of artesian basins.

Table 6.4. *Groundwater classification on a basis of conditions of deposition, N.I. Tolstikhin (Anon., 1959)*

Location	Zone	Groundwater of the hydrogeological regions				
		Outside the permafrost regions		Within the permafrost regions		
Groundwater in the crust	Zone of aeration	Soil water		Soil water		
		Suspended water	Fissure and karst water zone of descending movement of groundwater	Supraperma-frost water of seasonal taliks	Interpermafrost fissure and karst water of the zone of aeration	Subpermafrost fissure and karst water of the zone of aeration
	Zone of saturation	Water of the capillary fringe		Water of the capillary fringe		
				Seasonal and long duration supraperma-frost taliks	Interpermafrost taliks	Subpermafrost taliks
		Subsurface water	Fissure and karst non-pressure water of the zone of horizontal water movement	Supraperma-frost water of permanent taliks: subfluvial, sub-lacustrine, in talus cones below mountain fronts		
		Interstratal non-pressure water	–	–	Interpermafrost non-pressure water of alluvial and other deposits (fissure and karst)	Subpermafrost non-pressure water of alluvial and other deposits (fissure and karst)
		Stratal pressure water of artesian water	Fissure and karst pressure water of the zone of descending–ascending move-ment of groundwater	–	Interpermafrost pressure water: stratal and artesian; fissure, and karst; fissure-vein	Subpermafrost pressure water: stratal and artesian; fissure and karst; fissure-vein
		Pressure fissure water of the basement of artesian basins	Fissure and karst pressure water of the zone of the deepest migration of groundwater	–	–	Subpermafrost pressure water; artesian; fissure and karst; fissure-vein; zones of deepest migration of groundwater
	Hot vapour	1. (*a*) Hot water vapour of the deep parts of artesian basins and the underlying basement; (*b*) hot water vapour of the hydrogeological folded regions 2. Hot water vapour of the sial layer (granitic) 3. Hot water vapour of the sima (composed of basic and ultrabasic rocks)				
Water in the zone of plasticity at subcrustal depths and of intra-crustal magma chambers		Subcrustal and magmatic (interior) water				

as a basis for constructing general coverage hydrogeological maps of the USSR.

In Western Europe, particularly in France, the classification of the French scientist Schoeller (1962) distinguishes beneath the surface the following zones, starting from the surface: (1) the evaporation zone; (2) the infiltration zone or zone of aeration; (3) the capillary fringe; and (4) the zone of saturation or zone of groundwater accumulation. In the saturation zone horizons with free surface of groundwater and horizons with a pressure surface are distinguished.

The Austrian hydrogeologist A. Thurner approaches the classification of groundwater from a different position (Thurner, 1967). He groups the classes as follows:

(1) water and springs in hard rocks (including: (a) water in fissures or karst voids, (b) water in the zones of tectonic fracturing);

(2) water in springs and in unconsolidated deposits (including: (a) water which saturates the pores completely, (b) water which does not saturate the pores completely);

(3) water which moves along various paths.

The American hydrogeologists S.N. Davies and R.J.M. De Wiest (1967) adopt a combined approach. As has already been noted, they keep to Meinzer's classification with minor changes and additions, distinguishing two zones (aeration and saturation) and dividing the groundwater of the zone of saturation into non-pressure and pressure water. But in the description the groundwater is divided into: (1) water of igneous and metamorphic rocks; (2) water of hard sedimentary rocks; (3) water of unconsolidated sediments; (4) water of regions with extreme climatic conditions (excessive atmospheric precipitation, the arid zone, or permafrost regions).

The above review of the principles which form the basis of a classification of groundwater according to the geological conditions under which they are deposited shows that in the classifications which are the most widely used, the basic subdivisions of groundwater, i.e. gravitational water and liquid-droplet water are: (1) water of the zone of aeration (suspended water); (2) subsurface water (as non-pressure water of the water-bearing horizon that is nearest to the surface, and which has a free surface); (3) artesian water (also called interstratal pressure water). These types of water are to be found in practically all classifications by Soviet hydrogeologists and by the majority of foreign hydrogeologists too. In many cases water in porous sedimentary deposits is also distinguished, and also in fissured and karsted rocks, this may be both non-pressure and pressure water; both of these are closely connected hydraulically. Sometimes interstratal non-pressure water is distinguished. Furthermore, groundwater is often described separately in specific rock conditions (in the frozen zone for example).

Which scheme is to be preferred? Is it adequate for the present day level of knowledge?

All subdivisions of natural objects must be based upon the most important classifying features, which ensure the maximum of derivative features. In this way completeness, simplicity, and constancy of the classification schemes are obtained. In their time many of the groundwater classifi-cations examined satisfied these demands. However, new facts, discovered in recent years, give rise to the necessity of making even the most widely used of them more precise.

All the above classifications are concerned with the hydrogeology of the land masses (continents). Groundwater below the oceans and the seas is not considered, although there is now no doubt that it is quite plain that this category of groundwater should be recognized. Furthermore, it is urgently necessary. Neither will there be completeness in a classification if only artesian water, the pressure properties of which are caused by hydrostatic pressure, is recognized, and not deep-lying water which is pressured by the action of the geostatic pressures, or other internal forces. At the present time groundwater of such a type is established both in deep horizons of sedimentary formations, and in faults or other deep-lying tectonic zones.

Simplicity is an important aspect of any classification scheme. It was the clarity and accessibility of the classifications of S.N. Nikitin, O.E. Meinzer, A.M. Ovchinnikov, and O.K. Lange which ensured their wide use. As soon as the question arises of introducing additions concerning the way in which groundwater is formed into groundwater nomenclature, then it must be said that the amendments must, as far as possible, not complicate the classification unduly.

Finally, one must discuss the necessity of maintaining consistency in a classification scheme. This concerns combined schemes, the subdivisions of which are mutually exclusive, or do not have strong correspondence between each other. Thus one can sometimes find in the same classification scheme (Klimentov & Bogdanov, 1977) such subdivisions as 'water of the zone of aeration', 'subsurface water', and 'artesian water'. It is not entirely correct to compare 'water of the zone of aeration' with 'subsurface water' and 'artesian water' because in the zone of aeration the totality of the different types of water is distinguished. At the same time 'subsurface water' and 'artesian water' are compared according to the way in which they are formed, and also according to the pressure of this category of water. Hence, when differentiating the 'water of the zone of aeration', 'water of the zone of saturation' must also be distinguished; this is logical and consistent.

Regarding the above remarks, the scheme proposed by the authors of the present monograph, in which groundwater is subdivided according to the way in which it is formed, is shown in Table 6.5. In this scheme *groups* are distinguished (depending upon the position of groundwater in the main elements of the Earth's crust and its surface), *sections* (according to the degree of saturation of the rocks with water), *types* (on the basis of their hydraulic features), *classes* (as basic varieties of groundwater according to the way in which they are formed), *subclasses* (on the basis of the water-collecting properties of the rocks), and *special conditions* (defined by the specific nature of the surroundings). This classification scheme is based upon the subdivision of the best known groundwater classifications, especially those of O.E. Meinzer and A.M. Ovchinnikov. At the same time it is in accord with the level of present day knowledge of the way in which groundwater is formed.

The following categories of groundwater are described separately: (1) water of the zone of aeration; (2) subsurface water; (3) artesian water; (4) groundwater in fissured and karsted rocks; (5) deep-lying water; (6) groundwater of the permafrost regions; (7) groundwater of recent volcanic activity; and (8) groundwater under great expanses of water, i.e. under the floor of the oceans or seas. Thus, together with the subdivisions of groundwater, which are usually expounded in textbooks and manuals of hydrogeology, newly recognized categories of groundwater also figure, which have not previously been described.

6.2 The zone of aeration in the subsurface water

The interstices of soils and rocks situated above the water table contain air, water vapour, hygroscopic water, and pellicular and capillary moisture. Periodically in the spring when the snow melts, or during rains, the soil moisture content increases to a point when it is greater than the maximum molecular moisture content, and then subsurface water appears in the zone of aeration. This water, percolating under the influence of gravity, may accumulate in the soil and locally above impermeable or semi-permeable regions, or reach the water table.

In the broad sense all free (gravitational) water in the zone of aeration, including that lying closest to the surface, is called *suspended water* (Ovchinnikov, 1955). However, this term is also used in the narrower sense which has become widely accepted. According to this definition, suspended water is water which is formed in the aeration zone on lenses of relatively impermeable materials and is called perched groundwater. With this approach groundwater may be divided into (1) soil water, (2) infiltrating water, and (3) perched water proper (Fig. 6.1).

Soil water. In the soil there are different forms of the so-called soil moisture. Broadly speaking, 'soil moisture' is taken to mean 'soil water' and only free (gravitational) water comes into this category.

Not all soil moisture is available to plants: water vapour and the hygroscopic (tightly bound) forms of moisture cannot be utilized by plants. Pellicular water can participate only partly in the nourishment of plants — its exterior layer can dissolve and transport nutrients. By contrast, capillary water, the actual amount of which depends upon the structure of the soil, plays an extremely important role. In regions in which the water table is deep-lying, capillary water is the main source of moisture for vegetation (see Fig. 6.1).

Structured soils, which are distinguished by their ability

Table 6.5. *Groundwater classification according to the manner in which it has been formed (Pinneker, 1979)*

Group	Section	Type	Class	Subclass: Water in strata of porous rocks (pore and stratal water)	Subclass: Water in fissured cavernous rocks (fissure and vein-fissure water)	Special conditions: Water in permafrost regions	Special conditions: Water in volcanically active regions
Continental groundwater	Groundwater of the zone of aeration	Suspended water	Perched water (in the broad sense)	Salt water and infiltrating water, perched water		Active layer	Upper part of lava cover
	Groundwater of the zone of saturation on continents	Mainly non-pressure water	Groundwater	Aquifer nearest to the surface on stable impermeable layer	Upper parts of the zone of intensive fissuring and karst massif	Suprapermafrost — Interpermafrost and intrapermafrost	Lower part of lava cover — Water of hydrothermal systems under hydrostatic pressure
		Pressure water	Artesian water	Industrial water under hydrostatic pressure	Buried fissured zone under hydrostatic pressure	Subpermafrost	
			Deep-lying	Sedimentary layers, which are subjected to the action of geostatic pressure and endogenic forces	Water of deep-lying faults within the sphere of activity of endogenic forces	Absent	Water of volcanic structures and hot spring systems, connected with a rising stream from the magma chamber
Groundwater below seas and oceans	Groundwater of the submarine zone of saturation	Mainly pressure water	Water connected with the land mass	Shelf and marine deposits	Karsted rock of the shelf and fault zones	Subpermafrost shelf of the northern seas	Submarine volcanic structures and marine hot spring systems
			Water not connected with the land mass	Water of deep basins	Trenches and mid-oceanic rifts	Absent	

to accumulate and retain capillary water, because of the presence of large pores and their distribution, contain it in an immobile and therefore slowly evaporating form (Kovda, 1973).

The presence of free (gravitational) water in the soil is caused by the penetration of significant quantities of atmospheric precipitation and the condensation of water vapour. The presence of a significant quantity of soil water is indicative of excess moisture which leads to marshy conditions in the ground. A deficiency of soil water also decreases the fertility of the soil. Drainage also includes the regulation of the free water content in soils.

Soil water is fed both from above and below — by the transfer of free capillary rise water, especially in the case when the subsurface water lies near the surface of the ground. A considerable amount of organic material passes into this water from the soil, together with a large number of micro-organisms.

The soil water regime is determined by the ratio of water inflow to outflow. A *circulating*, *compensated*, or *evaporating* regime exists according to which of these two factors predominates. The first of these regimes is observed when the quantity of water arriving in the soil exceeds the sum of the amount taken up by the vegetation cover and that lost by subsurface evaporation. A compensated regime is said to exist when the inflow and outflow are equal. Finally, for the evaporation regime, the evapotranspiration rate exceeds that at which the water is added to the soil from the atmosphere. This last leads to saline water encroachment. If the evaporation regime persists, the mineral content of the soil water and groundwater increases, passing through the following stages (Kovda, 1973): silicate (0.02–0.1 g/l), calcium bicarbonate (0.2–0.5 g/l), sodium bicarbonate (0.5–3 g/l), sulphate (5–20 g/l) and chloride (> 30 g/l).

Infiltration water. Below the soil cover, in a zone of aeration of considerable thickness (usually greater than 5 m), between the suspended and the capillary rise (capillary fringe) water, is situated the so-called *intermediate layer*, or 'dead

horizon' (see Fig. 6.1). Here, periodically, free (gravitational) water arises which moves downwards or sideways.

Surface water, entering the zone of aeration, is expended to a considerable extent in moistening the soil, and in the formation of physically bound and capillary water. Only after this does the infiltration water, which moves downwards under gravity, appear below the soil cover. In those layers in which the pores are completely saturated, movement proceeds more rapidly because it is no longer necessary to moisten the soil. On the other hand, in dry soils, infiltration falls off with increasing depth.

Infiltration is the primary means by which free (gravitational) water moves in unsaturated soils. G.N. Kamenskii (1943) has distinguished between *free percolation* and *normal infiltration*. In free percolation the water moves accompanied by the partial filling of the pores in the form of isolated streams (Fig. 6.2). Normal infiltration takes place over a significant area with a continuous flow (all the interstitial spaces of the soil are filled with water). Sometimes 'inflowing', the influx and movement of water through large interstices, is referred to.

The rate at which moisture is transferred in the zone of aeration fluctuates. Infiltration opposes evaporation and processes intermediate in character between these two also occur (Table 6.6).

As V.N. Chubarov (1973) showed, the groundwater is fed through the zone of aeration even in desert dune regions (about 7 mm per year), which is almost exactly the same rate

Table 6.6. *Main types of continuous processes in the zone of aeration (after V.N. Chubarov)*

Direction of movement of moisture	Name of process	Relationship between inflow and outflow of moisture	Changes in moisture content with time
Descending	Infiltration (absorption)	Inflow > outflow	Increasing
	Downflow	Inflow < outflow	Decreasing
Ascending	Evaporation (desorption)	Inflow < outflow	Decreasing
	Renewal	Inflow > outflow	Increasing

Figure 6.1. Characteristics of water movement in the upper part of the Earth's crust.
I, zone of aeration; II, zone of saturation.
1, soil with soil water and suspended water; 2, sandy gravel deposits; 3, saturated rocks; 4, impermeable rocks; 5, capillary fringe (capillary rise water); 6, water table; 7, direction of movement of percolating water; 8, direction of groundwater flow.

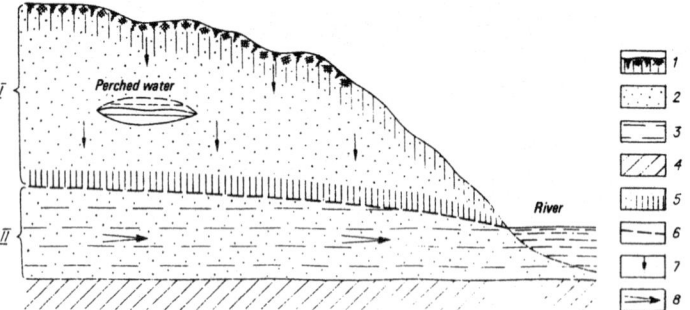

Figure 6.2. Diagram of free water penetration through rocks of the zone of aeration (after Klimentov & Bogdanov, 1977).
1, coarse-grained and gravelly sand; 2, medium-grained sand; 3, direction of penetration of free water; 4, capillary fringe; 5, atmospheric precipitation; 6, direction of water movement.

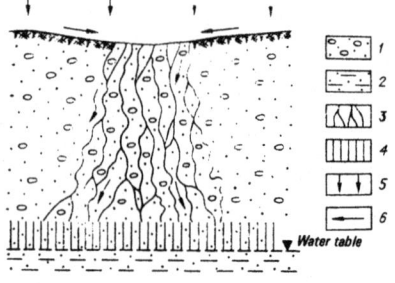

as that of those regions in which the moisture content fluctuates, or which have excess moisture (10 mm per year). This is because in the deserts, owing to the absence of vegetation and the presence of a deep zone of aeration (20–30 m and more), no water is removed by plants and evaporation within the rocks falls off sharply with increasing depth. In general, with a zone of aeration of more than 5 m thickness, the intensity of underground evaporation does not exceed 0.5 mm/year (Chubarov, 1973). Infiltration is greatest if the zone of aeration consists of soils with large interstices and in particular, fissures. On watersheds with poor powers of retention it reaches, according to the same author (Chubarov, 1973), 20–80 mm/year, but decreases to 1.4–7.6 mm/year on slopes and in valleys where a layer of loam and the soil and vegetation cover hinder infiltration.

Conditions in permafrost regions are favourable to the penetration of infiltration water. Here the zone of aeration consists of soils in which condensation of water vapour proceeds more rapidly in summer and which are distinguished by a high moisture content. It therefore remains frozen for long periods (Koldysheva, 1975). In these circumstances melt water, flood water, and rain water disperse mainly as surface runoff, only a small fraction being able to infiltrate through the zone of aeration in summer.

Perched water. This arises when infiltration water meets a relatively impermeable horizon between water-saturated soils. The impermeable layer may be a clay lens of loamy soil in a layer of sand, parts of the weathered rock over bedrock, frozen layers, or even alluvial soil horizons (when the surface water is soil water). The water-bearing layer formed on such an impermeable layer is distinguished in the first place by its seasonal character because surface water arises only during rains and, secondly, by local geology. Consequently it is not a continuous water-bearing horizon. Perched water is situated above the water table and is therefore suspended water (Lange, 1969). It usually has no hydraulic continuity with ground or river water, although in some areas such a possibility cannot be ruled out.

The thickness of a perched water body is more often than not 0.4–1.0 m, rarely reaching 2–5 m. It is usually found in sandy-loamy soils. Homogeneous very permeable beds and those which have poor water-retaining properties (for example coarse-grained sand or fissured hard rock) are very unfavourable for the formation of perched water. In just the same way it is not formed in clays: as a result of the swelling of the colloids the fine upper layer of clays is saturated comparatively quickly and then is no longer permeable.

The formation of perched water is strongly influenced by the relief. Slopes, especially steep slopes, where the surface runoff exceeds the infiltration, do not contain perched water; in an extreme case, it forms a thin water-bearing layer for a short time. The best conditions for the formation of perched water arise on flat water-divides and steppes with local depressions into which rain water runs and where melt water is held. Perched water is often found in sections of river terraces. In cities and industrial areas the formation of perched water is assisted also by the depressions or old hollows left over from previous construction work which are covered with fill: atmospheric precipitation and effluent water infiltrate these easily.

The perched water regime is completely determined by the amount of infiltrating atmospheric precipitation and in cities and industrial areas, by the volume of effluent discharged into the zone of aeration. As P.P. Klimentov & G.Ya. Bogdanov pointed out (1977), the length of time for which it lasts is dependent upon the size and thickness of the impermeable bed, the ability of the bed to hold water, and the recharge conditions. When the impermeable layer is small and thin, the perched water exists only for a comparatively short time: after a while the water manages to pass through the semi-permeable rocks or flows away beyond the margins of the lens. As the size and thickness of the containing bed increase there is an increase in the duration of the existence of the perched water. In arid regions where perched water is formed close to the surface, some of the water may evaporate.

Perched water is not only rapidly exhausted, but is also easily polluted. Hence in built-up areas (for example in towns, and industrial and housing estates), where leakage of domestic and industrial waste occurs, the perched water is very polluted. In natural conditions its quality varies greatly. In regions of high rainfall it is polluted to some extent with a relatively high content of organic material. In those places in which the rainfall is deficient, the subsurface evaporation gives rise to saline water and brines.

6.3 Subsurface water

The characteristics of subsurface water. The definition of the term 'subsurface water' given by S.N. Nikitin in 1900, with certain additions, has been adopted by the majority of Russian and Soviet researchers (F.P. Savarenskii, G.N. Kamenskii, A.M. Ovchinnikov, N.I. Tolstikhin, O.K. Lange, G.V. Bogomolov, N.A. Marinov, P.P. Klimentov, and others). In the USSR the term subsurface water is taken to mean free (gravitational) water of the permanent water-bearing layer situated nearest to the surface, which changes with time and is contained in unconsolidated deposits, or in the upper fissured region of the bedrock, and which lies above the containing bed nearest the surface. N.A. Marinov (1979) also places the groundwater which lies directly beneath the sea floor, reservoirs, and beneath river beds in this category. Incidentally, groundwater is not only formed 'by absorption of atmospheric precipitation' as S.N. Nikitin proposed; it can be of marine origin (in places which have been reclaimed from the sea), be derived from condensation, or be of mixed origin.

Upwards, the subsurface water is not covered by impermeable rocks and therefore is closely connected to the atmosphere via the zone of aeration. The surface of the subsurface water is mainly free: the term 'subsurface' has become practically synonymous with the term 'pressureless' or 'nonpressure' water. When subsurface water is met in boreholes or wells its level is established at exactly the same level at which it is found. Only in places where lenses of impermeable rock lie above the level of the subsurface water does it acquire a small local abnormal pressure: in boreholes the level of the subsurface water under an impermeable cover, and in neighbouring areas (which do not have such a level), is established at the same level. In Russia the surface of the subsurface water is also called a *mirror*, and occasionally a *tablecloth*.

Depending on the geomorphological conditions, geological structure, and hydrogeological features, there can be a *subsurface current*, a *subsurface basin*, and a *combination* of subsurface current and a subsurface basin.

A subsurface water-bearing horizon (or a part of it), in which the movement of water takes place under the action of gravity in the direction of the water table slope is called a subsurface stream (see Fig. 6.1).

A subsurface water basin is usually confined to the downwarping in the impermeable bed which is filled with permeable rocks which are saturated with water; the water table is horizontally situated (see Fig. 6.3). These basins are formed in recharge conditions in which the infiltrating or condensing waters do not fill the layer completely. Certain geomorphological elements or geological bodies, in which the subsurface water is widely distributed (subsurface water basins of fluvio-glacial deposits, ancient valleys, etc.) are also often called subsurface water basins.

Under natural conditions a combination of currents and subsurface water basins is often found. In this case there are close hydraulic connections between them; therefore it is difficult to establish the boundaries between the current and the basin. The character of the currents and the subsurface water basins of Uzbekistan described by N.N. Khodzhibaev (1970) is very interesting.

Comparatively shallow deposits of subsurface water of the impermeable stratum nearest the surface and the connection of the water with the surface and meteoric water determine the following of its features:

(1) it is non-pressure water, usually having a fres surface and the pressure on it is equal to atmospheric pressure, and it moves under gravity in the direction of the water table slope;

(2) it is recharged mainly by the infiltration of atmospheric precipitation and the condensation of moisture in the zone of aeration, the recharge region coincides with the region of distribution;

(3) discharge takes place at the base of the slope or into surface reservoirs, and currents of water such as streams, to which the subsurface water is hydraulically connected;

(4) as a consequence of the direct influence of surface factors, the level, water inventory, temperature, and other parameters associated with subsurface water, it is subjected to great time variations;

(5) shallow deposits and intensive underground runoff give rise mainly to fresh (i.e. non-saline) subsurface water, but when there is a deficiency of moisture in an area they become mineralized.

The recharge and distribution conditions of subsurface water (Klimentov & Bogdanov, 1977). The subsurface water is recharged basically through the zone of aeration by infiltration of atmospheric precipitation (rain, melt water, and flood water) over its whole area of distribution, and by the condensation of water in the zone of aeration. In any particular region there may be other recharge sources (river water, the entry of water from irrigation canals, inflows of artesian water from deeper-lying water-bearing horizons, etc.).

The magnitude of infiltration depends on the duration and intensity of the precipitation and also on the lithologic composition and permeability of the rocks in the zone of aeration. Prolonged steady rain, falling when the humidity is about 100%, is the most significant factor in the recharge of subsurface water. This ensures the maximum infiltration through the zone of aeration.

Winter precipitation as snow can serve as a source of recharge for subsurface water only in spring after the soil which has been frozen during the winter has thawed and the transition of the solid precipitation to liquid has occurred. The permeability of the soil, the local relief, the type of vegetation, and some other factors play an important role here. For instance, on a level plateau the conditions for the infiltration of the spring melt water are more favourable than on a steep slope; on places covered with vegetation infiltration will also be intensive because the vegetation reduces the rate at which the snow melts and reduces the surface runoff.

In the steppes, where the snow cover has insignificant thickness and the strong winter winds carry away great quantities of snow into ravines and river valleys, only a very small fraction of the solid winter precipitation succeeds in percolating into the soil layer during the rapid spring thaw, and then only to an insignificant depth. More intensive recharge takes place in low-lying areas, for example in ravines, in sink-holes, 'saucers' or 'dry' limans (Fig. 6.4), where melt water accumulates. Naturally, in such places the subsurface

Figure 6.3. Diagram of a groundwater basin.
Arrows show movement of moisture: downwards, infiltration; upwards, evaporation.
aa, water table; bb, surface of impermeable bed.
1, sand; 2, water-bearing sand; 3, clay.

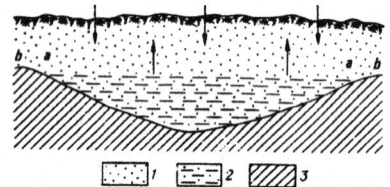

Figure 6.4. Diagram showing how a lens of fresh water is formed on saline water under limans in summer in the pre-Caspian lowland (after G.Ya. Bogdanov).
1, loams and sandy loans; 2, water table; 3, infiltration of surface water and atmospheric precipitation; 4, direction of movement of local surface runoff; 5, direction of groundwater movement.

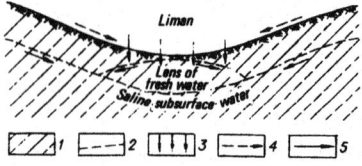

water level rises considerably in the spring and the water itself is made less saline. In the steppes, wells to the subsurface water, obtained here at a depth of several metres, are often the only source of fresh water.

The rise in the level of the subsurface water and the increase of the flow of the springs that are fed from the subsurface water, depend on the quantity of atmospheric precipitation and the local conditions (the water-conducting properties of the rocks, the depth of the water table, etc.), and take place over different periods of time after intensive rainfall. The sharp increase in the flow of springs which is observed a month or more after the maximum rainfall occurs is a well known example. There is also the fact which is particularly characteristic of fissured and karsted rocks in which groundwater moves at greater velocities than in sandy water-bearing strata, that the greatest spring flow occurs several days or even hours after rain.

In many cases, especially in deserts and semi-deserts, the subsurface water is fed from condensation, i.e. it is replenished as a result of the condensation of water vapour from the air, which settles in those parts of the rock which have cooled. For example in the Kara-Kum desert moist sand can be found after three or four months of drought. The underground moisture is formed from condensed water vapour also on the shores of the Caspian Sea and the Gulf of Kara-Bogaz-Gol, i.e. where atmospheric precipitation in an extremely dry climate cannot provide any noticeable source of groundwater recharge.

As has already been pointed out, subsurface water is replenished from undercurrents of artesian water from lower-lying strata. This form of recharge is possible in places where the impermeable cover of artesian water-bearing horizons has been pierced (through geological 'windows'), only when the pressure head exceeds the level of the piezometric surface of the subsurface water.

Discharge of groundwater takes place through dispersed and concentrated outlets, stratal seepage, or in marshy areas. Springs are usually situated in places where the water-bearing horizons have been exposed by erosion, or by the wedging out of the water-bearing beds. If the discharge consists of water-bearing strata made of fine- or medium-grained sands, then minor water outlets concentrate in small local hollows. Sometimes stratal seepage takes place near the outlets, usually on a slope which is wet over the whole area of the outlet of the water-bearing strata. Such places often continue along the slope as marshy strips where marsh vegetation grows, and water accumulates in hollows. In arid regions, as a result of the evaporation of water, a fine layer of salt is formed on the surface of such places in the form of a white bloom.

Springs from alluvial deposits are often observed in the steep sides of terraces or river bends where a river undercuts its bank, where the cross-section of the subsurface flow is reduced. Water outlets can also be connected with the changes in the composition of the alluvial deposits. For example, when the composition of the alluvium changes from sandy to sandy-clayey the velocity of the subsurface water is reduced and this causes its surface to rise and produces springs in depressions in the ground.

The flow of springs is usually small (less than 1.0 l/s). It is strongly time-dependent. The greatest flow is found in fissured, and especially karsted rocks; it often amounts to several hundreds of litres and even several tens of cubic metres per second.

The way in which subsurface water is formed depends upon many factors (the recharge conditions, the permeability of the rocks, the configuration of the banks of the reservoirs, and the streams with which the subsurface water has hydraulic connections, the position of the confining bed, etc.). The water table of the subsurface water is shown on a map of water table contours (*hydroisohypses*), which are lines joining points of equal water table level. Water table maps are constructed from data on the water table levels taken simultaneously (because they vary with time) in boreholes, excavations, and wells. Apart from this, data on the position of springs, swampy hollows and bogs, which arise as a result of the exit of groundwater on to the surface, upon the levels of the water in the rivers, lakes, and other bodies of water, and streams with which the subsurface water is hydraulically connected, are also used.

With a water table contour map one can find the direction of movement and the magnitude of the gradient of the subsurface current, the depth and way in which the surface of the groundwater is formed, and its relationship with the surface relief. With a stratoisohypse map of the surface of the impermeable bed it is also easy to determine the thickness of the water-bearing bed at any time. It is important to note that there is a marked dependence of the gradient of the water table on the permeability of the deposits and the thickness of the water-bearing horizon (see Fig. 6.5).

The depth of the water table often depends on the relief of the locality. As geological surveys of vast areas show, the top surfaces of the subsurface water-bearing horizons are for the most part uneven, undulating and often following the

Figure 6.5. The position of the water table (after Ovchinnikov, 1955).
(a) where there are changes in the permeability of rocks;
(b) where there are changes in the magnitude of groundwater flow.
1, water table; 2, sand; 3, gravel; 4, confining bed; 5, direction of water flow.

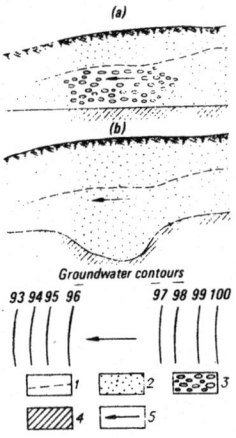

curved shape of the relief, although in certain places, for various local reasons (the draining of the subsurface current by a river valley, a sharp increase in the thickness of the water-bearing stratum, etc.) such a relationship between the ground surface and the water table may be absent. In river valleys, gorges, ravines, and other low points in the relief, groundwater is comparatively near the surface, but at water-divides the depth to groundwater may reach several tens of metres. The movement of groundwater, with rare exceptions, is from higher to lower areas.

The connection between subsurface water and river water is illustrated in Fig. 6.6. In mountainous regions, on plateaus and high ground with a broken relief, groundwater, as a rule, drains towards the river. The European part of the USSR is an example where the majority of the rivers (the Volga, Oka, Kama, Dnieper, Don, the Western Dvina and others) drain the groundwater. On the other hand, on the plains, especially where the climate is arid, and where the water table is deep, the rivers often feed the groundwater, i.e. the water table sinks away from the line of the river bank (see Fig. 6.6(*b*)). In the USSR many rivers of the plain of Central Asia are of this type. There can also be a more complex inter-relationship between subsurface water and river water. For example in mountainous regions subsurface water may flow from one slops of the river valley into the river bed, and the other slope may at the same time absorb river water (see Fig. 6.6(*c*)).

During periods of high water and floods, when the river water level is high, the water table near river banks rises (Fig. 6.7). The curve of the head of water extends outwards into the land between the rivers for several hundred metres and even for several kilometres in rare cases. When the water level falls, quite a sharp fall of the water table results, close to the bank.

In large hydraulic engineering works, as a result of the considerable head of river water, on the banks of the reservoir the water table rises (see Fig. 2.4) over a greater distance. The new position of the water table in the pressure zone becomes established for large rivers over a period of several months, and sometimes continues for many years.

The subsurface water regime, i.e. the processes of changes in its quantity and quality with time, is extremely unstable and is influenced by natural or artificial factors. A regime which is determined by natural factors alone is called a *natural* or an *undisturbed* regime, but if artificial factors play the major role in its formation, then it is called a *disturbed*

regime (Kamenskii, 1953; Kovalevskii, 1973). Natural factors have the greatest influence on shallow groundwater. As the depth increases the influence of many factors weakens or disappears completely.

The recharge of the subsurface water at any given place is determined by the infiltration of meteoric water, the condensation of water vapour, and subcurrents or side currents from neighbouring areas, and it is discharged as a result of outflows, by evaporation from the surface of the subsurface water, and the transpiration by vegetation. The quantitative determination of the elements of inflow and outflow in the subsurface water balance enables the basic factors forming the hydrogeodynamic and hydrogeochemical regimes of the subsurface water to be distinguished, i.e. all those objective causes which determine the seasonal and long-term variation in its level, volume, and direction of the process of formation of the chemical composition of subsurface water (Kovalevskii, 1973; Lebedev & Yartseva, 1967).

Subsurface water is differentiated by the way in which it is formed, the peculiarities of its distribution, recharge, and its regime into: (1) interfluvial and watershed areas, (2) river valleys, (3) alluvial cones and lines of foothills, and (4) seashores. The conditions of distribution and formation and especially of the temporal and spatial variations in these conditions are peculiar to each of the types.

For water of the interfluvial and water-divide areas, an infiltration–runoff or watershed regime is characteristic, despite the fact that it is formed in rocks which differ in composition and collecting properties: the water increment is determined by the infiltration of atmospheric precipitation and surface water, and the decrement by outflow and local drainage.

The subsurface water regime in valleys, which is closely associated with streams, is completely different. The recharge of such water is complex (atmospheric precipitation, river

Figure 6.7. The different cases of the relationship between subsurface and river water during flood (after M.A. Veviorovskaya). (1) the river usually drains the subsurface water horizon, but during flood feeds it; (b) the river always feeds the subsurface water; (c) the hydraulic connection between the subsurface and the surface water is absent even during floods; (d) there is hydraulic connection between the subsurface water and the surface water only during periods of low river levels; (e) the river influences the groundwater level only over a narrow belt near the banks.
1, permeable rocks; 2, impermeable rocks; 3, subsurface water level.

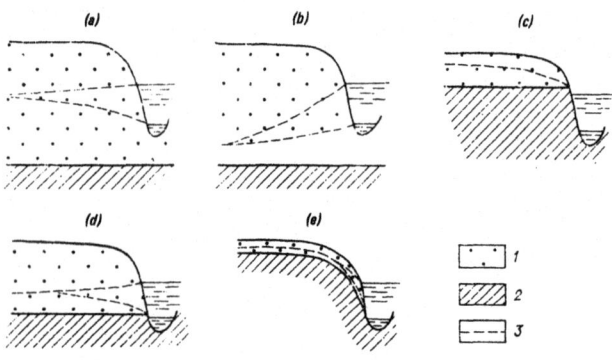

Figure 6.6. Relationship between subsurface and river water. (a) the river drains the subsurface water horizon; (b) the river feeds the subsurface water; (c) the river drains the subsurface water horizon on one side and feeds it on the other.

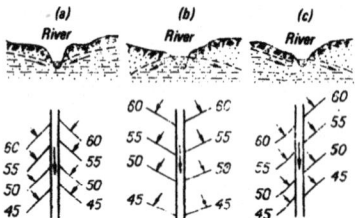

water, groundwater streams from watershed areas). On the decrement side surface rivers and outflows predominate, sometimes, if the water table of the aquifer is at a depth of less than 5 m, evaporation is the dominant factor.

In alluvial cones and lines of foothills there are great accumulations of subsurface water. The thickness of the permeated layer here reaches tens of metres. For the increment part of the balance of this water, along with the infiltration and the atmospheric precipitation, condensation and the absorption of the surface runoff are of great significance. In the decrement part underground runoff dominates and in favourable conditions evaporation also occurs.

The subsurface water of coastal areas has specific features. Atmospheric precipitation percolating from the surface and the condensing moisture accumulate in the porous strata at a level higher than the saline water of marine origin, and so saline water is usually expelled. The thickness of the fresh water varies within considerable limits; mixing with saline water takes place extremely slowly, but is accelerated by an irregular withdrawal regime.

One ought especially to distinguish dry steppes, semi-deserts and deserts, which are characterized by arid climates, which also determine the essential features of the subsurface water of these areas, i.e. the dominance of evaporation in the decrement part of the water balance, and, for deep-lying deposits (more than 5–8 m), underground outflow, which is combined with insufficient moisture. Because the depth of the deposit determines the amount of evaporation, the relief and the drainage of the territory become the basic factors controlling the regime.

The zonation of subsurface water. A wide zonation is characteristic of subsurface water. The latitudinal zones correspond to certain landscapes and succeed one another from north to south. As V.S. Il'in noticed as long ago as 1923, the zonation of subsurface water is influenced mainly by climatic factors, in particular the degree of wetting of the territory; such factors as the depth of the down-cutting by erosion, the collecting properties and composition of the water-bearing rocks are of no small importance. V.S. Il'in contrasted zonal water, which is differentiated according to the totality of the above-mentioned three factors with azonal water, i.e. water which does not possess all the zonal characteristics. Only one of these enumerated factors plays a leading role in the life of such water.

Of the innumerable schemes for the latitudinal zonation of subsurface water, that put forward for the territories of the USSR by O.K. Lange gives the most complete picture. It is based on the existence of three latitudinal macrozones of subsurface water, which succeed one another from north to south (Lange, 1960).

(I) A macrozone (a province) of permafrost with the following zones: (1) stone ice, (2) dense continuous permafrost, (3) thawed permafrost, and (4) island permafrost. These zones are clearly defined in Eurasia and North America.

(II) A macrozone of surplus moisture with the following zones: (1) subsurface water of the tundra type, in the unfrozen tundra and the dwarf woodland belt, (2) the high waters of the North, the basins of the rivers Pechora, the Northern Dvina, the Onega, the Western Dvina, the Pripet, the Leman, etc., (3) the drainage of shallow ravines – the basin of the Upper Volga, parts of the Oka, etc., (4) the drainage of deep ravines – the Upper Oka, the Tsna, the Moksha, the Inzar, etc., (5) the drainage of the transition region between the ravine system proper and the ravine–gulley system.

(III) A macrozone (province), semi-arid and arid, which contains the following zones: (1) developed drainage by way of evaporation with subzones (a) the deep gulleys of the Black Sea region, and (b) the wide gulleys of the Caspian region, (2) the equalizing of the inflow of subsurface water and its evaporation, which stretches from Central and Middle Asia through the Near East, and North Africa to the south of North America.

Although O.K. Lange takes features of the relief into account, and the lithology of the rocks in addition to the climate, the zones which he recognizes are subordinated exclusively to the climatic factor. Subsurface water, in the life of which the other factors predominate, is categorized as azonal or intrazonal water. Thus it is the lithologic factor which is predominant in the formation of azonal water. Karst water, fissure water, and alluvial water come into this category and are samples of azonal water. The saline water of salt-marshes (solonchaks), and salt lakes and springs (these are solonetzes), or lenses of fresh water which are deposited above saline water-bearing horizons are examples of intrazonal water.

G.N. Kamenskii (1949) believed that for subsurface water independent geographical zones could be established according to hydrogeological features, mainly according to the conditions in which they were formed, and the composition of the subsurface water. A.A. Konoplyantsev (1960) noticed the presence of zonal phenomena in the subsurface water regime, i.e. changes in temperature, level, and others. In his opinion, karst-fissure water, or alluvial water are by no means azonal: they are formed in response to the climatic regime of the subsurface water zone to which they are related. It is impossible to regard intrazonal water as an anomaly. For example, the water of solonchaks is a zonal phenomenon because it is formed under the influence of the general climatic conditions.

A hydrogeological indicator such as the degree of mineralization and the ion-salt composition of the subsurface water, is a concrete parameter which reflects not only the climate, but also the whole complex of the hydrogeological regime. Taking these parameters as a basis, I.V. Garmanov (1948) established within the European part of the USSR a progressive change from north to south of ultra-fresh hydrocarbonate water (with silicic acid), through fresh hydrocarbonate water to saline sulphate and chloride water. Consequently, from the northernmost latitudes to the arid regions, subsurface water not only becomes 'mineralized', but its ion-salt composition also changes.

The processes of formation of the subsurface water find expression in the variation of the degree of mineralization and the composition. G.N. Kamenskii (1949) distinguished two major zones, which correspond respectively to the two genetic types of subsurface water — *leaching* and *continental salination*. A third could probably be added — subsurface water of *frozen regions*, in the formation of which leaching and salination combine. Each of these types corresponds to macrozones of shallow water considered with the nature of zoning of subsurface water (see Section 3.4) — humid, arid, and ice.

A combination of the principles for distinguishing latitudinal zones of subsurface water, used by O.K. Lange on the one hand and also by G.N. Kamenskii on the other, is used in the scheme proposed by I.K. Zaitsev and M.P. Raspopov (Fig. 6.8). In this scheme the macrozones are called provinces and zones are called belts (Zaitsev, 1960).

6.4 Artesian water

The features of artesian water. Subsurface water which is situated in water-bearing horizons (complexes) which are covered with, or underlain by, confining beds of relatively impermeable material, and which have hydrostatic pressure which causes the water level to rise above the upper confining bed when these horizons are met in boreholes and excavations, is known as artesian water. Sometimes it is called interstratal pressure water (Lange, 1969). Usually subsurface water in sedimentary strata is put into this category, although, as has already been said, it can also be found as fissure water in crystalline rocks.

In favourable structural—geological and hydrogeological conditions boreholes produce spouting water. At one time subsurface water and aquifers, in which water is under excess pressure and which gushes out when met in boreholes, were classified in this way. However, not everywhere, and certainly not always, does an artesian aquifer, as was considered initially, yield gushing water: this does not occur when the pressure head is below the surface of the ground. Furthermore, earlier, (Keilhack, 1935), artesian water was associated with ideal basin-shaped downwarping of the Paris Basin type (see Fig. 5.6). Later it was found in a much more varied and complex arrangement (monoclinal slopes, saturated fault zones, etc.). This water is found in Precambrian rocks, although it is also found in Quaternary deposits.

Figure 6.8. Groundwater zoning in the USSR (after I.K. Zaitsev and M.P. Raspopov).
Subsurface water of the stable permafrost province: I_1, belt of solid permafrost with rare thawed areas (thickness of permafrost is 200–500 m and more); I_2, belt of solid permafrost with widespread development of thawed areas (thickness 100–200 m, in the Viluiski syneclise 400–600 m); I_3, belt of scattered patches of permafrost (thickness 25–100 m and less); I_4, belt of high mountain ice and glaciers.
Subsurface water of the non-permafrost province: II_1, belt of predominant development of leaching and salt removal processes (humid); II_2, belt of predominant development of continental salt accumulation in the subsurface water and in rocks (arid); II_3, belt of vertical zonation of the processes of leaching and salt accumulation in mountainous regions.

Degree of mineralization and chemical composition of subsurface water: 1, very fresh (up to 0.2 g/l), mainly hydrocarbonate, often with a high content of silicic acid and organic materials; 2, fresh (up to 0.5 g/l), hydrocarbonated; 3, fresh to saline, mainly with 1–3 g/l degree of mineralization; 4, from fresh to slightly saline with up to 10 g/l mineralization, mainly sulphate and chloride; 5, from fresh to brines with up to 200 g/l mineralization, mainly chloride, rarely sulphate and hydrocarbonate (soda); 6, boundary between the climatic provinces; 7, boundary between belts.

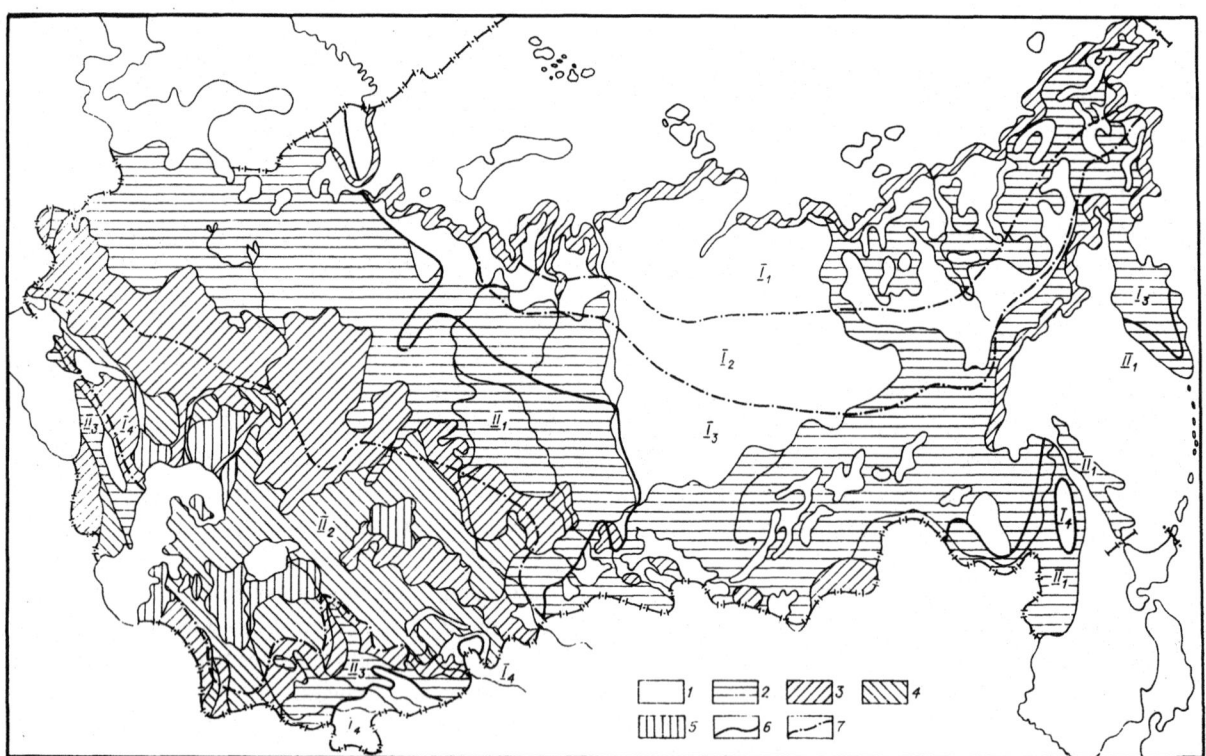

The following features are characteristic of artesian water:

(1) it is interstratal pressure water, the horizons and complexes of which are isolated both above and below by confining beds;

(2) the recharge region and the creation of the pressure of the artesian water and the region in which it is distributed do not coincide and are often separated from one another by great distances;

(3) when an artesian aquifer is met, the appearance of water in the borehole is always noticed deeper than the established level; often the level becomes established at a higher level than the ground surface, and then water gushes out of the borehole;

(4) the artesian water regime is more stable than that of subsurface water: surface waters have much less influence;

(5) artesian water is fresh in the upper part of the cross-section, but its degree of mineralization increases with depth and it becomes saline water and even brine.

The recharge and distribution conditions of artesian water. Among the natural reservoirs of artesian water, following A.M. Ovchinnikov (1955), the fundamental ones are: (1) artesian basins (basins of stratal water), and (2) artesian slopes (monoclinal formations of stratal water).

1. By an *artesian basin* (see Section 5.2) is understood the totality of pressure-water-bearing horizons or complexes, which are formed in synclinal structures where subsurface water issues under hydrostatic pressure. Water-bearing horizons (complexes) of the artesian type are distinguished by the relatively small dimensions of their recharge areas (the creation of head of pressure) compared with the area of the runoff region (the development of the pressure head). Within platforms the recharge region may be situated on various favourable forms of relief of the interfluvial areas and uplands (Fig. 6.9). These are characterized by the convexity of the pressure surface of the water-bearing horizons and the fall of the head with depth. Such a correlation of levels gives rise to the possibility of the flow of water from the upper horizons to the lower through relatively impermeable rocks.

Sometimes the recharge of artesian water is brought about by an undercurrent from the lower-lying horizon to the higher one; this takes place when the pressures of the alternating aquifers increase with depth. An undercurrent usually occurs along tectonically weakened zones or through the relatively impermeable confining beds.

Some researchers (A.M. Ovchinnikov, P.P. Klimentov and others) do not subdivide the regions of recent recharge into internal and external types because the term 'external' indicates that the recharge area where the pressure is created does not come into the composition of an artesian basin. In their opinion, to separate artificially in the same way such a basic element as the recharge area from the artesian basin would be to violate the principle of distinguishing a unified hydrogeodynamic system. The argument can hardly be considered convincing because in artesian basins it is necessary to include all the hydrogeological massifs, i.e. in general (as does N.A. Marinov) not to distinguish the latter. Many hydrogeologists (Kamenskii *et al.*, 1959; Zaitsev & Tolstikhin, 1972) hold the contrary opinion.

Artesian water is hydraulically connected to other subsurface water in places where the upper confining beds are eroded, or where there is a lateral change in lithology into permeable rock. Depending on the relationships of the level of the subsurface water in such places, either recharge or discharge of the artesian water will occur. The close connection of artesian with other subsurface water can be noted both in the recharge areas and in the discharge areas of artesian aquifers (Fig. 6.10).

Fig. 6.10(*b*) shows the case of the transition of subsurface water (i.e. non-pressure water) into artesian water (i.e. interstratal pressure water) which also has an intermediate stage — that of interstratal non-pressure water. This has already been noted in Section 6.1. The present occurrences are distributed mainly in the recharge areas of artesian basins; they are also found, for example, when cross-currents flow from one river to another (see Fig. 6.11), and in the coastal belt of streams or basins when they drain artesian aquifers.

The area of distribution of pressure (runoff) is situated within the basic area of the artesian basin; within this area the pressure levels, which are generally known as *piezometric surfaces*, are typical of artesian aquifers. The vertical distance from the upper confining bed of the aquifer to the piezometric

Figure 6.10. Interrelationship between artesian water and groundwater (after Ovchinnikov, 1955).
(a) recharge of artesian water by subsurface water; (b) transition between artesian water and subsurface water; (c) recharge of subsurface water by artesian water.
1, unconsolidated Quaternary aquifers; 2, water-bearing horizons in the basement rocks; 3, impermeable rocks; 4, water table; 5, direction of water flow.

Figure 6.9. Correlation of the Piezometric surfaces of aquifers in Cretaceous, Neocene, and Oligocene deposits and the water table dependence on the relief (after I.V. Garmonov, A.V. Ivanov, and V.M. Sugrobov).
1, ground level; 2, water table.
Respective piezometric surfaces of the aquifers: 3, Neocene; 4, Oligovene; 5, Cretaceous.

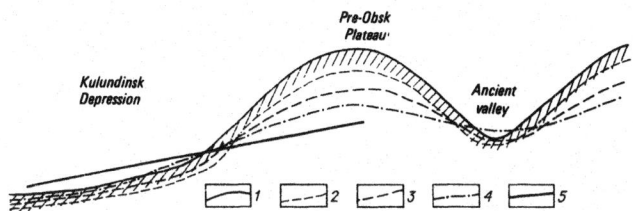

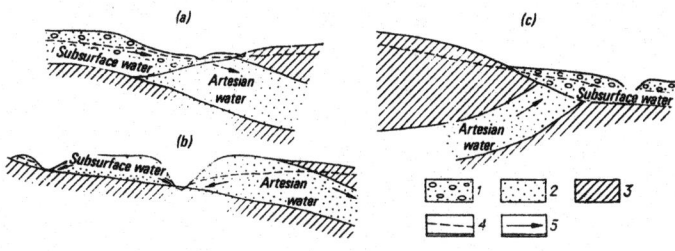

surface is called the *head* of the subsurface water. The head determines the potential energy of the water.

The piezometric surface is found by measurements taken in boreholes, by which the aquifer is tapped. The character of the piezometric surface in any pressure aquifer is usually shown on maps by hydroisohypses (Fig. 6.12). *Hydroisohypses* (sometimes called isopiestic lines or piezoisohypses) are lines joining points of equal level on the piezometric surface.

In the usual bowl-shaped formation of strata, it is the relief which has the greatest influence on the relationship of the piezometric surfaces of different artesian aquifers.

The intensity of the underground runoff is very varied, and depends upon the hypsometric position of the recharge area and the discharge area, and also on the location within the artesian basin. Artesian basins are often found in which water exchange in their deep parts is made difficult because the recharge area is at approximately the same altitude, and there are no visible discharge areas. The discharge of subsurface water takes place along tectonically weakened zones, and also

Figure 6.11. Formation of interstratal non-pressure water. 1, sands; 2, impermeable rocks; 3, water table; 4, level of the interstratal non-pressure water; 5, descending spring.

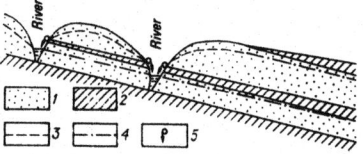

Figure 6.12. Groundwater contour map. 1, hydroisopiestic absolute level; 2, area of emerging water; 3, direction of artesian water movement; 4, borehole (the numerator gives the number of the borehole, the denominator gives the piezometric level in the borehole); 5, topographic contours.

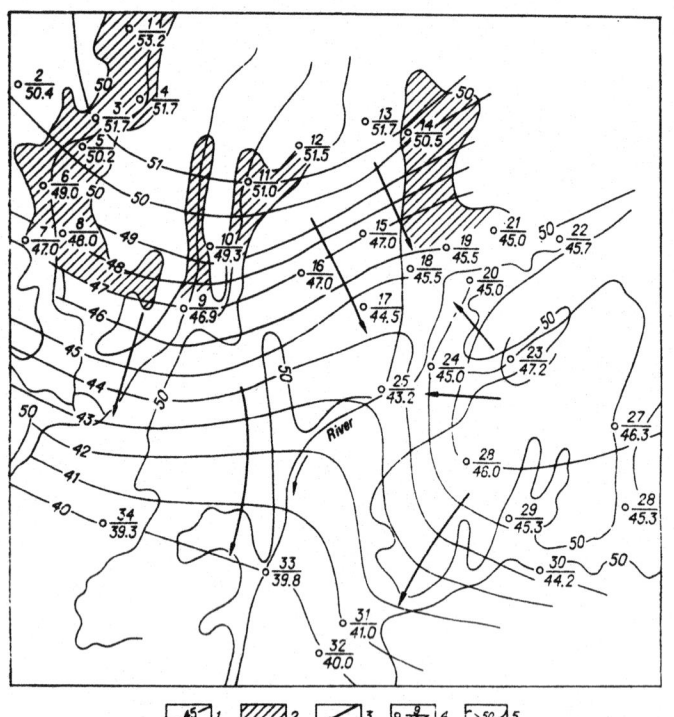

over the whole area through the rocks of the upper confining bed, which are relatively impermeable for a considerable range of pressures (Fig. 6.13).

2. A.M. Ovchinnikov (1955) considered an *artesian slope* to be a unique asymmetric basin of artesian water. Such basins are restricted usually to monoclinal deposits when the water-bearing rocks wedge out by sinking, or the facies changes into rocks which are relatively impermeable, and even into confining beds. The specific hydrogeodynamic features of such aquifers are determined by the geological conditions (see Fig. 6.14). The recharge areas (recent infiltration and creation of head) and discharge (drainage) within the artesian slope are usually situated in the immediate neighbourhood, and the region of distribution of head (runoff) is towards the lower levels.

Under natural conditions artesian slopes, just as artesian basins, are of common occurrence. They usually gravitate towards the marginal parts of foothill downwarps and inter-montane valleys, or towards the coast. They are also found within the slopes of synclines and depressions on platforms, especially when the layers of water-bearing rocks are thick. In this sense in particular, the Angara–Lena artesian basin in Eastern Siberia is an artesian slope.

The structure of artesian slopes, the conditions of recharge and discharge of the subsurface water can be extremely varied. This type of artesian reservoir has so far been but poorly studied.

Both in artesian basins and artesian slopes, subsurface water is discharged in the form of ascending, mostly concentrated springs; sometimes the discharge has a stratal character. The outflow is extremely variable: discharge centres are found which have discharge rates of up to 10 m^3/s and more, but often the outflow is in tenths or hundredths of a litre per second.

Some features of the dynamics of artesian water. One of the characteristic features of artesian reservoirs is the presence of a compression regime. According to V.N. Shchelkachev (1959), this is a regime in which the behaviour of the aquifer while it is being exploited is determined essentially by the resilience of the rocks and the liquids with which they are saturated (water and oil).

Figure 6.13. Artesian basin with an impeded water exchange. A, the limits of distribution of a weakly water-bearing rock complex; a, recharge area (and partial runoff), b, head and impeded recharge through a cover of almost impermeable rocks; 1, aquifer; 2, almost impermeable rocks; 3, impermeable rocks; 4, piezometric level; 5, water table; 6, boundary between fresh and mineralized water.
Arrows show direction of water movement.

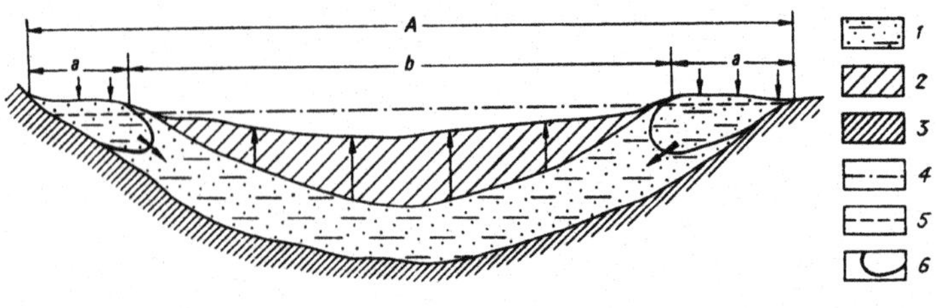

$12-15$ m^3/s, but in spring during snow-melt it increases to 50 m^3/s. It must be emphasized that the spring flow could supply the requirements of the population of Moscow.

On territories which consist of karst limestone streams are lost and even large rivers have reduced flow. For example the Angara and its tributaries lose a great amount of water. All this water goes to recharge the karst water.

Thus the interrelationship of the surface water and the groundwater in regions where rocks are undergoing karst formation is not only very close, but usually complex too. If the total flow of all the springs which emerge in the region is taken as the magnitude of the underground runoff, then it is possible to come to wrong conclusions, because some springs emerge on to the surface twice (Klimentov & Bogdanov, 1977).

D.S. Sokolov (1962) distinguished the following vertical hydrogeodynamic zones in karst rocks on the basis of the character of the movement and regime of the water (Fig. 6.17):

(1) the zone of aeration, in which infiltration and descending movement of water takes place mainly along vertical fissures;

(2) the zone of seasonal variation of groundwater level which in the period of increased recharge and with the rise in the water table merges with the lower, but in the period of falling levels joins to the upper, i.e. to the zone of aeration; when the level stands high in this zone the water flows horizontally, when the level is low it flows vertically;

(3) the zone of complete saturation which is situated in the sphere of the draining action of the hydrogeographic network with movement of water towards the river valley, which is cut into the karst rocks; the basement of this zone is lower than the horizons of the surface water and the movement of water near it is directed upwards; this zone contains the main stores of groundwater;

(4) the zone of 'deep' movement, where the current flows outside the immediate draining action of the local hydrogeographical network (not shown in Fig. 6.17) and the direction of groundwater movement arises as a result

Figure 6.17. Conditions of groundwater movement in various vertical zones of karst rocks (Sokolov, 1962).
I, zone of aeration; II, zone of seasonal variation of the water table level; III, zone of permanent saturation.
1, karst limestone; 2 and 3, upper and lower levels of groundwater.
Arrows show direction of water movement.

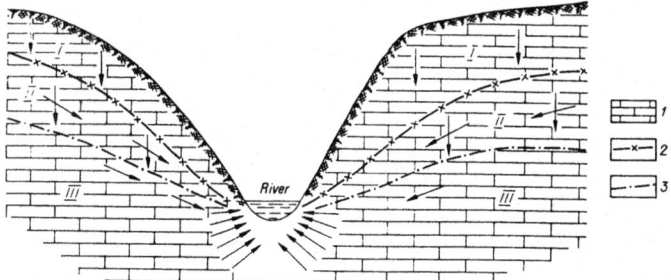

of structural–geological features and depth of deposition of the karst rocks; the groundwater moves slowly towards the discharge centres (tectonic depressions or deeper erosion incisions situated in neighbouring areas).

The chemical composition of karst water is very varied. High-flow springs from karsted limestones usually emit fresh water of a calcium hydrocarbonate composition. Calcium sulphate water is found in gypsum deposits. In some regions in which saline karst occurs, strongly mineralized water and sodium chloride brines are found in boreholes sunk into the deep-lying strata.

Groundwater in karst is subject to even more pollution from the surface than fissure water. In karst regions it is well known that several hours after rain falls cloudiness appears in the groundwater, not to mention the sharp increase in bacterial pollution. One more example: along the shore of the River Chusov springs are confined to the strongly karsted limestones and dolomites of Devonian and Carboniferous age, where flow during the rainy season increases considerably, and the karst water becomes cloudy as a result of the intensive penetration from the surface of atmospheric water and the rapid passage of this through the karst voids. On the southern coast of the Gulf of Finland in the water of the karst caverns in the Ordovician and Silurian limestones tiny fish have been discovered. Sometimes, together with the tiny fish there are also minute crabs, molluscs, and pieces of vegetation pumped to the surface with the groundwater from boreholes. Similar phenomena have been noted not only in the USSR, but in other countries too.

6.6 Deep-lying water (deep water)

In the existing classification it is usual to call all groundwater, apart from non-pressure water, artesian, subdividing it further into zones with different rates of water exchange. It is logical to categorize as artesian water which is under hydrostatic pressure alone, because there also exists groundwater with completely different sources of pressure and recharge. Certainly, the now established widespread occurrence of such water justifies its being recognized as an independent category under the name of 'deep water', or 'deep-lying water'.*

At present it is not possible to give universal criteria, especially qualitative criteria, for the recognition of deep water: it is necessary to take into account the totality of its features. Deep water is not a genetic concept. It is mainly the abnormal pressure which is significantly higher than the hydrostatic pressure and which is explained by the compacting of the deposits (elision recharge), the geotectonic forces, the outflow of fluids from the mantle (endogenic recharge), and a number of other reasons, which serve as the basis for the diagnosis of this water. A summary of the basic hypotheses on the origins of the abnormal pressure of such water can be found in the work of G.Yu. Valukonis & A.E. Khod'kov (1973).

*The conditionality of this term is obvious. The authors use it for lack of a better equivalent.

Deep water can be divided on the basis of the geological structure of the regions in which it occurs and the factors governing the pressure, as (1) stratal basin water, (2) crystalline basement water, including that of ancient shields, (3) water in the different kinds of faults of deep deposits in tectonically active regions. In the first case, and particularly in the second, passive causes (gravity, the plasticity of the rocks, and the presence of hydrocarbons, etc.) predominate; in the third case colossal endogenic forces predominate. Correspondingly, at the other end of the scale, elision water and endogenic sources of recharge will be found.

Basins of stratal water provide the most typical and characteristic occurrences. For example, the basic mass of water in the hydrogeodynamic zone of very impeded water exchange would be called deep water. Plastic sedimentary rocks have the best properties which are necessary on the one hand for the isolation of groundwater from the action of the hydrostatic pressure, and on the other hand for the creation and the maintenance of geohydrodynamic anomalies. Such properties are characteristic first and foremost of young sandy clay sediments. In these the magnitude of the abnormal pressure may reach extremely high values (Fig. 6.18).

It is precisely to the Cenozoic sedimentary downwarps of the Alpine fold regions that the classic appearances of the strained condition of fluids in the interior are confined. These lead to the formation of active mud volcanoes. The depth of formation of these volcanoes can reach 10–12 km. The oscillatory and fold-forming tectonic movements which are experienced by plastic and essentially clayey beds, which contain groundwater and gases, are the primary causes of the formation of these volcanoes. Gushers are characteristic of boreholes that meet such sedimentary formations. For example, in the Mirbashir oilfields of Azerbaijan one such borehole yields up to 10 000 m^3/day of hot water, while the average flow rate for the same deposits (the Maikopian Series) is no more than 10 m^3/day (Askerov & Durmish'yan, 1967). In another case, also in the region of mud volcanoes, the flow of gushers reached 3.5 m^3/s (about 300 000 m^3/day).

Hydrogeodynamic anomalies (but without mud volcano action) have not only been discovered in comparatively quiet tectonic conditions, but also in poorly lithified deposits. In the Western Siberian basin they are found both in gas-bearing and in water-bearing layers of Jurassic–Cretaceous age (Rosin, 1977).

Even on the ancient Siberian platform in the Lower Cambrian sediments abnormal pressure of the brines is found. The initial flow of the gushers from boreholes is anomalously high and is 1–2 orders of magnitude higher than the average value for the region, reaching hundreds of m^3/day (Balykhtinskii borehole 5–200 m^3/day, Omoloiskii borehole 13–1600 m^3/day, with a degree of mineralization of the brines of 600 and 620 g/l). The presence of rock salt in the series is not essential for the maintenance of abnormal pressures. Thus the Sukotungusskii borehole 2r, which is situated on the western border of the Siberian platform, met a horizon of strong brines (360 g/l) in the dolomites of the Lower Cambrian at a depth of about 1000 m and the flow from the gusher reached 4500 m^3/day. Sealing has been brought about by a thick stratified group of volcanic trap rock.

A fluctuating regime is typical of boreholes that tap deep water. Their flow is usually rapid, rarely gradual, and decreases as a result of the relaxation of the controlling geostatic pressure. Such a flow regime is typical of the brines, mentioned above (Balykhtinskii and Omoloiskii), in the boreholes of the Siberian platform. It is also typical of brines of the Jurassic halite–anhydrite beds in Central Asia; when these are tapped at depths of 2–3 km gushers occur with flows of up to 1000–8500 m^3/day (Sokolovskii & Sedletskii, 1970). It is usual with deep-lying brines for the flow to reduce rapidly and the period of the flow to be limited (from 3.5 months to several days).

The depth at which this water is found can vary greatly. In many saline basins 'sealed' lens-shaped brine deposits with anomalously high abnormal pressures are found at depths of 0.4–0.5 km. On the other hand, depending on the local

Figure 6.18. Hydrogeodynamic anomalies in sandy clay Cretaceous–Palaeogene deposits of the east Caucasus region artesian basin, based on the data of I.G. Kissin (1967). The arrows at the sides of the wells show the piezometric water levels (the numbers at the arrow heads give the magnitude of the head relative to datum).

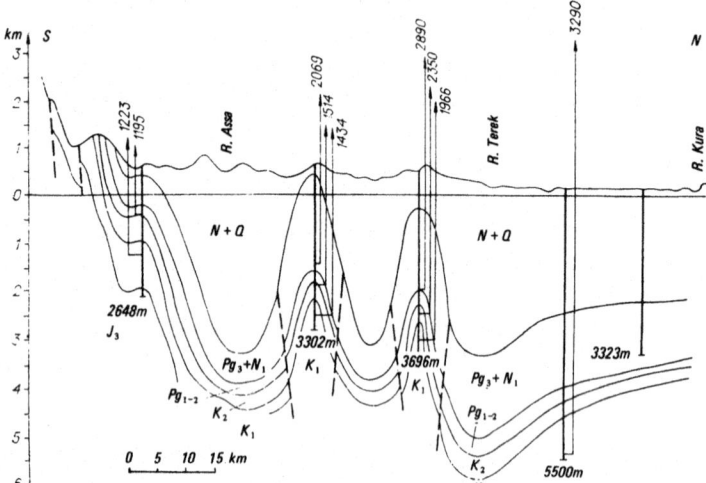

structural–geological conditions, and infiltration regime, depths of even 5.0–6.5 km are not excluded. This is shown in particular by the discovery of hydrostatic pressures at depths of more than 6 km in the U.K. Tippen No. 1 borehole in Texas, USA (Kissin, 1967).

High pressure deep water in the interior of the crystalline basement of platform and ancient shields has been found by very deep drilling. It has been found in the Minnibaev borehole 20 000 in the Tatar anticline in granite–gneiss at a depth of 4.7–5.1 km (more than 3 km below the level of the base of the sedimentary cover), and consisted of calcium chloride brines with a degree of mineralization of 333 g/l (Muslimov *et al.*, 1977). Interesting data have been obtained from the Kola borehole SG-3 on the Baltic shield. According to the data of V.D. Bezrodnov, L.V. Borevskii and others, the abnormal pressure of the aquifer at a depth of 6350 m reaches 1170 atmospheres, which is 1.9 times the normal hydrostatic pressure and is close to the geostatic. The piezometric surface at such pressures is 5.3 km above the top of the borehole. The degree of mineralization of the water is about 300 g/l.

The deep waters of the various deeply faulted regions of recent volcanic activity and regions which exhibit active endogenic energy are distinctive. They are usually hot springs (Fig. 6.19); in these circumstances water in the zone of intensive water exchange may also have a high temperature, but the deep water is distinguished by qualitative features and its hydrogeodynamic regime. For example, in the Salton Sea area (California, USA), the hot brine springs are met at a depth of about 1.8 km and are confined to the northern continuation of the graben of the Gulf of California. They are under a pressure which is 25% greater than the normal hydrostatic pressure (White, 1967). In the land parts of the Great Afro-Asiatic Rift hot springs with a degree of mineralization similar to that of the brines of the Salton Sea area are found, although these differ in their chemical composition and in places are mineralized to only a small degree (up to 1–3 g/l). The pressures (rather higher than hydrostatic) are found when hot springs are tapped in the faults of the Alpine fold zone which extends through Europe and Asia.

Figure 6.19. Diagram of the possible formation of deep water with the participation of magmatic gaseous–liquid solutions (after White, 1965).
1, liquid phase; 2, gaseous phase; 3, direction of movement of fluids (meteoric waters take part in the circulation).

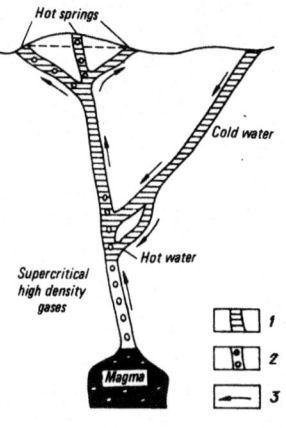

Deep water is composed of waters of different genetic origins. Water of one or other genetic category may predominate according to the features of the palaeohydrogeological development of the different regions, but more often than not it is polygenetic. Occupying the lowest horizons, deep water, following the normal pattern of hydrogeochemical zonation, is usually mineralized and its composition is more often than not of a chloride nature. Moreover, it may be saturated (degree of mineralization higher than 500 g/l) – the brines of the Angara–Lena basin for example. Deep water very often contains a number of macro- and microcomponents (alkaline earths, bromine, fluorine, strontium, calcium, heavy metals, etc.) in high quantities.

The period during which deep water remains in closed systems under abnormally high pressure is relatively short. The best studied anomalies, which are under study at the present time in the stratal water basins, appeared probably recently – not later than the Neocene–Quaternary. According to the calculations of V.F. Linetskii (1961) their balancing out takes several million years, but according to the data of other researchers, they level out in hundreds or even tens of thousands of years. Favourable conditions for the occurrence of, and the longest life of anomalies exist in relatively young stratal water basins. On the other hand with increasing age and degree of lithification of the sediments the possibility of this occurring reduces. However, under the action of neotectonic movements they may arise periodically in the same place, or in a different part of the same geological region. Essentially, deep water is permanently subject to the influence of increased forces in general.

6.7   Groundwater in permafrost regions

According to the definition of N.I. Tolstikhin and N.A. Tystovich, regions in which the temperature of the rocks is zero or below, which contain ice and which remain in such a condition for many years, even hundreds and thousands of years, are called *permafrost regions* (Dostovalov & Kudryatsev, 1967; Mel'nikov & Tolstikhin, 1974).

According to the calculations of Black (1954), permafrost regions ('the frozen zone of the lithosphere' according to N.I. Tolstikhin) cover more than 35 million km^2 (about 25% of the land area) and occupy the north of Eurasia and America, the islands of the Arctic, Greenland, Antarctica, and the high mountain belts of all the continents. Information on the area occupied in the USSR is available, but even taking the minimum values, it covers 10 million km^2 (45% of the country).

The basic features of permafrost regions are as follows: (1) the existence of H$_2$O in all three states – solid (ice), liquid (free and bound water), and vapour; (2) interaction between water in these states and with the frozen and thawed rocks which contain it; (3) the transformation of rocks which are permeable in the thawed state to relatively impermeable rocks when cemented with ice.

N.I. Tolstikhin (1941) proposed that the water be divided into suprapermafrost water, interpermafrost water, and subpermafrost water according to the relationships between the liquid phase, i.e. the groundwater proper and the layer of (impermeable) frozen rocks. This classification, which was devised as far back as the 1930s, has stood the test of time. All subsequent classifications have been based on it. Of these classifications that of N.N. Romanovskii (1966) subdivides the groundwater in permafrost into five types: suprapermafrost water, interpermafrost water, intrapermafrost water, subpermafrost water, and water in open talik zones (Fig. 6.20), and deserves mention.

Suprapermafrost water. This is water that is situated above the surface of the permafrost. The latter, in the majority of cases, is the lower confining bed of this type of water. This is subdivided (Tolstikhin, N.I. & Tolstikhin, O.N., 1974) into seasonally frozen, seasonally semi-frozen, and seasonally unfrozen, on the basis of the way in which it is formed and its regime.

The water of the active layer is of the seasonally frozen type. The most important feature of its regime is the temporary existence of the liquid phase: from two months (in the Arctic) to six months (at the southern boundary of the permafrost). This water occurs everywhere, and sporadically in layered deposits. The direction of movement is governed by the slope of the surface of the lowest confining bed (frozen rock), which usually conforms to the topography. Hence the discharge of seasonally frozen water occurs in the river valleys and other depressions of the land surface. The thickness of the thawing layer increases from north to south, and in mountain regions with decrease in altitude and degree of exposure of the slope. At depth thawing also influences the composition of the frozen ground. In the northern arctic regions of the USSR the maximum depth of thawing in the sandy ground is 1.0–1.5 m, but further south (in the Angara region and in southern Yakutia) it may reach 4 m and more. On the arctic slope of

Alaska (Barrow Point) the depth to which thawing occurs has been established by drilling to be from 1.2 to 1.5 m, and in the south of the Seward Peninsula up to 3 m (Hopkins *et al.*, 1955). In the Khangai–Khentei mountain region (Mongolian People's Republic) the maximum thickness of the seasonal thaw reaches 4.2 m (Zabolotnik, 1974).

L.N. Maksimova *et al.* (1966) distinguish three types of seasonally frozen water (Fig. 6.21) based on the character of the cross-section of the active layer and the permeability of its intercalated layers (in the thawed condition): (*a*) open, in contact with the upper boundary of the permafrost; (*b*) open, not in contact with this boundary; (*c*) closed, contacting. It is obviously expedient to distinguish yet a fourth type (*d*) closed, not contacting. In practice there are of course sections of the active layer which are more complex in character.

Within fissure water massifs, on parts of the water-divides and the slope of the range which are exposed to the sun during periods of heavy rainfall quite large ground streams are formed, which quickly dry up. In detrital deposits on those slopes that are exposed to the sun, and which contain a great amount of ice, springs fed from the melting ice flow for short periods (a matter of hours). Such places are of wide occurrence on the exposed sides of river valleys in the north of the Chitin region in the north-east of the USSR. Plains, river terraces, and low foothills are characterized by a more regular areal distribution of water in the active layer. On the margins of river valleys and lake hollows it emerges as a multitude of springs, the flow of which varies considerably and may in the rainy season reach 10 l/s. The water temperature is usually close to zero, but sometimes increases to 5 °C or more.

Seasonally frozen water is usually extremely pure, with a degree of mineralization which rarely exceeds 0.1 g/l. Its composition varies from chloride-hydrocarbonated sodium (with a degree of mineralization of up to 0.05 g/l) to hydrocarbonated calcium. In the overwhelming majority of cases the water is weakly acid (pH 5–6). In the north of the Chitin region for example, it has been found that a pH greater than 7 is a sign of the presence of open taliks. In the same way as surface water, they are easily polluted.

Seasonally semi-frozen water is found comparatively rarely in places where the active layer consists of very permeable rocks and covers water-bearing talik. In this case the

Figure 6.20. Diagram of the relationship between the groundwater and frozen ground (after N.N. Romanovskii, 1970).
A, water in the active layer above the permafrost; B, talik below the lake; C, water in open talik; D, water in open talik below river; E, interpermafrost talik; F, water in closed talik inside the permafrost; G, unfrozen water below the permafrost; H, pressure water not in contact with the permafrost.
1, sands; 2, pebble–gravel deposits; 3, loam; 4, detritus and grèzes litées; 5, limestone; 6, sandstone; 7, shale; 8, permafrost boundary.

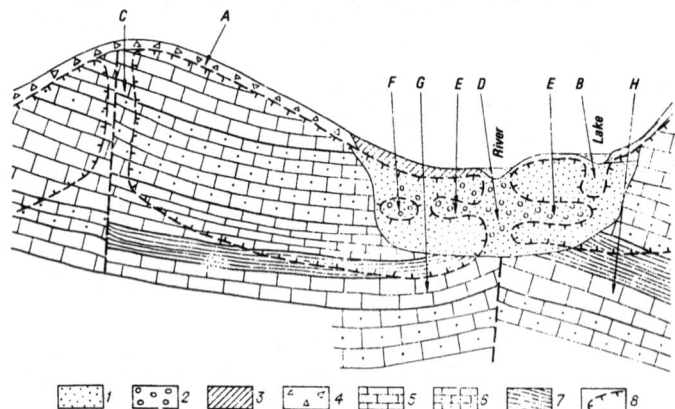

Figure 6.21. Basic types of seasonally frozen water for a section of an active layer (in a period of maximum thaw).
1, permeable water-bearing ground (sands); 2, impermeable ground (loams and clays); 3, surface of the water above the permafrost; 4, head of the water above the permafrost; 5, upper boundary of the permafrost.

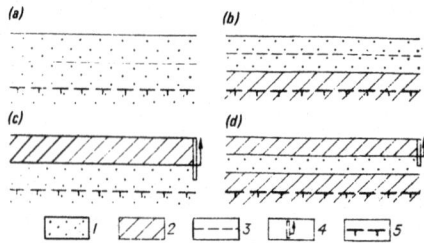

water of the active layer does not freeze throughout its thickness in winter and in some years the liquid phase is preserved in the lower parts. Such parts, being elongated in shape, usually are confined to the talus cones of the foothills, to river valleys, and lake depressions. The regime of this water differs from that of seasonally frozen water only by the head of liquid water which is preserved in the winter, and is very similar to it in composition.

Seasonally frozen water is associated with open taliks. Henceforward the term will be understood to be a layer of rock, the temperature of which is above zero or above the freezing point of the gravitational water contained, and which is situated inside the permafrost (Nekrasov, 1967). Closed taliks, the thickness of which, in contrast to open taliks, is less than the thickness of the permafrost layer, do not reach the horizons of interpermafrost and subpermafrost water and thus are not part of their interrelationship with the water above the permafrost. They are found in flood plains, and on the lower terraces of river valleys (subfluvial taliks), under lakes and lake terraces (sublacustrine taliks), and in the talus cones emerging from the foothills of mountain slopes (slope taliks), in very permeable deposits, mainly pebble gravel, and sandy detrital deposits.

Subfluvial taliks are of widespread occurrence in permafrost regions, especially in the valleys of the great rivers. They have been studied in some detail in Yakutia, and in the northeast of the USSR (Nekrasov, 1967; Vel'mina, 1970; Anisimova, 1971, and others), in Canada and in Alaska (Hopkins *et al.*, 1955). Their recharge regions (just as in the sublacustrine and slope taliks) often exceed the area of distribution considerably. They form both subfluvial streams of suprapermafrost water, and the basins of separated areas which are often combined into the same aquifer.

A talik in the form of a subfluvial stream stretching for 15 km was surveyed by I.A. Nekrasov (1967) in the flood plain of the River Ugol' (in the Anadyr river basin). Data on the existence of a 'tube-shaped' talik (stretching for more than 7 km) in the pro-alluvial and alluvial deposits of the valley of the River Naminga (in the north of the region) with a minimum flow of 0.3 l/s was put forward by Yu.G. Shastkevich (1966). Closed taliks of limited area have been described (Efimov, 1964; Anisimova, 1971; and others) in the River Lena basin. The thickness of such taliks, which occupy alluvial deposits and underlie the rocks at their base (bedrock), may reach 20–30, and even 60 m. They occur both under the bed of the river and on the adjacent terraces. The infiltration properties of the deposits of taliks vary greatly (the infiltration coefficient varies from tenths of metres to 50 m/day). The yield of wells sunk into taliks is from 10 m^3/hour (in Yakutsk) to 65 m^3/hour (a talik on the River Menda).

The regime of the water of the subfluvial taliks which lie above the permafrost depends directly upon their hydraulic connections with the surface water. The regime in the talik streams above the permafrost is characterized by a rapid cut-off in the autumn–winter period and a maximum flow in the middle of the summer. The degree of mineralization of subfluvial water varies from 0.1 to 0.5 g/l, and its composition

from hydrocarbonate magnesium-calcium to chloride-hydrocarbonate calcium-sodium.

Sublacustrine taliks are formed in the basins of large lakes, both in unconsolidated deposits and in the bedrock. According to the data of P.N. Anisimova (1971), in central Yakutia the thickness of such taliks may reach 40–60 m. The position of the surface of the water in taliks under the existing lakes within water masses is determined by the depth of the lake, and in the margins of lakes with overhangs they may be under pressure. In sublacustrine taliks, especially under lakes which dry out, there may be several intercalated aquifers, both pressure and non-pressure types. The quantities of water in open taliks are usually limited by the dimensions of the lake basins. The flow from wells is from 10–20 m^3/hour. The composition of both fresh, and brackish or saline water with mineralization from 0.2 to 6 g/l, varies with both area and cross-section; in relict sublacustrine taliks it may reach 60 g/l (the composition being sulphate-chloride to sodium-magnesium with a change of mineralization from hydrocarbonate-calcium to sulphate-magnesium).

Slope taliks are found at the foot of mountain ranges in the outfalls of temporary streams, gulches, and ravines, where talus cones are formed, made up of various detrital materials. The thickness may reach several tens of metres. They are fed from atmospheric precipitation, the condensation of water vapour, and undercurrents of subfluvial water from above the permafrost. Discharge takes place either via springs with rapidly fluctuating flows or via bogs and swampy hollows at the outside edge of the slope. In the autumn–winter period the flow quickly ceases, and a considerable amount goes to form icings. The degree of mineralization does not exceed 0.2 g/l; the composition is hydrocarbonate calcium and sodium.

Interpermafrost and intrapermafrost water. Liquid solutions in layers which are bounded above and below by layers of permafrost in which there is no liquid water, are classified as interpermafrost groundwater. Intrapermafrost water is liquid water in the form of lenses and intercalated water which is bounded on all sides by permafrost.

N.N. Romanovskii (1966) distinguishes two basic groups of interpermafrost–intrapermafrost water: those above zero temperature, and those below zero.

Interpermafrost water whose temperature is above zero is formed mainly as a result of different magnitudes of the thawing and freezing of the upper part of open taliks, and also the thawing action of the liquid water contained in the underlying frozen rocks, where there is no heat loss. As a result, after a certain period of time (from several years to tens of thousands of years depending on the climatic conditions and a number of other factors), the upper part of the lower frozen confining bed must thaw, even if only a layer with the liquid phase does not freeze completely. Furthermore, interpermafrost aquifers may arise and be preserved in thawing layers in which the permeability of the environment of semi-permeable and impermeable rock is increased with recharge and discharge across open taliks. Such conditions often occur under lake basins and on river terraces in central Yakutia (Fig. 6.22). Interpermafrost water is discharged as springs, which form huge icings. The flow from such springs can reach 40 and even 160 l/s (Anon., 1970a).

The degree of mineralization and the composition of the interpermafrost water whose temperature is above zero are determined by the recharging water and also by the dynamics of its movement. It may be fresh or brackish. Closed lenses of intrapermafrost water can form under lakes which dry out or which freeze; these, as a result of cryogenic metamorphism (Anisimova, 1971) are relatively highly mineralized (20–80 g/l).

Interpermafrost layers are also present as represented by water in the solid state as ice. As a result of human activity the ice may pass over to the liquid state. For example, in one pit on the island of Vaigach the workings were completely flooded as a result of the melting of underground glaciers (Mel'nikov & Tolstikhin, 1974). The melting of underground glaciers probably occurs under natural conditions as well.

Interpermafrost water whose temperature is below zero and intrapermafrost water (cryopeg, according to N.I. Tolstikhin) occur mainly in regions of continuous permafrost. The preservation of such supercooled water in the liquid state is due to its high degree of mineralization (from 35 to 320 g/l) which lowers the freezing point.

Shallow interpermafrost cryopeg (metres and tens of metres) is to be found on the shore of the Bering Sea and the Sea of Okhotsk (Anon., 1972a), and the shelf and islands of the Arctic Sea (Neizvestnov & Semenov, 1973). On the southern slope of the Anabarsk shield it is found at depths of 200–300 m (Ustinova, 1964).

Subpermafrost water. Liquid aqueous solutions which lie beneath the base of the permafrost come into this category. N.N. Romanovskii (1966) divides this water into contacting, non-contacting, and 'deep'.

Contacting subpermafrost water, for which the permafrost layer is the upper confining bed, takes part in the active heat exchange with the permafrost. Its temperature may be above or below zero depending on the thickness of the over-lying frozen layer, the presence or absence of water exchange with the surface, and the composition of the water-bearing deposits. Subpermafrost cryopegs are usually characterized by a high degree of mineralization. According to N.I. & O.N. Tolstikhin (1975), saline water and brines whose temperature is below zero constitute the lower zone of the belt of subzero temperatures. The concentrated chloride-calcium brines at a temperature of $-1.8\,^{\circ}C$ which were found in the Markhin boreholes in the north of the Siberian platform at depths of more than 1000 m (Mel'nikov, 1967) are an example of such water.

Contacting subpermafrost water usually forms under thin frozen strata which are recharged from atmospheric and surface water via taliks. The position of the base of the frozen layer does not change throughout the year. This water is generally under pressure.

Non-contacting subpermafrost water is separated from the frozen layer either by impermeable rocks (in this case it is pressure water), or its position in the permeable rocks is characterized by a free surface which is subject to seasonal variation. In places situated near the recharge zones subpermafrost water has a varying heating effect on the base of the permafrost. More often than not such water is formed in fissured crystalline rocks of hydrogeological massifs where freezing of the frozen layer at the base occupies a layer several metres deep, but may also be found in the sedimentary deposits (see Fig. 6.20).

'Deep' water is separated from the frozen layers by several hundred metres, probably isolated from interaction with the frozen rocks. If the arguments are based upon a short period of time then such an opinion is correct, but from the historical point of view (thousands and tens of thousands of years) then obviously the possibility of interaction between permafrost and deep water cannot be ruled out.

Open talik water. This occurs in taliks or talik zones which form conducting channels for gravitational water, which occupies either a considerable part of the frozen layers of rock outside the talik, or the whole of this layer in the area. These reservoirs are bounded by frozen rock on all sides (in contrast to interpermafrost water, which is bounded above and below) and for practically the whole of the permafrost region fulfil a role which is analogous to fault zones. The movement of water in such taliks may be both ascending or descending. All this allows the water of open taliks to be regarded as an independent category of groundwater in permafrost regions. Open taliks may be areas which consume or discharge water, and may connect the water above the permafrost only with the interpermafrost water, or only with the subpermafrost, or occupy the whole water system of the frozen ground – hydrogeological section.

There is a whole series of classifications of open taliks, which take account of their formation, situation, and relief, role in the water exchange, etc. (Vel'mina & Uzemblo, 1959; Romanovskii, 1966; Nekrasov, 1967; Vel'mina, 1970). Considering these, a classification (see Table 6.8) is proposed in

Figure 6.22. Diagram of recharge and discharge of interpermafrost water in the valley of the River Ulakhan-Taryn (Anisimova, 1971).
I, subfluvial talik; II, seasonally frozen layer; III, layer of interpermafrost water; IV, sublacustrine talik.
1, silty sand; 2, sand with pebbles; 3, water-bearing sand and pebbles; 4, semi-permeable sandstone; 5, boundary of the permafrost; 6, direction of water movement; 7, spring of interpermafrost water.

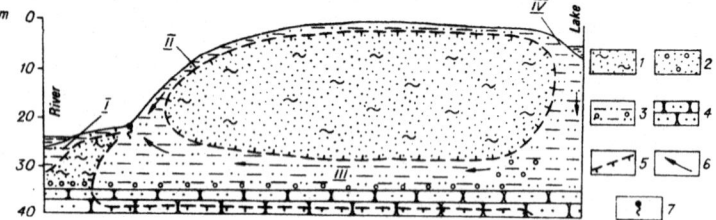

which three basic genetic types of open taliks are distinguished: meteorogenic, tectonogenic, and thermogenic.*

Meteorogenic open taliks are formed either as a result of the action of insolation or the conduction of heat by the surface water (conducting taliks). In the majority of cases they are consuming water by direct infiltration (influxes) of water whose temperature is above zero into the permeable deposits (convective taliks). These include subfluvial, sublacustrine, slope, and other taliks. In most cases they consume water and could be called swallet taliks. Karst taliks form in carbonate and gypsum-halite-bearing rocks which outcrop at the surface, or which are covered with a thin layer of unconsolidated deposits. The role of taliks in the recharge of subpermafrost and interpermafrost water is often more significant than the infiltration of atmospheric precipitation on water divides. An interesting swallet talik which recharges the interpermafrost water has been discovered on the Seward Peninsula (Alaska) near Lake Imurek in intercalated layers of basaltic lavas and pebbles (Fig. 6.23).

Tectonogenic open taliks are connected with the saturated faults in the zones of contact between rocks of different permeability. In the southern regions they occur on water-divides, slopes, and in the sides of river valleys; in the northern parts they are mainly in river valleys. Higher than the local

erosion base levels, and in the upper reaches of mountain rivers, tectonogenic taliks may be consuming water and serve as paths for recharge of interpermafrost and subpermafrost water, but in the overwhelming majority of cases they are emitting taliks, which are interconnecting and discharging channels on to the surface for subpermafrost and interpermafrost water. Their role in the water exchange and the water circulation of the whole permafrost region is extremely important.

Discharge of subpermafrost and interpermafrost water along tectonogenic taliks occurs in the form of springs, either into the beds of rivers or into sublacustrine taliks. The flow of the springs may be considerable (tens and hundreds of litres per second). The group discharge centre of subpermafrost water in the valley of the Bol'shaya Anyui river (north-east USSR) is one where V.Ya. Kovalenko (1964) described more than 20 springs in the zone of intersecting faults, with yields from 100 to 1500 l/s.

The appearance of open taliks along the zone of tectonic contacts is especially characteristic of the Daursk system of groundwater massifs and basins. Thus in the Namingsk talik (in the Udokan range) the specific yield of wells reaches 50 l/s.

Subpermafrost water can be discharged into the alluvium of river terraces along tectonogenic taliks and merge into a meteorogenic talik (see Fig. 6.24) enabling it to remain stable.

The quantities of water discharged along tectonogenic

Table 6.8. *Genetic classification of open taliks*

Basic genetic type	Subtype according to character of formation	Main processes of talik formation	Permafrost hydrogeological features
Meteoro-genic	Conductive	Insolation and conductive heat effect of surface water action	Mainly swallet taliks, which feed the inter- and subpermafrost water in the shallow (to a few tenths of a metre) thickness of frozen rock
	Convective	Infiltration and inflow	
Tectono-genic	Fault	Active tectonic movements	Both swallet and discharging, aid the interconnection of suprapermafrost water with the inter- and subpermafrost waters in frozen rocks of considerable thickness (tens and hundreds of metres)
	Contact	Tectonic movements at contacts of different permeability, gravitational movements	
Thermo-genic (hydro-thermal)	Conductive	Deep warm currents with a conductive heat-transfer regime	Discharging taliks which discharge subpermafrost water in great thickness of frozen rock (thousands of metres)
	Convective	Rising infiltration of thermal water and hot vapour	

*There are also chemogenic taliks, which form as a result of the heat effect of physical–chemical reactions taking place inside the Earth, but they are of rare occurrence.

Figure 6.23. Draining talik on the Seward Peninsula, Alaska (after Hopkins *et al.*, 1955).
1, silt with an abundance of basalt boulders; 2, layer of basalt lava; 3, pebbles and breccia; 4, permafrost; 5, direction of water movement.

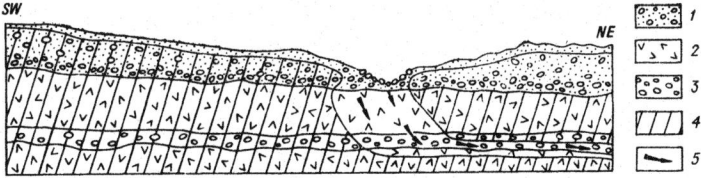

Figure 6.24. Open talik in the estuary of the Lower Larba river, located at junction of andesite and granite (after T.N. Kaplina *et al.*, 1975).
1, sands; 2, gravelly sand deposits with sand lenses; 3, andesite; 4, granite; 5, weakened and saturated contact zone; 6, permafrost boundary; 7, borehole; 8, vertical electric profile; 9, direction of groundwater movement.

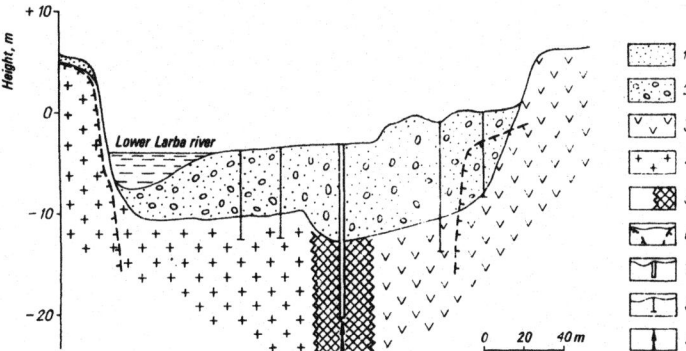

taliks in fault zones which affect limestones are especially significant. The interaction of the tectonic and the karst processes favour the formation of vast channels for the upwards or downwards movement of groundwater.

Thermogenic (hydrothermal) taliks are formed in places of elevated heat flow from inside the Earth, in regions of present day and recent volcanic activity, and of vigorous neotectonic activity. These are exclusively emitting taliks. The transfer of heat to the surface takes place either by conduction (for example the formation of a conductive talik in the frozen layers at some distance from an active volcano), or by the discharge onto the surface or into thin unconsolidated deposits of thermal water, hot vapours and gases (convective taliks). An obvious example of a talik which is connected with the discharge of thermal water is the Charsk hot spring in the margin of the Charsk intermontane basin in the north of the Chitin region. At different times of the year from 20 to 90 l/s of water at a temperature of 40–50 °C are discharged through this talik. The discharge point is circular and the radius of the zone of thawing rocks is 450 m; beyond that the permafrost layer is 500–600 m thick.

Icings are of particular significance in permafrost regions. These are bodies of ice accretion of various dimensions which are formed on the surface by the discharge of groundwater. The length of time for which the solid phase (ice) exists is limited by the duration of the cold period of the year; in spring the icings begin to melt, and by autumn they have either melted completely or a small amount of ice remains. Thus, temporarily, considerable quantities of groundwater accumulate in icings.

6.8 The groundwater of active volcanic regions

Territory in which there are volcanoes which are active or have been active in historical times are considered to be regions of volcanic activity. They are situated in the active fold belts, the Alpine, and Pacific Ocean belts, within island arcs, deep trenches, and the shores of inland seas (98% of all active volcanoes); only 2% of all volcanoes are associated with young faults of the ancient continental blocks in Europe, Africa, and central Asia (Fig. 6.25). In such regions the flow of heat from the depths to the surface is often 50–100 times greater than the average heat loss from the Earth. Here groundwater, being the active agent of heat and mass transfer, is involved in the water-bearing systems which have specific hydrogeological conditions. It has become the regular practice of hydrogeologists when referring to such water, which is characterized by a wide range of high temperature and a combination of different phase states, to use the term 'hot springs', and this term is also used in this collective sense in the present monograph. The term hot spring is used for jets of steam, hot aqueous solutions (thermal water), and mixtures of steam and water, i.e. all kinds of hot water and vapour which occur in the subsurface hydrosphere. Hydrothermal activity associated with volcanic activity has been thoroughly studied on Kamchatka and in the Kuriles (USSR), in America, Italy, Japan, and New Zealand.

6.8.1 Natural hot springs

The complex hydrogeothermal regime of the water-bearing systems of volcanically active regions gives rise to a diversity of hydrothermal phenomena on the Earth's surface.

Figure 6.25. World distribution of major hot springs.
I, hot spring locations; 2, active volcanic areas.

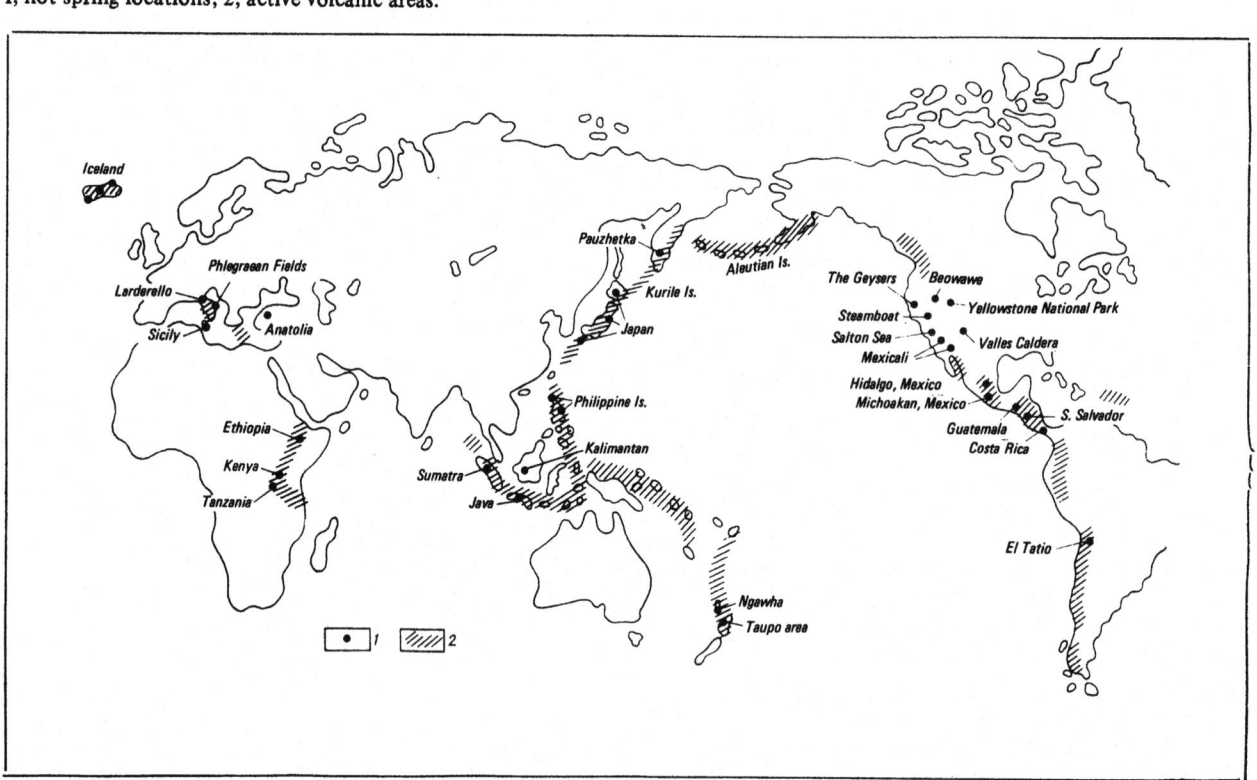

Their form depends on the structural–geological situation of the system, the permeability of the rocks near the surface, the development of zones of tectonic disruption, the heat potential of the system, and the ratio of the liquid and vapour phases so caused, and the possibility of the merging of steam jets with cold water of the near-surface layers.

The basic types of such phenomena are fumaroles, geysers, steam jets, and hot springs.

Fumaroles are volcanic emanations in the form of steam jets or quiet emissions from fissures and channels in the craters of and on the external slopes of volcanoes (primary fumaroles), or on the surface of hot lava flows and from the pyroclastic cover (secondary fumaroles).

The temperature and composition of the active gases of fumaroles depend upon the degree of volcanic activity.

D.E. White & G.A. Waring (1963) consider the following to be the fundamental factors which determine the composition of fumarole gases: (*a*) the initial amount of each volatile component in the magma; (*b*) the temperature at which the gaseous mixture is given off by the magma; (*c*) the time elapsed from the commencement of emission of gases; (*d*) the location of the emission; (*e*) the degree of mixing and reaction with the air and atmospheric water; (*f*) reaction with the rocks along the path of the emission.

Fumaroles are divided into: (1) fumaroles proper (mainly chloride-sulphide-carbon dioxide gases with temperatures of up to 800 °C); (2) solfataras (vapour and gas jets with mainly hydrogen sulphide or sulphurous gas and a temperature of 90–300 °C); (3) mofettes (mainly carbon dioxide gas jets with temperatures of up to 100 °C). A slightly different classification of volcanic gases has been put forward by V.V. Ivanov (Table 6.9).

Water vapour is clearly the predominant gas (up to 99%

by volume) in fumaroles. However, in some regions practically dry jets of volcanic gases are known. One example is the high temperature (300 °C) solfatara volcano Papandayan on the island of Java (Bemmelen, 1949). In addition to steam, mixtures of steam and very minor additions of gas arise as a result of the boiling of high temperature water when the hydrostatic pressure falls, and these escape via discharge points on the surface.

The discharge of steam and gases may reach enormous proportions. In this respect the most representative is the Valley of Ten Thousand Smokes in Alaska. Fumarole activity began here as a result of the eruption of the volcano Katmai in 1912. The rate at which steam at a temperature of up to 650 °C escaped was about 23 million l/s, and was produced from several tens of thousands of fumaroles. The steam jets reached heights of 150 m and some even 300 m. Usually it was pure steam with an addition of active gases of between 0.1 and 1.6% by volume. This region was considered as a possible source of geothermal energy (Mavritskii & Antonenko, 1967); in the last decade hydrothermal activity has fallen off sharply.

Geysers form hot springs which periodically discharge water and steam. Morphologically a geyser is a system which consists of a channel which conducts superheated water or hot steam to an underground reservoir (chamber) near the surface into which cold meteoric water enters from side channels or through fissures. From the chamber there is also a channel leading to the surface. This is crowned with a bowl-shaped crater.

A generally convincing explanation of the principles of geyser processes has been put forward by D.E. White (1967) based on the study of geysers, in particular those of the greatest geyser region in the USA, Steamboat Springs in Nevada. White introduced the concept of surface and subsur-

Table 6.9. *Major types of gas–steam emissions in volcanically active regions (after V.V. Ivanov, 1958)*

Emission	Temperature at the surface (°C)	Predominant composition on emerging at the outlet	Origin
I. *Volcanic gases*			
1. High temperature (chlorino-sulphuro-carbonic acids)	Hundreds of degrees	H_2O, HCl, H_2S, SO_2, CO, CO_2, N_2 (+ rare gases), O_2, sometimes NH_3, CH_4, and some others	Gases of magmatic and thermal metamorphic origin (emitted from magma chambers and strongly heated hard rocks), partly transformed as a result of the interaction with atmospheric gases
2. Low temperature (hydrogen sulphide – carbon dioxide)	100–150	CO_2, H_2S, H_2O, sometimes CH_4	Gases of magmatic and thermal metamorphic origin filtered through groundwater
II. *Steam*			
1. Deep	approx. 100	H_2O, insignificant additions of CO_2, N_2 (+ rare gases), traces of H_2S, sometimes CH_4	Conversion to steam 'underground boiling') of deep superheated pressure water
2. Surface	approx. 100	H_2O, insignificant additions of N_2 (+ rare gases), O_2, traces of CO_2	Conversion of descending cold water to steam in heated hard rocks

face (in the crater) discharges. On the basis of temperature measurements made in boreholes drilled into regions of geyser activity (in USA, Mexico, Salvador, New Zealand, Japan, Iceland, and USSR) he came to the conclusion that the temperature in the aquifers concerned must increase from the surface downwards for a geyser to exist, starting with a temperature of 150–170 °C. The highest temperatures were measured in the geysers of Cerro Prieto in Mexico. The geyser process itself, according to White, is connected with deep convection currents, caused in the first place by the temperature-dependent variations in the density of the water and its dissolving ability. The energy necessary for the eruption arises as a result of steam being emitted from the water at temperatures greater than 150 °C.

According to modern concepts (Nekhoroshev, 1959; Droznin & Razina, 1977; and others), the geyser process is based on the mixing of two currents with different heat contents (endogenic steam and infiltration water). The eruption of the geyser is like an explosion which occurs as a result of the rapid emission of energy from the superheated water. Laboratory models of this process have recently been made (Merzhanov *et al.*, 1973; Droznin & Razina, 1977), which treat geysers as a hydrothermal activity phenomenon, caused by a mechanism which most closely approximates to active volcanoes.

T.I. Ustinova (1955) distinguishes four stages of the geyser process (Fig. 6.26): (1) charging — when, after eruption, the dry upper part of the geyser again fills with mixing currents of hot and cold water; (2) overflowing (but not always) — the overflow of water via the edge of the channel, with a gradual increase in discharge; (3) eruption — the ejection of superheated, boiling water and steam; (4) the discharge of steam during the period of intensive boiling of the superheated water which is rising once more.

S.I. Naboko (1954) in a rather different fashion (in our view more successfully) subdivided the whole cycle of the geyser process into four phases: (1) quiescent — from the moment the emission of steam ceases to the appearance of water at the bottom of the channel, or until the water level starts to rise (if the channel does not empty completely); (2) preparation — from the time the water level starts to rise to the moment at which the spurts, which are connected with the boiling of the water, appear; (3) eruption — from the commencement of overflow of boiling water to the moment gushing ceases; (4) steam — from the moment of completion of gushing to the cessation of steam emission.

Geysers were named after Big Geysir in Iceland, where they were first studied. Springs with geyser regimes of activity are to be found in many of the regions of present day volcanic activity, but the number of such regions is comparatively small. The highest concentration of geysers is in Yellowstone National Park (USA) where there are 200 geysers — this comprises 10% of the total number of natural hydrothermal phenomena (Barth, 1950). In Iceland, out of 700 known hydrothermal phenomena only 30 are true geysers. In the USSR the Valley of the Geysers in Kamchatka is well known; this was studied for the first time by T.I. Ustinova. Here there are 12 large geysers and a large number of small ones. The largest of them, Velikan (The Giant) discharges a great jet of water to a height of 40 m and a column of steam to several hundred metres every 2–3 hours. However, the total number of geysers in Kamchatka is small (according to S.I. Naboko up to about 3% of the total number of hot springs). Geysers also occur in New Zealand, Chile, Guatemala, Costa Rica, Japan, Java, etc. A comprehensive summary of the main regions of geyser activity has been compiled by S.I. Naboko (1954).

The largest geyser in the world is Waimanga in New Zealand: in one discharge it has hurled 800 tonnes of water to a height of 450 m. In Yellowstone Park the height of discharge of the largest geysers (Giant, Giantess, and Old Faithful) is from 35 to 80 m.

The regime of a geyser is usually unstable: periodicity varies with time. The temperature of the superheated steam at the surface may reach 117 °C (Naboko, 1954), and the temperature of the water is close to the boiling point for the altitude concerned. Barth (1950) has published interesting data on the geysers of Tibet at an altitude of 4800 m, which discharge water at a temperature of 84 °C to heights of 12–15 m. The deposition of siliceous tufa (geyserite), formed in quite large quantities, is characteristic of almost all geysers.

The most widely occurring natural hydrothermal phenomena in regions of active volcanoes are hot springs, which have a wide temperature range. Among these are fluctuating hot springs (essentially these are small geysers with pauses between spurts of less than one minute), 'boiling' springs with temperatures which reach the boiling point, and hot water flows with temperatures greater than 20 °C. The flow of the hot springs may reach considerable magnitudes. One of the 'boiling' springs with the highest flows in the world is Deildartunga in Iceland which has a flow of 250 l/s (Kononov & Polyak, 1977). In Kamchatka 'boiling' springs are known with flows of 10–18 l/s.

Figure 6.26. Basic stages of the geyser process (after T.I. Ustinova, 1955).
Stages: (a) filling, (b) varying level, (c) ejection, (d) steam emission.
1, ash tuffs; 2, geyserite; 3, fissures with superheated steam; 4, fissures with cooled water; 5, superheated water at temperatures greater than 100 °C; 6, cooled water at less than 100 °C.

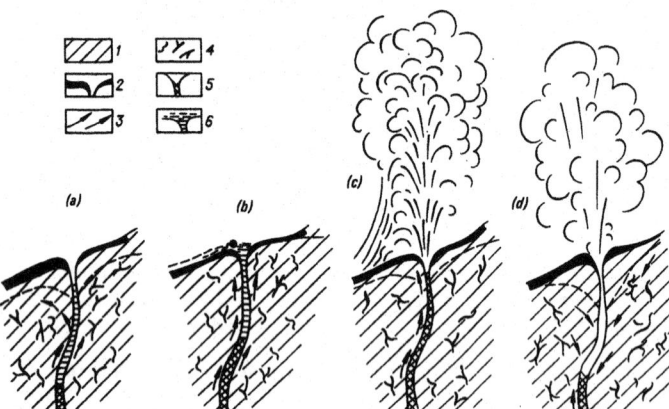

Of the other surface thermal phenomena, mud volcanoes and hot lakes may be mentioned.

The physical–chemical features and composition of hot springs. Hot springs which occur in volcanoes themselves (in craters, calderas, and on external slopes) form a group which is specific only to regions of volcanic activity. They are usually chloride or sulphate water with a complex cation composition and strongly acidic (pH < 3). CO_2 usually predominates in the gaseous phase of such springs; of the other gases H_2S, N_2, and H_2 are present. On the basis of the gases which they contain, the hot springs of active volcanoes are subdivided into fumarole, solfatara, and mofette types; these are found mainly in the craters of extinct volcanoes. V.V. Ivanov (1961) considers the first two as hot springs which are formed at depth, and the third as being of surface origin, but such a subdivision must be regarded as provisional.

Fumarole hot springs are of chloride, more rarely chloride-sulphate composition, and of very low pH (up to 0.2), consequently sulphate exists in them in both the forms SO_4^{2-} and HSO_4^-. Of the cations, H^+, Al^{3+}, and Fe^{3+} are contained in various compounds and Na^+, Ca^{2+}, and Mg^{2+} are present in low concentrations. A high concentration of H_2SiO_3 (100–800 mg/l) is characteristic. The degree of mineralization of this water is usually from 1–20 g/l but brines are known with a degree of mineralization of 36–65 g/l (in the volcanoes of Mutnovskaya Sopka in Kamchatka, Ebeko (in the Kurile Islands), Akeyama (Japan) and others), and even 100 g/l and more (in the crater lake of the Kawah Idjen volcano on Java, and Donald Mound fumaroles of White Island in New Zealand). The water temperature usually varies from 70 °C to the boiling point, and in the steam emissions more than 100 °C. In the gas phase of the majority of springs the fraction of CO_2 comes to 74–90% by volume. Hydrogen sulphide – carbon dioxide hot springs ($H_2S > 20$%) are found in Japan (Akinomiya-Minase, Tamagava); nitrogen (88%) in the Kurile Islands (Nizhne-Mendeleev and the island of Kunashir).

In solfatara hot springs the predominant anions are sulphates (in the forms of SO_4^{2-} and HSO_4^-), while the cations are represented by compounds of various concentrations (Na^+, Ca^{2+}, NH_4^+, H^{3+}, Al^{3+}, Fe^{3+}). The pH value is 0.3–4.0, the H_2SiO_3 content from 22 to 476 mg/l, and that of HBO_2^- up to 780 mg/l (the springs of Kastatiki in New Zealand). The degree of mineralization is usually 0.5–5 mg/l, less often 10–20 g/l (Gryaznyi spring, the islands of Iturup, Verkhne-Yur'evkie, and Paramushir in the Kuriles). The water temperature varies from 30 to 100 °C. They are, in the majority of cases, carbon dioxide springs (CO_2 content is 80–97% by volume). Hydrogen sulphide – carbon dioxide springs (H_2S content from 28–56% by volume) are found in Japan (the springs of Yunokanzava and Manza), nitrogen – carbon dioxide springs (N_2 content from 21 to 39%) on the island of Kunashir (Kuriles).

Natural hot springs in large calderas and tectonic–volcanic depressions (outside active volcanic regions) include geysers, 'boiling' springs, steam jets, and thermal phenomena with temperatures of 30–90 °C which are characterized by very variable salt and gas compositions: from pure sodium chloride, carbon dioxide, and hydrogen sulphide – carbon dioxide waters to hydrocarbonate waters of complex cation composition which emit nitrogen with consequent reduction in the degree of mineralization. Thus water of the most varied composition may be concentrated in the same thermally active area.

An excellent example of this is the Uzon Geyser depression in Kamchatka. Here, in the north-west part of the caldera of Uzon, localized zones have been charted in which are concentrated different surface thermal phenomena within an area of 15 km^2. G.F. Pilipenko (1974), as a result of a detailed hydrochemical investigation distinguished the following basic types of hot spring here:

(a) Alkaline and weakly acid water (pH 5–8), containing sodium chloride, and carbon dioxide (rarely methane – carbon dioxide), with a degree of mineralization of 1.5–4.5 g/l and temperature close to the boiling point. The H_2SiO_3 content is up to 400 mg/l and that of HBO_2^- is up to 265 mg/l;

(b) Weakly acid (pH 5–6) and acid (pH 2–3) water containing sodium sulphate and sodium chloride, with a degree of mineralization of 1.5–3.0 g/l and temperatures of 50–76 °C (there are also *griffons bouillants*). The H_2SiO_3 concentration is up to 260 mg/l, and that of HBO_2^- is up to 150 mg/l;

(c) Weakly acid (pH 5.5–6.5) water, containing sodium chloride and sodium sulphate, with a degree of mineralization of 0.5–1.5 g/l with temperature at the boiling point (95–102 °C), H_2SiO_3 content up to 250 mg/l and that of HBO_2^- up to 60 mg/l;

(d) Neutral (pH 6.5–7.0) water containing sodium chloride and hydrocarbonate, with degree of mineralization of 1–2 g/l and temperatures 64–70 °C (H_2SiO_3 up to 250 mg/l, and HBO_2^- up to 12 mg/l);

(e) Water with sulphate, hydrocarbonate, calcium, sodium, and carbon dioxide is found in only one spring – the Uzon narzan No. 100 (degree of mineralization is 1.3 g/l, temperature 20 °C, pH 6.1, H_2SiO_3 is 104 mg/l);

(f) Weakly acid (pH 5.2–7.0) water with a complex cation composition and hydrocarbonate and degree of mineralization less than 1 g/l, temperature 50–80 °C, and H_2SiO_3 concentration up to 250 mg/l;

(g) Acid water (pH 1.5–3.0) with specific cation concentration (Al^{3+}, NH_4^+, Fe^{3+}, H^+), sulphate as anion and a degree of mineralization of 1.5–7.5 g/l and temperature 30–100 °C.

As G.F. Pilipenko remarks, there is clear zonation in the spatial distribution of these types of thermal waters. Sodium-chloride-containing water gravitates to the centre of the thermal anomaly, and is surrounded by sodium-sulphate–sodium-chloride water, and the other types of water form outer thermal zones. Judging by the chemical composition of the water in the Uzon caldera all the basic types of hot spring found in other volcanically active regions are represented.

Of the hot springs with specific gas compositions the hot springs at Kanmer (New Zealand) ought to be mentioned. The predominant gas is methane (65% by volume); they also contain nitrogen (18%) and hydrogen (16%).

6.8.2 Aquifers in volcanically active regions

In volcanic regions there are two types of aquifer (groundwater reservoirs), which are connected with volcanic structures (the initial stage of the cycle of volcanic activity above the magma chamber), calderas, and tectonic–volcanic depressions (the final stage). Naturally this subdivision does not exclude the possibility of the renewal of the activity of the magma chamber, i.e. the appearance of a new cycle. E.A. Vakin & V.M. Sugrobov (1972) divide, in our view successfully, these aquifers into (1) 'positive', and (2) 'negative'.

(1) 'Positive' volcanic aquifers are volcanic structures which are elevated above the surrounding area to a considerable height and which are characterized by a high rate of water exchange. A great quantity of atmospheric precipitation, falling on the permeable rocks of which these structures are made, determines their hydrogeological significance as parts of the intensive recharge of the groundwater and surface water. The groundwater of 'positive' volcanic structures is divided according to the correlation of the volcanogenic rocks (the accumulations of pyroclastic material and sheets of lava), the conditions under which it is deposited, and the way in which it moves and is discharged.

On the huge lava plateaus formed by shield-shaped volcanoes as a result of discharges from fissures over large areas, groundwater collectors consist mainly of volcanogenic cover whose permeability is of fissure-stratal type and the congealed lava flows at the base serve as a relatively impermeable confining bed. Here there occur streams of subsurface water which are discharged at the bottom of the slope. In places made up of lava blocks the vertical permeability of the cover rises sharply and in localized parts (tongues) the discharge of subsurface water takes place as springs with high flow rates (tens and hundreds of litres per second) and stable regimes. The external slopes of shield volcanoes are characterized by similar conditions of movement and discharge of groundwater. But on the summits of many of them small craters form (small bowl-shaped hollows with steep walls and uneven floors) with surfaces made of semi-permeable igneous rocks which ensure an underground runoff into the interior of the volcano.

The interbedded pyroclastic deposits and lava sheets of stratovolcanoes are distinguished by a high but uneven permeability. Usually on the surface lava sheets, in spite of the abundance of atmospheric precipitation, there is no surface runoff. Depending on whether unconsolidated deposits or lava predominate here a complex system of poro-stratal water and fissure-stratal water is formed.

Subsurface water which is discharged at the foot of volcanic structures and on their external slopes is usually cold (1–4 °C) and very weakly mineralized (less than 100 mg/l). This forms the upper storey of aquifers which, in the regional system, is not connected with deep-lying water.

The presence of impermeable and semi-permeable layers of lava and pyroclasts, the hydrothermal alteration of rocks close to the volcanic vents, and the thickening of the walls of the vents by injections of lava ensure the isolation of the interior parts of the volcanic structure from the percolating atmospheric water. Such isolation, naturally, is destroyed during eruption and active tectonic movements, when groundwater of the upper storey, actively participating in the volcanic process and its consequences, penetrates along the fissures to the vent zone.

Hydrothermal activity, in the form of steam and gas jets, (fumaroles, solfataras, and mofettes) in the craters and on the slopes of volcanoes occupies small areas, which are directly in contact with gas discharge channels.

The heating effect of water vapour and gas near the surface is localized to areas of a few hundred square metres; the surrounding areas remain cold. E.A. Vakin & V.M. Sugrobov (1972) give examples of large fumaroles which have temperatures of 700 °C in the crater of the volcano Mutnovskaya Sopka (Kamchatka); at a distance of 50 m from the exit point in 1963 the ground was still covered with snow. Fumaroles of the same kind are to be found on the shores of ice-covered lakes in the crater of the volcano Gorelyi.

Apparently some part of the meteoric (or marine) water enters into the volcanic vents along large tectonic fractures, but how great a portion this is or how it penetrates is not yet clear. When meteoric water enters directly into the volcanic crater along permeable zones it may penetrate into the gas-conducting vents and take part in the formation of the emissions of water vapour.

Thus, in 'positive' aquifers conditions favourable to the formation of significant quantities of hydrothermal resources are absent.

(2) 'Negative' aquifers in volcanically active regions are represented by complex reservoirs which contain hydraulically connected horizons of poro-stratal and fissure-stratal water of artesian basins, and vein water of tectonic faults. They are confined to calderas (vast circular hollows and volcano-tectonic depressions), which are formed as a result of the sinking of large blocks of volcanogenic deposits of differing permeability into the surrounding rocks. The formation of such depressions is accompanied by intensive comminution of the rocks and increase in the fissure permeability which assists both ascending and descending groundwater movement. It is considered that the very formation of a caldera indicates the presence of a magma chamber at a relatively shallow depth (from 1 to 7 km), and the hydrothermal phenomena (fumaroles, geysers, thermal springs, etc.) within the volcano-tectonic depressions indicate continuing magmatic activity, the differentiation and crystallization of the magma and the emission of volatile fractions, etc. At the same time the multitude of fractures and fissures form pathways for the penetration of infiltration water into the zone of heating and chemical reaction of the magma chamber.

The presence of a roof or upper confining bed of dense impermeable rock which covers the underlying permeable layers is characteristic of all volcanically active regions. As many researchers have noted, a similar character for the cross-section is especially typical of volcano-tectonic depressions (Anon., 1970b; Anon., 1977; etc.).

In the upper parts of such depressions, which are filled with unconsolidated volcanoclastic material, there is an abundance of water-bearing layers of poro-stratal water which is characterized by intensive water exchange. The discharge of groundwater from the upper layers takes place basically along the fault zones and the regional erosion exposes the drainage area. In places which consist of interlayered pyroclasts and lavas, horizons of poro-stratal and fissure-stratal water alternate, often separated by impermeable intercalations which lead to the creation of heads of pressure and a slower rate of movement.

Accumulations of hot masses of water and vapour, which are fed by meteoric water along major tectonic zones, both transverse and circular along the perimeter of the depressions can be formed in deep-lying regions where the heating effect of the magma chamber is felt and which are isolated from the cooling effects of meteoric water.

It is specially necessary to distinguish, among negative aquifers, active hot spring systems with great heat energy potential and significant groundwater resources.

6.8.3 Active hot spring systems

Large subsurface reservoirs in which hydrothermal activity occurs and which are associated with anomalously high heat transfer from the interior of the Earth by means of a fluid heat carrier are called hydrothermal systems of volcanically active regions (Vakin & Sugrobov, 1972). From the hydrogeological point of view they are treated as reservoirs of hot water and intensely heated rock which often discharge water or water vapour on the surface (White, 1967). The rising current of hot aqueous fluids is regarded as the supplier of heat from the depths.

Hot spring systems of the active volcanic regions are of wide occurrence both on continental and marine depressions. The most important of these, which contain the largest stream spring resources, are shown in Fig. 6.25.

The majority of hot spring systems are associated with volcano-tectonic depressions or huge calderas (Uzon Geyser, Pauzhetka in Kamchatka, Onikobe in Japan, Wairakei in New Zealand, etc.), more rarely with superimposed trenches of the fold mountain surrounding the Paratunskii system in Kamchatka. Areas of direct hot spring activity within depressions are often confined to local structural elevations (complex horsts), and are bounded by steeply inclined faults. The Wairakei system is a typical example (Fig. 6.27). A block volcano-tectonic structure is characteristic also of the Pauzhetka (Aver'ev, 1961) and the Onikobe (Nakamura & Sumi, 1961) systems.

The hydrogeological structure of hot spring systems is extremely complex. They are reservoirs which contain poro-stratal, fissure-stratal, and vein cold and hot water between which there is a hydraulic connection, the nature of which is different for the various parts of the system, but which is determined in principle by the stratal pressure of the reservoir.

This is confirmed by the correspondence of the piezometric surface of the cold and hot pressure water which varies spatially with the geomorphological environment of the area concerned. The absolute level usually falls in the direction of the point of discharge. This has been established for example in the Wairakei system (Studt, 1958) and in the Pauzhetka system (Sugrobov, 1964). Furthermore, high temperature water, remaining in the liquid state as a result of considerable geostatic pressure, forms localized rising currents. The position of the piezometric surface of such currents is determined not only by the stratal, but also by the so-called *thermoartesian pressure* which arises as a result of thermal expansion (Studt, 1958).

The high temperatures affect the physical composition of the water and may cause changes in its gas phase (Kononov & Polyak, 1977). Thus, with increase in temperature up to 300 °C the viscosity falls by factors of tens of times; this enables water at high temperature to percolate through strata which are impermeable to cold water. The change in temperature from 5 to 250 °C (at a constant pressure of 200 atm.) reduces the density by 20%. This also produces a rise in the level of the hot water, and there discharges on to the surface as a result of both the thermoartesian pressure, and especially as a result of *steam-lift*, boiling hot water with reduction of hydrostatic pressure (Sugrobov, 1964). It is precisely the formation of steam which predetermines the differences in the surface hydrothermal phenomena ('boiling' springs, geysers, steam jets, steaming ground, etc.).

Tectonic dislocations and large fissures play an important role in the dynamics of hydrogeothermal systems. In the upper horizons they are sites of increased permeability and saturation, and intersecting the impermeable layers, they often determine the position of the discharge points of hot springs. At depth they are the main heat-conducting channels along which the deep water vapour or fluid reaches the upper parts of the hydrothermal systems.

In present day hydrothermal systems considerable resources of hot steam and hot water are concentrated. In many of these systems deep drilling reveals huge sources in which the quantities of a steam−water mixture are very considerable. Geothermal power stations, community heating systems, hot-houses and the production of chemicals are based on such resources. The resources of steam hydrothermal deposits are determined by two main factors: (1) the volume of the steam−water mixture; this depends primarily on the

Figure 6.27. Geological section across the hot springs system at Wairakei, New Zealand (The thermal regime . . . , 1961).

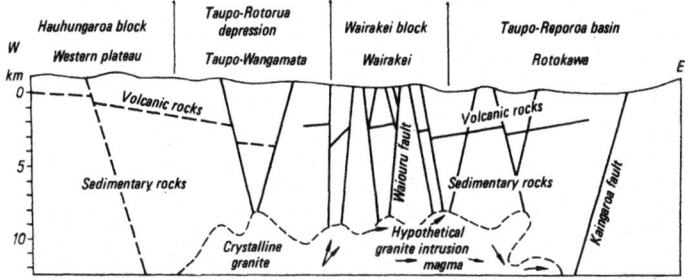

recharge of the hydrothermal systems by infiltration water; (2) the quantity of deep heat supplied from the magma chamber; this determines the total amount of heat and, in the final analysis, whether the steam resource can be exploited and how it will be used.

At the present time energy is being extracted from hydrothermal systems in the USSR (Kamchatka), USA, Italy, Iceland, and New Zealand (see Table 6.10). In Reykjavik (Iceland), the vertical column is characterized by changes in temperature to a depth of 2200 m. The maximum temperature (388 °C) is found in Cerro Prieto in Mexico at a depth of 1500 m (Mercado, 1969); in the same resource the highest yield from the wells is 680 tonnes/hour and the heat content of the steam−water mixture is about 1000 kcal/kg.

As can be seen in Table 6.10, the temperature in boreholes at depths down to 500 m is usually less than 200 °C; this figure is exceeded only in the resources at The Geysers (207 °C at a depth of 230 m). Higher temperatures, (250−390 °C) are characteristic of depths greater than 1 km. In terms of yield, together with the Cerro Prieto resource, those of Larderello

(300 tonnes/hour), and in terms of heat content (on present data), The Geysers resources (680 kcal/kg), are the highest.

In the upper parts of hydrothermal systems, tectonic faults and fissures provide paths along which the steam−water mixture moves, and these form, in the zone of increased permeability (of a porous or fissured nature), aquifers of varying area, through which the hot water rises. Hence, in a whole series of hydrothermal resources lower temperatures can be found in impermeable or semi-permeable layers between or below those layers that are saturated with water. A similar temperature inversion is observed in the Pauzhetka, Onikobe, and Cerro Prieto resources. Thus the 'productive' horizons of the resources, particularly those closed from the surface by impermeable rocks, are localized in the vertical sense, and in area, and are controlled by the permeability of the rocks. A graphic example of such a localization of steam spring zones is the Cerro Prieto site (Fig. 6.28).

In the majority of sites of steam springs (Pauzhetka, Phlegrean Fields, Khengil, Krisuik, Onikobe, Wairakei, etc.), high temperature (200−306 °C) steam jets in deep boreholes

Table 6.10. *Information from boreholes drilled into the steam and hot water deposits and in areas of potential geothermal energy**

Sites of potential, or producing, geothermal energy	Maximum temperature (°C)	Depth of maximum temperature point in borehole (m)	Maximum yield of the well of steam and water mixture (tonnes/hour)	Maximum heat content of the steam and water mixture (kcal/kg)	Usage
Pauzhetka, Kamchatka, USSR	199	350	120	260	Experimental industrial geothermal power stations
Bol'she Bannoe, Kamchatka, USSR	171	480	32	162	Developed for community heating use
Goryachii Plyazh, Kunashir Is., USSR	170	400	108	300	Developed for construction of geothermal power stations
Larderello, Italy	220	1660	300	no data	Geothermal power station and by-product chemicals
Phlegrean Fields, Italy	300	1840	100	no data	Geothermal power station
Reykjavik, Iceland	146	2200	140	no data	Community heating system
Reykjanes, Iceland	292	1700	100	no data	Community heating system
Matsukawa, Japan	250	1050	2	no data	Geothermal power station
The Geysers, California, USA	207	230	no data	681	Geothermal power station
Steamboat Springs, Nevada, USA	172	295	57	no data	Exploratory work in progress
Cerro Prieto, Mexico	388	1500	680	1000	Experimental geothermal power station
Wairakei, New Zealand	265	1200	100	no data	Geothermal power station

*Data used (Mavritskii & Antonenko, 1967; Ellis, 1979; Anon., 1972b; Dvrov, 1972; Pampura, 1977).

contain chloride-sodium alkaline water (pH up to 9.6) with a degree of mineralization of 1–5 g/l, with a high H_2SiO_3 content (up to 1000 mg/l) of boron, alkaline earths and some microcomponents. The vapour phase contains an insignificant amount of gas (up to 1% by volume), among which CO_2 is by far the dominant component (67–99% by volume, not including water vapour and air), H_2S is present (up to 11% by volume), H_2 (up to 10% by volume), N_2 (6% by volume) and CH_4 (5% by volume).

The temperature of the steam–water mixture falls as it approaches the surface (to 130 °C) and metamorphosis of the hydrothermal waters occurs, and they become chloride-sulphate (Bol'she Bannoe), hydrocarbonate-chloride (Reykjavik, Steamboat), sulphate-chloride (Ngawha, New Zealand), and even hydrocarbonate-sodium (Wairakei). The dominant gas in the majority of resources is still CO_2, but nitrogen-carbon dioxide springs are known (Krisuik, Reykjavik). The hydrocarbonate-sodium hot springs of Wairakei are distinguished by the reduction in the degree of mineralization and an increase in the gas composition in the vapour phase (Al'binskii, 1967).

'Coastal' steam springs (Kononov & Tkachenko, 1974; Kononov & Polyak, 1977), which are located on the periphery of hydrothermal systems near the sea, are of especial interest. By composition they are chloride, mainly sodium salts and brines with a degree of mineralization of 10–77 g/l, and are alkaline or weakly acid. The gas content is characterized by different combinations of carbon dioxide and hydrogen. It is thought that entrained sea water as a result of steam-lift plays an active part in its formation. The most highly mineralized of the coastal steam springs are the brines of the Reykjanes (Iceland) and Arima (Japan) hot spring systems.

Steam-spring brines on the Reykjanes peninsula have a temperature of 292 °C at a depth of 1700 m, and have a chloride-sodium composition with a degree of mineralization of 48 g/l. The water is weakly acid (pH 6.2), with a very high H_2SiO_3 concentration (700 mg/l), Li (3 mg/l), Sr (6.3 mg/l), Zn (0.88 mg/l), and other microcomponents. CO_2 is by far the largest component in the gas phase (93.6%), nitrogen is also present (4.5%).

The brines of the Arima system are also chloride-sodium, but are mineralized to a higher degree (77 g/l), a temperature

Figure 6.28. Hot spring at Cerro Prieto, Mexico (Svyatlovsky, 1975).
1, trap (centres of superheated steam); 2, cover impermeable to heat transfer for liquids and gases; 3, aquifer; 4, volcano-tectonic type of horsts; 5, faults which conduct hot spring water.

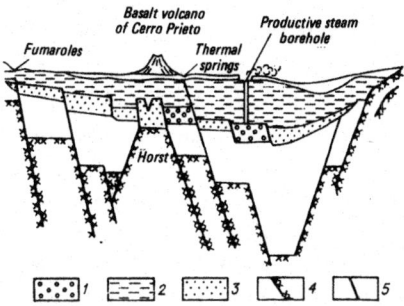

of 133 °C at a depth of 168 m has been recorded, and the water is weakly acid (pH 5.8). The gas composition of the steam phase is represented by hydrogen (51.4%) and carbon dioxide (46.7%).

V.I. Kononov and B.G. Polyak consider the presence of considerable quantities of hydrogen in the gas phase to be a distinguishing feature of marine rift hydrotherms. In Iceland such steam and water mixtures are found in the Namafjall, where siliceous (H_2SiO_3 – 770 mg/l), sulphate-hydrocarbonate potassium-sodium water with a degree of mineralization of 1.3 g/l and a temperature of 289 °C are found, in which H_2 (32.8%) is the main component, and which contain CO_2 (31.4%), H_2S (24.1%), and N_2 (4.5%). In other regions there are hot springs with a high hydrogen content, in the Arima system (which has the greatest value), and Akuchanan in Salvador (up to 40%).

6.9 Groundwater under seas and oceans

The water-bearing capacity of rocks which have been deposited on the bottom of large marine areas (oceans, seas, and lakes) has, up to now, received insufficient study. Moreover, the prospects of the exploitation of the world ocean demands a knowledge of hydrogeodynamics, hydrogeochemistry, and the hydrogeothermics of the submarine aquifers. With this in mind, in recent years there have been suggestions put forward on the necessity of dividing marine hydrogeology into independent branches at the junction of marine geology and hydrogeology (Kissin, 1974; Dzhamalov *et al.*, 1978; and others).

The total area of the world ocean is 361.3 million km² (71% of the Earth's surface), the maximum depth is in the Pacific Ocean (11 km) and the average depth varies from 1.1 km in the Arctic Ocean to 4 km in the Pacific (Anon., 1974a).

As E.A. Baskov (1971, 1975) has proposed, under the floor of the seas and oceans there exist aquifers which are analogous to those of the land mass — basins of stratal and fissure-stratal water, and also the so-called *thalassobasins*. He distinguishes two groups of submarine artesian reservoirs: (1) those forming single impermeable systems with the artesian basins of the land mass (coastal and shelf zones); (2) the deep water marine basins and trenches, margins of the land masses, and abyssal plains of the ocean bed, which have no connection with the land masses. In the first of these both the entry of infiltration water into the bottom deposits and the migration of sedimentary water from the submarine basins into the deposits of the land take place. Water of oceanic (marine) origin is distributed in the second.

I.G. Kissin (1974) divides the submarine hydrogeological reservoirs into: (1) those distributed in the sphere of influence of the infiltrating water — pressure systems (with a continental recharge region); (2) those outside this sphere of influence. He considered that the hydrogeodynamic environment in the deep parts of the Pacific Ocean has special features and it is possible to compare such reservoirs with the artesian basins of the land masses.

There is also a third point of view (Dzhamalov *et al.*,

1978), according to which the groundwater of the land mass has a connection with the atmosphere (via the unsaturated zone in the upper part of the rock formations), and with the submarine groundwater which is subject to the overriding influence of the marine and ocean water (although the infiltration of water from the continents into the submarine reservoirs cannot be excluded).

Finally, apart from the above, submarine hydrothermal systems are probably a special type of hydrogeological reservoir formed in conjunction with the action of endogenic forces. Their existence and the nature of their water-exchange processes have been quite objectively explained from the position of the new global tectonics (see Section 3.2).

A summary of these opinions indicates the complex laws of formation of groundwater under the floor of large water masses (especially oceans). These are as yet far from fully explained.

Thus the available limited factual material on the various large water masses of the Earth and modern theoretical concepts allow the submarine hydrogeological reservoirs to be regarded as different from the purely continental reservoirs and, depending on the causes and nature of the water exchange, to divide them into:

(1) aquifers (of the shelves, continental slopes, and inland seas) which exchange water with the continents;
(2) aquifers of the deep ocean basins (marine abyssal plains), which are not connected with the continents;
(3) submarine hydrothermal systems.

Submarine aquifers that exchange water with the continents. Essentially there exist submarine parts of the continental hydrogeological reservoirs, i.e. a transitional variety; they occupy the coastal shelf zone and the continental slope of the large marine regions. It is in these reservoirs that the underground runoff from the land takes place, and this has been the subject of a number of publications devoted to its establishment and evaluation over the last decade, including two compendiums by R.G. Dzhamalov, I.S. Zektser, & A.V. Meskheteli (1977, 1978), in which both Soviet and foreign experience on the investigation and evaluation of underground runoff into the seas and the world ocean is summarized. According to the calculations of the authors the total magnitude of the underground runoff from the land to the world ocean is about 2460 km^3/year. This includes the Atlantic Ocean − 850 km^3/year, the Pacific Ocean − 1340 km^3/year, the Indian Ocean − 220 km^3/year, the Arctic Ocean from the territories of Europe and Asia − 50 km^3/year.

The discharge of the underground runoff into the coastal and shelf zones takes place via submarine springs of the stratal type in karst and fissured rocks, often by dispersion in permeable unconsolidated deposits or in the form of overflow via semi-permeable bottom sediments.

Submarine springs are found in many marine regions, but the largest discharge centres are described in some detail in the publications of G.A. Maksimovich (1956, 1957), F. Kohout (1966), I.S. Zektser *et al.* (1972), I.F. Glazovskii *et al.* (1973), etc. As an example we quote the well known Greek

discharge points of karst water at the bottom of the Mediterranean, and other inland seas.

On the Adriatic coast there are 700 such springs and groups of springs. Their outlets are in the form of large ascending streams in the karst funnels and channels which are exposed in the Mesozoic−Cenozoic limestone beds by the sea. They are usually found at distances of from 1 to 75 km from the shore and at depths of 1 to 30 m. At greater depths on the Côte d'Azur at a depth of 162 m, and near San Remo at a depth of 190 m, and off St Martin's Point on the Adriatic coast of Yugoslavia (700 m), there are fresh water springs. The flow of the karst springs is usually very great. Thus in the Levant the total flow of the group of submarine springs at a distance of 2 km from the shore changes from 6 m^3/s in the dry season to 50 m^3/s in winter. The ascending streams of fresh water from the karst funnels are under considerable pressure and form cupolas and 'boiling cauldrons'.

On the Black Sea coast of the Caucasus, submarine discharge centres have been investigated in some detail in the regions of Gagri and Gantiadi (Maksimovich & Kiknadze, 1967; Yurovskii, 1975; and others). Here outlets of fresh water from the karst funnels are found in Cretaceous limestones 10−100 m from the shore and at depths of 5 to 10 m and more. Karst deposits on the bottom are characteristic of areas with outlets with high permeability (infiltration coefficient of 40−300 m/day). The total flow of the springs in the region of Gantiadi has been calculated by Yu.G. Yurovskii at 1.35 m^3/s. The water in the ascending streams (a mixture of fresh and sea water) is chloride-sodium in composition and has a degree of mineralization of 15.2 g/l (Buachidzhe & Meliva, 1967).

Submarine springs with fresh and saline water are well known in many regions, on the shores of the Atlantic, Indian, and Pacific Oceans.

On the Atlantic coast major discharge centres are found on the Florida peninsula. Here the best known spring is at Crescent Beach (4 km from the shore not far from the town of Jacksonville). The water flows from a kratogenic funnel 15 m in diameter and 21 m deep. The pressure is so great that at the surface (with a depth of water of 38 m) the water is quite turbulent. The flow of the spring has been calculated to be 1.1 m^3/s (Dzhamalov *et al.*, 1977).

A considerable part of the underground runoff infiltrates from the land to the seas and oceans across a wide front along the coastline and penetrates to various distances into the shelf with the wedging out of aquifers. A graphic example of this penetration of fresh water into the sea is given by the infrared picture of the coastal zone of the island of Luzon in the Philippines (Fig. 6.29), where the penetration of tongues of surface and groundwater into the sea can be clearly traced.

The continental slopes have received considerably less study than the shelves. Nevertheless, the possibility of water entering from the land and being discharged in them is confirmed by data from marine drilling at considerable distances from the shore. The coarse-grained nature of the bottom deposits on the surfaces of many continental slopes and the presence of a multitude of faults in the consolidated deposits and the crystalline rocks create excellent conditions for ascending discharges. Aquifers of fresh and brackish water are found in the shelf and slope deposits at quite considerable depths. In the South China Sea north of the island of Hainan a borehole at a depth of 200 m (below the sea-bed) revealed a horizon of water-bearing sands 100 m thick in the sandy-clay Neocene–Quaternary deposits (Anon., 1974b): the water is pressure water (the surface being 10 m above sea level), hydrocarbonate-sulphate-chloride-sodium with a degree of mineralization of 1.5 g/l. Off the Atlantic coast east of Jacksonville (Florida) 43 km from the shore boreholes at depths of 130 and 255 m revealed the presence of brackish pressure water (piezometric surface 9 m above sea level) and 100 km off Florida stratal water with a degree of mineralization of 0.7 g/l was found (Dzhamalov *et al.*, 1977). The occurrence of water with a degree of mineralization of about 1.5 g/l in the Cretaceous deposits of the Australian shelf at a depth of 1200 m below the sea-bed (at a depth of 48 m) has been noted. And, on the continental slope off the Gulf of Mexico 70 km from the Mississippi delta, a borehole 140 m below the sea-bed revealed saline water with a degree of mineralization of 18 g/l.

The underground water exchange between the land and the sea is not limited to movement of water from the continents to the sea and its discharge from the sea-bed. A reverse movement of marine water into the deposits of the land and its discharge on the surface is also observed. The considerable variations in sea level lead to the recurrence of an underground discharge from the land areas to the sea and from the sea to the land. Some researchers (Ulanov, 1965; Pavlov, 1968) consider that the scale of the penetration of marine water into the land areas is so great that this process must be taken into account in the reckoning of the water balance of coastal regions. According to Ulanov's calculations, the infiltration of sea water into the bottom and shore of the Caspian Sea constitutes about one third of the underground runoff from the land. There are a multitude of examples of the discovery of saline water in the wells and boreholes of coastal zones (the Baltic and Caspian coasts, USA, Morocco, etc.). In the Netherlands sea water is found in the coastal belt at a depth of 200 m under fresh water. A.N. Pavlov considers the seasonal penetration of Black Sea water to be one of the main causes of the formation of the Matsesti mineral waters.

Many coastal saline springs are fed from sea water including the coastal hot springs described earlier (see Section 6.8). Shallow inland seas are in practically all cases under the influence of land water and take part in the general hydrologic cycle. On the floor of such seas the discharge of groundwater occurs, which, for example in the Caspian Sea, is found along the hydrogeological anomalies in the bottom deposits (Brusilovskii, 1971).

The hydrogeological section of the bottom deposits of the Caspian Sea have been studied in more detail than those of any other inland sea. As is well known (Anon., 1974b), it has a Precambrian basement, and Palaeozoic, Mesozoic, Paleocene, Neocene, and Quaternary deposits. The water-bearing capability of Precambrian rocks has not been studied. In the podsol terrigenic and carbonate Palaeozoic deposits of the northern part there occur strong chloride-sodium-calcium and calcium-sodium brines which have high temperatures and which contain methane. Above this the Permian and Mesozoic formations of saline deposits are characterized by a great variety of water-bearing capacities and by the presence of chloride-magnesium and calcium-sodium brines with a degree of mineralization of up to 450 g/l, and in the Neocene deposits 10–80 g/l. Methane and carbon dioxide are the dominant gases. In the Cenozoic deposits, which are represented mainly by sandy-clayey formations, groundwater occurs which down to a depth of 2–4 m below the floor is no different in composition and degree of mineralization from sea water.

The underground runoff into the Caspian Sea has been evaluated by many researchers, but is very provisional: its magnitude, calculated on the basis of different data is from 0.3 to 49 km^3/year. The most reliable figure is a volume of underground runoff of 1.0–1.4 km^3/year (Kortsenshtein, 1962; Dzhamalov *et al.*, 1977).

With such a significant amount of groundwater entering the sea, the question naturally arises of the ways in which it is discharged. It is mainly accomplished by means of ascending filtration and subsequent discharge into the sea through the sea-bed. The most likely channels of such a discharge are considered by V.N. Kortsenshtein (1962) to be zones of large regional faults (Fig. 6.30).

The aquifers of deep water basins which are not connected with the continents. In all oceans the continental slopes and deep trenches gradually give way to flat abyssal plains (depressions). They are isometric in form and several thousand

Figure 6.30. Discharge mechanism of deep groundwater in the Caspian depression (Kortsenshtein, 1962).
1, sea water; 2, clays; 3, sands; 4, limestones; 5, sandstones; 6, crystalline basement; 7, faulting; 8, direction of groundwater movement.

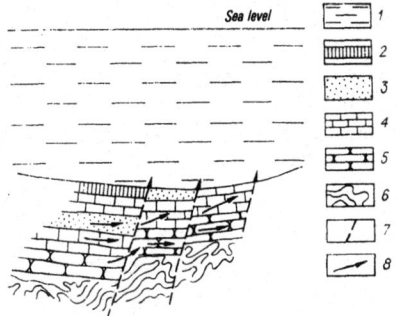

Figure 6.29. Diagram of the infrared picture of the coastal zone of the island of Luzon, Philippine Islands (Zektser *et al.*, 1972).

kilometres in diameter. The depth of the ocean here reaches 5–6 km. The sedimentary layer of the ocean floor consists of alternating layers of clay, and siliceous and carbonate silts. The thickness of the layer over the whole area rarely exceeds 1 km (Zonenshain *et al.*, 1976). The basement is a basalt layer about 2 km thick and below this there is a layer of serpentine which is 6 km thick (Monin, 1977). Thus we have here a typical complete section of the oceanic crust. The absence of a granite layer which predetermines the small thickness of the oceanic crust in comparison with that of the continental, is an important hydrogeological, and also hydrogeochemical and hydrogeothermal feature of the deep part of the ocean bed.

Unfortunately from the hydrogeological point of view, the deep submarine aquifers remain completely uninvestigated. They are characterized by the specific conditions under which groundwater is formed in the sedimentary layer which is subject on the one hand to considerable pressure of the oceanic water, and on the other hand to the cooling effect of this water which is felt at a depth down to 2–3 km (Kortsenshtein, 1978). Furthermore there is no doubt that the sedimentary nature of such aquifers is characterized by an elision regime (Kissin, 1974).

In the upper parts of the vertical section of the bottom deposits pore solutions are represented by trapped sea water. In the deeper layers of the compacted sediments there occurs water which has been squeezed out from the beds and laminations. It is thought that in these conditions active lateral migration of sedimentary water is not possible and so vertical movement predominates, which ensures the continuous interaction with the water of the ocean (Dzhamalov *et al.*, 1978). However, in layers of increased permeability lateral movement is definitely possible. In the peripheral regions of the mid-oceanic ridge in the fissured consolidated rocks the entry of oceanic water into the sedimentary layer and its mixing with the solutions which have been squeezed out of the pores may take place.

As a result of vertical (both descending and ascending) movement and of the transport of materials by diffusion, over geological time there has been an equalization of the concentration of stratal and oceanic water. The components of these and of other waters have become very close, albeit not in all components. It is true that this effect, obviously, is peculiar to the upper parts of the vertical section of submarine aquifers only.

Submarine hydrothermal systems, in contrast to continental systems, (see Section 6.8) have no connection with the atmosphere; they are closed layers of sea or oceanic water. Such systems may be divided into two types according to the structure of the hydrogeological reservoirs, hydrogeodynamic, and hydrogeochemical features: (1) systems of deep trenches connected with the Benioff–Zavaritskii zone; (2) systems of oceanic (along the axes of mid-oceanic ridges), and marine rifts.

(1) Submarine hydrothermal systems in deep trenches: these are deep and narrow 'gashes' in the ocean floor which run along the coastline and often border island arcs. They are formed as a result of the dehydration of the sinking ocean crust and its interaction with the material of the mantle. The lateral movement of the oceanic lithosphere in the Benioff–Zavaritskii zone leads to the dehydration of the serpentine and kaolinite and the production of great volumes of water (see Section 3.2). In conditions of high temperature, steam saturated with silica and alkalis, and volatile components, is liberated and because of excess pressure which accompanies the superheating this steam migrates into the overlying layers of rock and takes part in the processes that bring about its metamorphism (Monin, 1977). The bulk of active volcanoes are confined to the Benioff–Zavaritskii zone and a considerable quantity of groundwater and steam is discharged through these into the oceans.

Water exchange in the hydrothermal systems of the deep trenches is very complex, and fluid currents flow from the mantle; renewed water from the ocean crust, oceanic water, and also groundwater from the continental crust take part in it. As has already been noted, the mechanism of the process is by no means clear, but the water exchange itself and the formation of hydrothermal systems are a fact.

Submarine hydrothermal systems of deep trenches are of widespread occurrence on the periphery of the oceans. This is shown by the fact that the regions of volcanic activity, both on land and on islands, are confined to these areas. The major continental hydrothermal systems are also connected with such contact zones (see Section 6.8). The widespread occurrence in the upper horizons of the sedimentary layer of such zones of iron–manganese nodules, which are found for example in the Marianas Deep at a depth of 7700 m (Menard, 1964) is indirect confirmation of the discharge of hydrothermal currents into deep trenches. It is interesting that nodules are found more often than not on the surface of the ocean floor, but at some depth in the silt sediments too.

(2) Submarine hydrothermal rift systems, confined to the mid-oceanic ridges or the continental rift zones, are characterized by the discharge of deep water.

The mid-oceanic ridges are areas of the origin of new oceanic crust and its spreading out. Rift valleys, which are distributed along the axes of the ridges, are places where the material of the mantle comes into contact with the surface of the ocean floor (Monin, 1977). The hydrogeological information is extremely fragmentary in this respect. However, the very high values of heat flow, the widespread occurrence of volcanoes both submarine and on the surface, submarine hydrothermal springs, and the presence of hydrothermally altered rocks in such zones provide a sufficient basis for the belief that oceanic rifts are specific submarine hydrothermal systems (see Section 3.2).

Hydrothermally altered rocks and ores are found, for example, on the Eastern Pacific ridge, in a region in which the heat flow is high. Sediments on the sea-bed in this region are enriched with iron and many other metals (Lisitsyn, 1974). The regions of active volcanic and hydrothermal activity which are confined to zones of transform faults, and with which the volcanic activity of the Hawaiian Islands is connected, are rather different in their mechanism of formation. It is possible that active hydrothermal systems are also confined to these zones, but as yet this is only a supposition because there is no reliable data on these regions.

The rift depressions of some internal seas also belong to the category of submarine hydrothermal systems. The Red Sea rift, which is characterized by its great ease of study has already become a classic example of such systems. Here, within the deepest parts of the sea-bed, thermal springs of chloride-sodium brines have been found, and anomalously high temperatures and degree of mineralization of the water have been reported. Analysis of the materials obtained as a result of drilling boreholes, shows that the discharge of these brines, which arises as a result of active heat transport (temperatures up to 56 °C), occurs periodically along localized channels and is at times accompanied by earthquakes and the ejection of basalts (Anon., 1978). The various metallic deposits of the Red Sea, which are formed by hot springs, occupy an area of about 100 km^2 (Lisitsyn, 1974). According to the data of American oceanologists, the discovery of a hot spring discharge in the rift of the Galapagos mid-oceanic ridge is evidence of continuing hydrothermal activity in rift systems.

The above information on submarine aquifers is clearly inadequate. Many of the positions outlined require strengthening and also refuting by factual material which no doubt (because of the great emphasis on research in the seas and oceans) will be obtained in the near future. Furthermore, it must be borne in mind that in this section the first attempt has been made to summarize the data on the groundwater under the Earth's great marine areas.

References

Al'binskii, N.V. (1967). The chemical composition of the thermal waters and gases of New Zealand. *Dokl. Geograf. ob-va SSSR*, No. 2, 35–41.

Al'tovskii, M.E. (ed.) (1962). *Reference book of hydrogeology.* Moscow: Gosgeoltekhizdat. 615 pp.

Anisimova, N.P. (1971). *The formation of the chemical composition of the groundwater of taliks (for example those of Central Yakutia).* Moscow: Nauka. 186 pp.

Anon. (1959). *Manual of hydrogeology.* Leningrad: Gostoptekhizdat. 836 pp.

Anon. (1966). *Methods of hydrogeological research and the groundwater resources of Siberia and the Far East.* Moscow: Nauka. pp. 61–8.

Anon. (1968). *The hydrogeology of the USSR, Vol. 11, Armyanskaya SSR.* Moscow: Nedra. 352 pp.

Anon. (1970a). *The hydrogeology of the USSR, Vol. 20, Yakutskaya ASSR.* Moscow: Nedra. 383 pp.

Anon. (1970b). *The thermal regime in the USSR.* Moscow: Nauka. 171 pp.

Anon. (1972a). *The hydrogeology of the USSR, Vol. 26, The northeast of the USSR.* Moscow: Nedra. 296 pp.

Anon. (1972b). *The hydrogeology of the USSR, Vol. 29, Kamchatka, the Kuriles, and the Komandorskie Islands.* Moscow: Nedra. 364 pp.

Anon. (1974a). *The world water balance and the water resources of the Earth.* Leningrad: Gidrometeoizdat. 638 pp.

Anon. (1974b). *The hydrogeology of Asia.* Moscow: Nedra. 575 pp.

Anon. (1977). *Iceland and the mid-oceanic ridge: deep structure, seismicity, and geothermics.* Moscow: Nauka. 195 pp.

Anon. (1978). *The hydrogeology of Africa.* Moscow: Nedra. 372 pp.

Askerov, A.G. & A.G. Durmish'yan (1967). The varieties of deep thermal water discovered during drilling for oil and gas. In *Regional geothermics and the distribution of hot springs in the USSR*, pp. 175–7. Moscow.

Aver'ev, V.V. (1961). The conditions of discharge of the Pauzhetka hot springs in the south of Kamchatka. *Trudy Laboratorii vulkanologii AN SSSR*, No. 19, 80–98.

Babushkin, V.D., E.P. Lebedyanskaya, L.E. Levi, *et al.* (1972). *The forecasting of water inflows into mines and the intake of groundwater in fissured and karst rocks.* Moscow: Nedra. 194 pp.

Babushkin, V.D. & Yu.V. Ponomarenko (1972). Features of the development of fissuring and water-bearing capacity of rocks of various types. In *The forecasting of water inflows into mines and the intake of groundwater in fissured and karst rocks*, pp. 7–15. Moscow: Nedra.

Barth, T.F.W. (1950). Volcanic geology. Hot springs and geysers of Iceland. *Carnegie Institute, Washington, Publ.*, Vol. 587, 174 pp.

Baskov, E.A. (1971). The main types of hydrogeological structure of the world ocean. In *Problems of hydrogeological mapping and zonation*, pp. 22–5. Leningrad.

Baskov, E.A. (1975). Structural–hydrogeological zonation. In *The hot springs of the Pacific*, pp. 21–37. Moscow: Nedra.

Bemmelen, R.W. van (1949). *The geology of Indonesia.* The Hague: Government Printing Office.

Black, R. (1954). Permafrost – a review. *Bull. Geol. Soc. Am.*, 65, 839–55.

Borevskii, B.V., M.A. Khordikainen & L.S. Yazvin (1976). Prospecting for, and evaluation of, exploitable groundwater resources in fissure-karst strata. Moscow: Nedra. 247 pp.

Brusilovskii, S.A. (1971). The possibility of evaluating the submarine runoff according to its geochemical features. In *Complex investigations of the Caspian Sea*, pp. 68–74. Moscow: Izd-vo MGU.

Buachidzhe, I.M. & A.M. Meliva (1967). On the question of the discharge of groundwater into the Black Sea in the Gagri region. *Trudy NIL gidrogeol. problem. GPI, Tbilisi*, pp. 17–24.

Castany, G. (1963). *Traite pratique des eaux souterraines.* Paris. 657 pp.

Chubarov, V.N. (1973). *Investigation of moisture transfer in the zone of aeration for the solution of hydrogeological problems.* Moscow: Nedra. 70 pp.

Daubrée, A. (1887). *Les eaux souterraines.* Paris. 375 pp.

Davies, S.N. & R.J.M. De Wiest (1967). *Hydrogeology*, 2nd edn. New York: Wiley. 463 pp.

Dostovalov, B.B. & V.A. Kudryatsev (1967). *General permafrost studies.* Moscow: Izd-vo MGU. 398 pp.

Droznin, V.A. & A.A. Razina (1977). The nature of the geyser regime. In *The hydrothermal process in regions of tectono-magmatic activity*, pp. 96–103. Moscow: Nauka.

Dvorov, I.M. (1972). *The deep heat of the Earth.* Moscow: Nauka. 205 pp.

Dzhamalov, R.G., I.S. Zektser & A.V. Meskheteli (1977). *Underground runoff into the seas and the world ocean.* Moscow. 93 pp.

Dzhamalov, R.G., I.S. Zektser & A.V. Meskheteli (1978). The basic laws governing the formation and distribution of underground runoff into the seas and the world ocean. *Vodnye resursi*, No. 6, 32–47.

Efimov, A.I. (1964). The frozen hydrogeological features of the bank and river bed of the Lena near Yakutsk. In *The geocryological conditions of Western Siberia, Yakutia, and Chukotka*, pp. 97–110. Moscow: Nauka.

Ellis, A.J. (1979). The chemistry of some explored geothermal systems. In *Geochemistry of hydrothermal ore deposits*, 2nd edn, ed. H.L. Barnes. New York: Wiley Interscience. 798 pp.

Garmonov, I.V. (1948). The zonation of subsurface water of the European part of the USSR. *Trudy gidrogeologicheskikh problem*, 3, 131–8.

Glazovskii, I.F., V.A. Ivanov & A.V. Meskheteli (1973). The study of submarine springs. *Okeanologiya*, 12 (2), 249–54.

Gogoleva, N.P. & Yu.V. Ponomarenko (1971). The laws governing the changes in the infiltration properties of fissured rocks with depth. *Sov. geologiya*, No. 12, 104–7.

Gordeev, D.I. (1954). The main stages in the history of Russian and Soviet hydrogeology. *Trudy Lab. gidrogeol. problem im F.P. Savarenskii*, 7, 381 pp.

Haas, H.I. (1895). *Quellenkunde.* Leipzig. 220 pp.

Hopkins, D.M. *et al.* (1955). Permafrost and ground water in Alaska. *US Geol. Survey Prof. Paper*, 264 F, 113–46.

Ivanov, V.V. (1958). The main stages in hydrothermal activity of the volcanoes of Kamchatka and the Kurile Islands and the types of thermal water associated with them. *Geokhimiya*, No. 5, 473–85.

Ivanov, V.V. (1961). The steam springs of the Kurile–Kamchatka volcanic zone. In *Problems in geothermy and the practical use of the heat of the Earth*, Vol. 2, pp. 43–65. Moscow: Izd-vo AN SSSR.

Kamenskii, G.N. (1943). *The fundamentals of the dynamics of groundwater*. Moscow: Gosgeotekhizdat. 248 pp.

Kamenskii, G.N. (1949). The zonation of groundwater and soil-geographical zones. *Trudy Lab. gidrogeol. problem. im F.P. Savarenskii*. Moscow, 6, 5–21.

Kamenskii, G.N. (1953). The hydrodynamic principles of study of the groundwater regime. In *Problems of hydrogeology and engineering geology*, pp. 4–12. Moscow: Gosgeolizdat.

Kamenskii, G.N., M.M. Tolstikhina & N.I. Tolstikhin (1959). *The hydrogeology of the USSR*. Moscow: Gosgeoltekhizdat. 366 pp.

Kaplina, T.N., O.P. Pavlov, V.P. Chernoryad'ev & I.L. Kuznetsova (1975). *New global tectonics and the formation of permafrost and groundwater*. Moscow: Nauka. 123 pp.

Keilhack, K. (1912). *Lehrbuch der Grundwasser- und Quellenkunde*, 1st edn.

Keilhack, K. (1935). *Lehrbuch der Grundwasser- und Quellenkunde*. 3rd edn. Berlin: Borntraeger. 575 pp.

Keller, G. (1931). *Pressure water*. Moscow: Gostekhizdat. 79 pp. (Translated from the German.)

Khodzhibaev, N.N. (1970). *Natural currents of groundwater in Uzbekistan*. Tashkent: Izd-vo FAN. 175 pp.

Kissin, I.G. (1967). *Hydrodynamic anomalies in the subsurface hydrosphere*. Moscow: Nauka. 135 pp.

Kissin, I.G. (1974). Marine hydrogeology – a new branch of research into the bottoms of seas and oceans. *Sov. geologiya*, No. 11, 41–52.

Klimentov, P.P. & G.Ya. Bogdanov (1977). *General hydrogeology*. Moscow: Nedra. 357 pp.

Kohout, F. (1966). Submarine springs. In *The Encyclopaedia of Oceanography*. Encyclopaedia of Earth Sciences Series, pp. 878–83. New York: Rheinhold Publ. Corp.

Koldysheva, R.Ya. (1975). Features of the zone of aeration of the cryolithic zone. In *Regional and thematic geocryology research*. pp. 133–6. Novosibirsk: Nauka.

Kononov, V.I. & B.G. Polyak (1977). Geothermal activity. In *Iceland and the mid-oceanic ridge (deep structure seismicity and geothermics)*, pp. 8–82. Moscow: Nauka.

Kononov, V.I. & R.I. Tkachenko (1974). Coastal hot springs and the features of their formation. In *Hydrothermal mineral-forming solutions of the active volcanic regions*, pp. 38–45. Novosibirsk: Nauka.

Konoplyantsev, A.A. (1960). The zonation and azonation of groundwater. *Sov. geologiya*, No. 12, 86–97.

Kortsenshtein, V.N. (1962). The mechanism of discharge of deep groundwater in the Caspian depression. *Dokl. AN SSSR*, 142 (3), 667–9.

Kortsenshtein, V.N. (1978). The specific nature of conditions influencing the formation and preservation of marine gas deposits. *Dokl. AN SSSR*, 240 (6), 1426–9.

Kovalenko, V.Ya. (1964). The subpermafrost water of the right bank of the upper basin of the Bol'shaya Anyui. In *Material on the geology and minerals of the N.E. USSR*, No. 17, pp. 190–9. Magadan.

Kovalevskii, V.S. (1973). *The conditions of formation and the forecasting of natural groundwater regimes*. Moscow: Nedra. 152 pp.

Kovda, V.A. (1973). *Fundamentals of soil studies*, Book 2. Moscow: Nauka. 468 pp.

Kriger, N.I. (1951). Fissuring and the methods of studying it in geological surveying. In *Material on engineering geology*, No. 2, p. 139. Moscow: Metallurgizdat.

Kühne, V. (1932). *Groundwater studies*. Moscow, Leningrad: Gosstroiizdat. 196 pp. (Translated from the German.)

Lange, O.K. (1960). *Problems in hydrogeology*. Moscow: Gosgeoltekhizdat. pp. 15–25.

Lange, O.K. (1969). *Hydrogeology*. Moscow: Vyshaya shkola. 363 pp.

Lebedev, A.A. & E.N. Yartseva (1967). *An evaluation of the subsurface water recharge and balance*. Moscow: Nedra. 172 pp.

Linetskii, V.F. (1961). The hydrogeological significance of large abnormal pressures in closed structures. In *Proceedings of the First Ukrainian Hydrogeological Conference*, Vol. 1, pp. 236–47.

Lisitsyn, A.P. (1974). *Sediment formation in oceans*. Moscow: Nauka. 435 pp.

Maksimova, L.N., G.E. Perl'shtein & N.N. Romanovskii (1966). The influence of suprapermafrost water on the seasonal thawing of sediments. In *Methods of hydrogeological research and groundwater resources of Siberia and the Far East*, pp. 61–8. Moscow: Nauka.

Maksimovich, G.A. (1956). Freshwater springs on the seabed. *Priroda*, No. 4, 89–91.

Maksimovich, G.A. (1957). *Submarine karst springs*. Publication of the University of Perm, Vol. 11, No. 2, pp. 83–5. Izd-vo Khar'kovskogo yn-ta.

Maksimovich, G.A. & T.E. Kiknadze (1967). The submarine springs of the Black Sea and some karst regions of the Mediterranean. *Soobshcheniya AN GSSR*, 17 (3), 643–6.

Marinov, N.A. (1979). Some problems in the study of subsurface water. *Vodnye resursi*, No. 2, 83–94.

Marinov, N.A. & N.I. Tolstikhin (1973). The large karst springs of Eurasia. In *The hydrogeological significance of karst in the Leningrad combustible shale deposits*, pp. 3–15. Leningrad.

Matthess, G. (1970). *Beziehungen zwischen geologischen Bau und Grundwasserbewegung in Festgesteinen*. Wiesbaden. 105 pp.

Mavritskii, B.F. & G.K. Antonenko (1967). *The prospecting for, investigation of, and practical application of, hot springs in the USSR and abroad*. Moscow: Nedra. 176 pp.

Meinzer, O.E. (1923). The occurrence of groundwater in the United States: *US Geol. Survey Water Supply Paper*, No. 489, 321 pp.

Mel'nikov, P.I. (1967). The influence of groundwater on the deep cooling of the upper zone of the crust. In *Hydrogeothermic and hydrogeological research in the permafrost of the east of the USSR*, pp. 24–9. Moscow: Nauka.

Mel'nikov, P.I. & N.I. Tolstikhin (eds.) (1974). *General permafrost studies*. Novosibirsk: Nauka. 240 pp.

Menard, H.W. (1964). *Marine geology of the Pacific*. New York: McGraw-Hill. 271 pp.

Mercado, S. (1969). Chemical changes in geothermal well M-20, Cerro Prieto, Mexico. *Bull. Geol. Soc. Am.*, 80 (12), 2623–30.

Merzhanov, A.G., A.A. Razina, A.S. Shteinberg & G.S. Shteinberg (1973). A laboratory model of a geyser. *Dokl. AN SSSR*, 211 (3), 584–7.

Monin, A.S. (1977). *The history of the Earth*. Leningrad: Nauka. 228 pp.

Muslimov, R.Kh., T.A. Lapinskaya, I.Kh. Kaveev & V.I. Filippovskii (1977). The results from borehole 20 000 in the crystalline basement of the TASSR. *Sov. geologiya*, No. 6, 110–16.

Naboko, S.I. (1954). The geysers of Kamchatka. In *Trudy Lab. vulcanologii AN SSSR*, No. 8, pp. 126–209. Moscow: Izd-vo AN SSSR.

Nakamura, H. & K. Sumi (1961). Geothermal investigation of Matsukawa hot springs area, Iwate Prefecture. *Bull. Geol. Surv. Japan*, 12 (2), 1–12.

Neizvestnov, Ya.V. & Yu.P. Semenov (1973). Underground cryopegs of the shelf and islands of the Soviet Arctic. In *Groundwater of the cryolithosphere*, pp. 103–6. Yakutsk: Kn. izd-vo.

Nekhoroshev, A.S. (1959). The theory of geyser action. *Dokl. AN SSSR*, 127 (5), 1096–8.

Nekrasov, I.A. (1967). *The taliks of river valleys and the laws governing their distribution*. Moscow: Nauka. 137 pp.

Nikitin, S.N. (1900). *Subsurface and artesian water on the Russian plain*. Sib. 71 pp.

Ovchinnikov, A.M. (1938). Methods of studying fissuring. *Razvedka nedr*, Nos. 4–5, 32–41.

Ovchinnikov, A.M. (1955). *General hydrogeology*. Moscow: Gosgeoltekhizdat. 383 pp.

Pampura, V.D. (1977). *The formation of minerals in hot spring systems*. Moscow: Nauka. 203 pp.

Pavlov, A.N. (1968). The interdependence of the water of the land and the sea. *Sov. geologiya*, No. 12, 67–78.

Pilipenko, G.F. (1974). The hydrochemical features of the Uzon thermoanomaly. In *Volcanic activity, hydrothermal processes and ore-formation*, pp. 83–109. Moscow: Nedra.

Pinneker, E.V. (1979). A new groundwater classification based on the manner in which it is formed. In *Proceedings of the All-Union*

Conference on the Waters of the Eastern Regions of the USSR, Irkutsk—Petropavlovsk—Kamchatskii, pp. 6–7.

Pinneker, E.V., B.I. Pisarskii, I.S. Lomonosov *et al.* (1968). *The hydrogeology of the Baikal Region.* Moscow: Nauka. 168 pp.

Pisarskii, B.I. & S.I. Sherman (1967). Fissuring parameters and their significance in hydrogeological research. In *The formation and geochemistry of the groundwater of Siberia and the Far East*, pp. 25–9. Moscow: Nauka.

Plotnikov, N.I., M.V. Syrovatko & D.I. Shchegolev (1957). *The groundwater of ore deposits.* Moscow: Metallurgizdat. 614 pp.

Rats, M.V. & S.K. Chernyshev (1970). *The degree of fissuring and the properties of fissured rocks.* Moscow: Nedra. 160 pp.

Romanovskii, N.N. (1966). A scheme for the subdivision of the groundwater of permafrost regions. In *Methods of hydrogeological research and the groundwater resources of Siberia and the Far East*, pp. 28–41. Moscow: Nauka.

Romanovskii, N.N. (1970). Aspects of the study of groundwater in permafrost regions. In *Methods of permafrost hydrogeology and engineering geology surveys*, pp. 175–218. Moscow: Izd-vo MGU.

Romm, E.S. (1966). *Infiltration properties of fissured hard rock.* Moscow: Nedra. 283 pp.

Rosin, A.A. (1977). *The groundwater of the Western Siberian artesian basin and its formation.* Novosibirsk: Nauka. 100 pp.

Savarenskii, F.P. (1939). *Hydrogeology.* Moscow: ONTI. 113 pp.

Schoeller, H. (1962). *Les eaux souterraines.* Paris: Masson. 643 pp.

Shastkevich, Yu.G. (1966). Permafrost in rocks of the highland parts of the Udokan Range and the conditions of formation of the temperature regime. In *Geocryologic conditions of the North Transbaikal*, pp. 24–43. Moscow: Nauka.

Shchegolev, D.I. & N.I. Tolstikhin (1939). *Groundwater in fissured rocks.* Moscow: Gostoptekhizdat. 76 pp.

Shchelkachev, V.N. (1959). *The exploitation of oil-bearing strata in an elastic regime.* Moscow: Gostoptekhizdat. 468 pp.

Smekhov, E.M. (1962). The laws governing the development of fissuring in hard rock and fissured reservoir rocks. *Trudy VNIGRI*, No. 172, 145 pp.

Sokolov, D.S. (1962). *Fundamental conditions for the development of karst.* Moscow: Gosgeoltekhizdat. 321 pp.

Sokolovskii, L.G. & V.I. Sedletskii (1970). Geochemical properties and the origin of highly mineralized brines of the south of Central Asia. *Sov. geologiya*, No. 7, 101–12.

Steuer, A. (1907). *Die Entstehung des Grundwassers in hessischen Ried.* Festschrift zum 70 Geburtstage von A.V. Koenen. Stuttgart.

Studt, F.E. (1958). The Wairakei hydrothermal field under exploitation. *N.Z. J. Geol. Geophys*, No. 1.

Sugrobov, V.M. (1964). Pauzhetka hot springs of Kamchatka as an example of a high temperature impermeable system. In *Hydrothermal conditions of the upper parts of the Earth's crust*, pp. 72–86. Moscow: Nauka.

Svyatlovskii, A.E. (1975). *Regional volcanology.* Moscow: Nedra. 223 pp.

Thurner, A. (1967). *Hydrogeologie.* Vienna, New York: Springer Verlag. 350 pp.

Tolstikhin, N.I. (1941). *Groundwater of the frozen zone of the lithosphere.* Moscow, Leningrad: Gosgeolizdat. 201 pp.

Tolstikhin, N.I. & O.N. Tolstikhin (1974). Underground and surface water of permafrost zones. In *General permafrost studies*, pp. 192–229. Novosibirsk: Nauka.

Tolstikhin, N.I. & O.N. Tolstikhin (1975). Groundwater of the permafrost zone of the USSR (major results and problems of research). In *Second International Conference on Permafrost Studies. Reports and papers*, No. 8. pp. 73–88. Yakutsk: Kn. izd-vo.

Ulanov, Kh.K. (1965). Underground runoff into the Caspian Sea and the infiltration of its water into the sea bottom and the coast. *Dokl. AN SSSR.* 162, 166–8.

Ustinova, T.I. (1955). *Kamchatka geysers.* Moscow: Gosgeografizdat. 120 pp.

Ustinova, Z.G. (1964). On the hydrochemistry of the kimberlites of the Yakut pipes. *Trudy VSEGINGEO*, No. 9, 237–52.

Vakin, E.A. & V.M. Sugrobov (1972). The hydrogeological features of volcanic structures in active hydrothermal systems. In *Hydrogeology of the USSR, Vol. 26. Kamchatka, the Kuriles, and the Komandorskie Islands.* pp. 169–95. Moscow: Nedra.

Valukonis. G.Yu. & A.E. Khod'kov (1973). *The geological laws of movement of groundwater, oils, and gases.* Leningrad: Izd-vo LGU. 304 pp.

Vel'mina, N.A. (1970). *Features of the hydrogeology of the frozen zone of the lithosphere.* Moscow: Nedra. 324 pp.

Vel'mina, N.A. & A.V. Uzemblo (1959). *The hydrogeology of the central part of South Yakutia.* Moscow: Izd-vo AN SSSR. 179 pp.

White, D.E. (1965). Hot springs of volcanic origin. In *The geochemistry of recent post-volcanic processes*, pp. 78–100. Moscow: Mir.

White, D.E. (1967). Some principles of geyser activity, mainly from Steamboat Springs, Nevada. *Am. J. Sci.*, 265, 641–84.

White, D.E., E.T. Anderson & D.K. Grubbs (1963). Geothermal brine well: mile-deep drill may tap ore-forming magmatic water and rocks undergoing metamorphism. *Science*, 139, 919–22.

White, D.E. & G.A. Waring (1963). Volcanic emanations. In *Data of geochemistry: US Geol. Survey Prof. Paper*, No. 440-K.

Yas'ko, V.G. (1978). *Hydrogeology of the mineral deposits of Siberia.* Moscow: Nedra. 200 pp.

Yurovskii, Yu.G. (1975). Evaluation of the magnitude of the submarine discharge of groundwater. *Izv. Vsesoyuz. geogr. ob-va*, 105 (2), 174–9.

Zabolotnik, S.I. (1974). The seasonal freezing and thawing of ground. In *Geocryological conditions in the Mongolian People's Republic*, pp. 49–73. Moscow: Nauka.

Zaitsev, I.K. (1960). The regional laws governing the hydrochemistry of the groundwater of the USSR. In *Problems of hydrogeology*, pp. 24–41. Moscow: Gosgeoltekhizdat.

Zaitsev, I.K. (1961). *The methods and principles of making hydrogeological maps. The hydrogeological map of the USSR. Scale 1 : 2 500 000, explanatory note*, pp. 7–42. Moscow: Gosgeoltekhizdat.

Zaitsev, I.K. & N.I. Tolstikhin (1972). *The laws of distribution and formation of mineral (industrial and medicinal) groundwater in the USSR.* Moscow: Nedra. 279 pp.

Zektser, I.S., V.A. Ivanov & A.V. Meskheteli (1972). The discharge of groundwater into the seas. *Vodnye resursi*, No. 3, 125–46.

Zonenshain, L.P., M.I. Kuz'min & V.M. Moralev (1976). *Global tectonics, magmatism, and metallogenesis.* Moscow: Nedra. 229 pp.

7

Hydrogeothermics

Hydrogeothermics, the very important branch of hydrogeology, came into being at the interface between hydrogeology and geothermics. It is the science which is concerned with the 'thermal properties of the Earth, the origin of the Earth's heat, and the manner in which it is distributed' (Cheremenskii, 1972, p. 5).

The subject of hydrogeothermics, as a branch of hydrogeology, consists of the thermal properties of the subsurface hydrosphere, and primarily, the water of the interior of the Earth which is the active agent of heat and mass transfer, the heat interaction (heat exchange) with hard rocks which contain it and the changes of both water and the rock as a result of this interaction.

At the same time, as a branch of geothermics, hydrogeothermics must concern itself with the role of the water of the interior of the Earth in the formation and distribution of heat, i.e. the thermal regime of the Earth and the convective transfer of heat.

7.1 The hydrogeothermal regime of the interior of the Earth

The heat regime of the Earth has been and is being formed under the influence of a multitude of sources of heat energy, among which two main groups can be distinguished: external (cosmic), and internal (planetary).

The external (cosmic) sources of heat. First among these is solar radiation. According to M.I. Budyko (1964), taking the reflection into the atmosphere into account, the world as a planet absorbs on average 6.9×10^9 J/cm². year of radiant heat. Of this 35.7% is lost as long-wave radiation, 10.8% disappears into the atmosphere, and 53% is expended on evaporation. The heat balance of the surface of the Earth is positive, i.e. the input exceeds the output. This predetermines the influence of insolation (the irradiation of the Earth's surface by the sun's rays) on the hydrogeological conditions of the upper layers of the crust.

Furthermore, the intensity of insolation has changed and continues to change throughout the geological history of the Earth. This is connected with the variation in climate both at different latitudes, and in certain intervals of time — the climatic cycles and rhythms which have been evaluated for the USSR by A.N. Afanas'ev (1967). The large masses of surface water, in which active heat exchange leads to the cooling of the layers of rock under them to a considerable depth, play a significant role in the redistribution of solar heat. Hypotheses have also been put forward on the effect of the solar heat on the physical and chemical processes which take place in the interior of the Earth (Shvetsov, 1974; and others).

Cosmic rays of various kinds have an effect on the heat regime of the planet, but because this radiation is so small this has so far not been amenable to study.

Internal (planetary) sources of heat. Some hypotheses about the origin of the Earth from the primeval molten material connect the emission of heat in the interior with the cooling of the planet. Another group of hypotheses proposes that the Earth was formed from cold material and subsequently became hot.

Radioactivity, which was discovered in 1896, has played a very important role in the establishment and development of hypotheses about the heating of the Earth. Subsequently the idea was formulated, which held sway for a considerable period of time, that the heat emitted during radioactive decay was the main source of the Earth's heat energy. Calculation of the energy of a radioactive body showed that the subcrustal shell of the Earth contained practically no radioactive elements. Consequently in the oceanic crust (because of the absence of a granite layer) the generation of heat must be smaller than in the continental crust. However, measurements of the actual flow of deep heat through the ocean floor (Bullard, 1954; Smirnov, 1966; and others) showed the continental heat flow and that through the ocean floor to be practically equal. Modern concepts of the heat regime of the lithosphere (Polyak, 1966; and others) are influenced to a far greater extent by the inflow of heat from the interior of the Earth than follows from the radiogenic hypothesis.

The results of calculations of the heat energy emitted during the various processes accompanying the formation of the Earth and its development showed that it is completely comparable with the energy of a radioactive body. Among the sources of deep heat are (Kropotkin, 1948; Lyubimova, 1966):
(1) Gravitational energy:
    (*a*) the strain energy of the compressed planet — 2×10^{31} J,
    (*b*) the energy of gravitational differentiation — $1.5-2.0 \times 10^{31}$ J.
(2) Rotational energy:
    (*a*) the slowing down of the rate of rotation of the Earth — 0.36×10^{31} J,
    (*b*) the variations in the rate of rotation of the Earth — 2.0×10^{31} J.

Taking the energy of a radioactive body ($1.6-2.8 \times 10^{31}$ J), any of these sources can fully meet the present day deep heat losses.

V.V. Kesarev (1967) put forward the hypothesis of the important geothermal role played by physical and chemical processes in the core and the mantle. Considering that all the hydrogen of the Earth was formed as a result of exothermic reactions such as hydrolysis, he calculated the energy of the heat emitted over the whole period since the formation of the Earth to be about 1.6×10^{31} J.

The importance of the role played by geological processes in the energetics of the crust is beyond doubt. For example J. Goguel (1976) has calculated the possible effect of tectonic processes to be 3×10^{31} J/year. But the main significance of tectonic, metamorphic, and magmatic processes is in the changing of the heat energy, the transport of deep heat to the surface and its redistribution between different regions.

Conductive heat transfer. The overwhelming majority of researchers consider it firmly established and a fact which is well supported by innumerable calculations that the chief mechanism for the redistribution of heat in the crust is conduction by hard rocks which transmit the energy directly from particles of high energy (molecules, atoms, or electrons) to those with lower energy.

A quantitative parameter is the coefficient of conductivity (λ), which is numerically equal to the quantity of heat which passes in unit time through a unit area under the temperature gradient of 1 °C per unit length (Cheremenskii, 1972).

According to the experimental data of S.V. Timareva and others (Anon., 1970), the thermal conductivity of rocks in the USSR varies quite considerably: from 0.1×10^{-3} to 26.5×10^{-3} cal/cm.s.deg. In 80% of cases it is $3-7 \times 10^{-3}$ cal/cm.s.deg. The authors note the increased thermal conductivity of igneous rocks in comparison with sedimentary rocks, and that of the more ancient rocks compared with younger.

In practical geothermal research the inverse of the thermal conductivity is used; this is the thermal resistance E (cm.s.deg/cal):

$$E = \frac{1}{\lambda} \tag{7.1}$$

In the layer of the crust which is nearest the surface there are well defined daily and annual periodic temperature variations together with a variation over a period of many years. These are connected with the different intensities of insolation.

The amplitude of the temperature variation decreases sharply with depth. The constant temperature is at a depth of $1-2$ m depending on the winter temperature and the thermal properties of the rocks. The thickness of the constant temperature layer, which has been called the 'neutral layer', varies from 10 to 40 m.

The depth of the neutral layer (H_n) is found approximately from the relationship

$$H_n \approx 19\, H_c \tag{7.2}$$

where H_c is the depth of the constant temperature layer (D'yakonov, 1958).

The influence of exogenic heat sources on the deepest layers of the crust is not in dispute; however, its value, with rare exceptions, is insignificant in comparison with the internal heat of the Earth, even over a period of many years. This is borne out by observations made in various parts of the Earth, which show that below the neutral layer the temperature rises continuously with depth.

The main form of geothermal research consists of temperature observations made in specially prepared boreholes from which geothermograms (see Fig. 7.1) and geothermal sections are prepared. Data from thermometric observations and laboratory determinations of the conductivity of the rocks from samples are used for the calculation of the parameters of the heat flow. In a similar fashion temperature observations are made in boreholes drilled for groundwater, but in these calculations a number of corrections must be made.

Among the parameters which characterize the heat flow from the interior of the Earth are the geothermal gradient, the geothermal degree, and the thermal flux density.

The geothermal gradient determines the rate of increase of temperature with depth (usually for intervals of 1 m, 100 m and 1 km). The geothermal degree is the depth interval across which the temperature varies by 1 °C. Taking the depth of the neutral layer as 25 m, N.M. Frolov (1968) proposed the following formula for determining these parameters:

$$G = \frac{t_2 - (t_0 \pm 0.006\,h)}{h_2 - 25} \qquad (7.3)$$

$$G' = \frac{h_2 - 25}{t_2 - (t_0 \pm 0.006\,h)} \qquad (7.4)$$

where G is the geothermal gradient (°C/m), G' is the geothermal degree (m/°C), t_0 is the temperature of the neutral layer (25 m), t_2 is the measured temperature at a depth h_2, and h is the height above sea level of the point of observation.

For those places where direct observations have been made, the average value of the geothermal gradient is found to be 3 °C/100 m, and that of the geothermal degree is 33 m/°C.

The amount of heat flow, q, is a very important parameter. This can be determined from the formula (Ovchinnikov, 1963):

$$q = \frac{G}{\epsilon} = \frac{1}{\epsilon G'} \; (\text{cal/cm}^2.\text{s}) \qquad (7.5)$$

Figure 7.1. Geothermograms of deep boreholes in the Siberian platform (Lysak, 1968).
1, for the localized marginal areas of artesian basins (deep penetration of infiltration water); 2, central part of an artesian basin; 3, for the fault zone (discharge of thermal water).

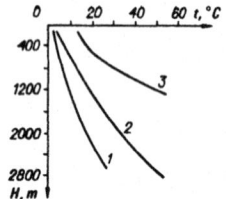

The average value of the heat flow of the Earth has been found to be 1.5 ± 10% cal/cm².s.*

Heat flow is unevenly distributed over the surface of the Earth (Table 7.1). The lower values of the flow are characteristic of the ancient folded regions and crystalline shields, the high values for regions of Alpine folding and especially for volcanically active regions.

Convective heat transfer. This form of heat transfer, the agents of which are groundwater, water–steam mixtures, and magmatic melts, is usually regarded as playing a secondary role. However, in recent years other points of view have appeared. In particular, N.M. Frolov (1976), as a result of analysing experimental data on heat transfer in the lithosphere came to the conclusion that the quantity of heat carried by groundwater is not only comparable with the molecular heat flow, but in some circumstances exceeds it.

N.M. Frolov's arguments are extremely convincing and are supported by data from regional hydrogeothermal research. The processes of convective heat transfer take place most actively in the upper parts of the continental crust (to a depth of 3–5 km), where the groundwater moves as an infiltration current, and water exchange with the surface hydrosphere takes place. In the deeper parts of the crust the dominant form of movement is diffusion and the significance of convective transfer reduces sharply. Only in deep tectonic zones, where H_2O moves as steam, which has a high migratory capability, does the role of convective heat transfer remain quite high.

Thus in the upper layers of the crust the role of groundwater in the redistribution and transport of heat from the interior is important. The first to pay serious attention to the role of groundwater in the redistribution of heat from the interior were N.A. Ogil'vi, A.M. Ovchinnikov, and especially F.A. Makarenko, whose work 'The geothermal conditions of the Caucasian mineralized water region' (Makarenko, 1948) may be regarded as a classic, because it sets out practically all the present day directions of hydrothermal research.

Only recently have the first attempts been made to evaluate the convective heat transfer quantitatively. Such an evaluation was carried out for two kinds of groundwater movement: stratal, in aquifers and ascending, in centres of discharge.

A scheme for the calculation of the possible heat exchange between the flow in the aquifer and the water-bearing rocks (see Fig. 7.2) has been proposed by N.M. Kruglikov (1963):

$$g_w = G_w H V \tan \alpha . c . \rho \qquad (7.6)$$

where g_w is the quantity of heat lost by the water in moving through the stratum, G_w is the geothermal gradient in the limiting conditions of the aquifer, H is the thickness of the aquifer, V is the rate of infiltration taking into account the dip

*The above data characterizes the total heat flow. The magnitude of its conductive component (the conductive heat loss of the Earth), according to Ya.B. Smirnov is 1.18 cal/cm².s.

of the stratum, α is the dip of the stratum, c is the heat capacity of the water, and ρ is the density of the water.

Effective heat exchange between water and the rocks containing it only takes place in conditions of an inclined aquifer. N.A. Ogil'vi (1956) calculated that at rates of infiltration greater than 4 m/year ($\approx 1.3 \times 10^{-7}$ m/s) groundwater plays a decisive role in the redistribution of heat through water-bearing rocks. The same author (1959) proposed a strict mathematical basis for the heat effect when groundwater moves vertically (ascending or descending):

$$G_H = G_{H=0}\,l\,\frac{\pm\,V \cdot \rho \cdot c}{\lambda} \qquad (7.7)$$

where G_H is the geothermal gradient at a depth H, $G_{H=0}$ is the initial gradient at $H = 0$, V is the rate of infiltration (for both ascending and descending water), λ is the coefficient of thermal conductivity of the rock in which the movement takes place.

The basis of the solution of the problem of the convective transfer of heat has been given by J. Goguel (1976). In his opinion convection predominates over conduction only in the presence of a temperature gradient in the direction of the water flow. His main conclusion (Goguel, 1976) was that if the difference in density and temperature of the water in the hard rocks is the cause of the circulation, the heat transfer in them occurs as a result of convection.

The transfer of heat by groundwater. The movement of groundwater, together with the redistribution of water horizontally and vertically, leads to the discharge of considerable quantities of water on to the surface of the Earth. The penetration of meteoric water, which is usually at a low temperature, into the interior of the Earth causes a gradual rise in the temperature of this water. With a rising current of groundwater, and also when it moves along a stratum, part of the heat is lost in heat exchange between the water and the rocks con-

Table 7.1. *Hot currents in different parts of the world (Kutas, 1978; Ovchinnikov, 1963)*

Structural elements	Region	Value of the heat flow (cal/cm^2.s)		
		Minimum	Maximum	Average
Shields	Baltic	0.6	1.3	0.9
	Ukrainian	0.6	1.1	0.9
	South African	0.7	1.3	1.0
	Russian	0.7	1.7	1.1
Kratons	Siberian	0.7	1.6	1.0
	North American	0.9	1.7	1.1
Palaeozoic fold regions	Altai–Sayan	0.9	1.2	1.1
	Western Siberian	1.2	1.6	1.4
	North Ural	1.1	1.7	1.5
	Appalachian	0.7	1.4	1.1
Alpine fold regions	Caucasus	1.4	2.8	1.9
	Alps	1.3	2.1	1.8
	Cordillera	0.7	2.2	1.7
Rifts	Rhine Graben	1.7	3.4	2.3
	Baikal rift	1.2	3.4	2.2
Oceans	Eastern part of the Atlantic Ocean	–	–	0.7
	Pacific Ocean	–	–	1.1
Inland seas	Black Sea	–	–	0.85
	Southern Caspian basin	0.9	1.2	–

Figure 7.2. Diagram of the calculation of the theoretical heat exchange in an aquifer (Kruglikov, 1963).
1, aquifer of thickness T (m); 2, impermeable (clay) beds.
$q_{cond.}$, heat-conducting current; $q_{conv.}$, amount of heat carried by the water; V, velocity of infiltration; α, dip of beds.

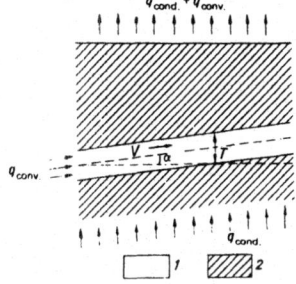

taining it. The physical, chemical, and biochemical reactions, and other processes in the subsurface hydrosphere are accompanied by heating or cooling effects. Hence groundwater is discharged at the surface with so-called excess (in comparison with the temperature in the neutral layer) temperature which enables the magnitude of the convected heat loss, i.e. the removal of heat by groundwater, to be calculated.

An attempt to determine the heat lost from the Earth as a result of water circulation was carried out by E.N. Lyustikh (1959): assuming an excess temperature of 1 °C and a discharge of two-thirds of the quantity of the meteoric precipitation which falls, he very tentatively evaluated it as close to that caused by the conductive heat loss.

F.A. Makarenko and others (Anon., 1970), on the basis of data on the magnitude of the underground recharge of the rivers of the USSR, and the same value of excess temperature of 1 °C, determined a tentative value for the specific heat transfer by groundwater as 0.16 μcal/cm^2.s, i.e. one order of magnitude less than that of the conductive heat transfer.

A more accurate solution to this problem, but still depending on the accuracy of the value of the excess temperature, was given by N.M. Frolov (1976). The formula he proposed is based on the value of the modulus of the underground runoff m and the magnitude of the geothermal gradient G for any particular conditions

$$g = 0.3 \, G(h - 2500) \, mc_0 \qquad (7.8)$$

where h is the thickness of the zone of underground runoff, 2500 is a correction term for the zone above the netural zone and which has no gradient, c_0 is the specific heat capacity of water.

This formula served as the basis for the calculation to determine the transfer of heat by groundwater in the USSR. This turned out to be 0.24 μcal/cm^2.s. Obviously this figure includes both the internal convective heat transfer and the heat received as a result of insolation, and, thus, neither refutes nor makes more precise the earlier value of 0.16 μcal/cm^2.s, although N.M. Frolov claims his value to be the minimum transfer of heat by groundwater.

The available, very tentative data on convective heat loss for some regions of the USSR and some other countries (Table 7.2) show that the average transfer of heat by groundwater for large regions varies from 0.09 to 0.25 μcal/cm^2.s, this confirms the validity of the average figure for the USSR. Higher values of this parameter are characteristic for hot spring systems, and local hot water discharge points (1.3–3.7 μcal/cm^2.s). There is doubt only about the high value of convective heat loss for the Crimea (0.6 μcal/cm^2.s), which is comparable with that of Iceland (1.0 μcal/cm^2.s).

There are hot spring systems in the Mongolian People's Republic (Khangai and Kentei-Daursk – the latter is also in Soviet territory), for which the scale of heat transfer by groundwater is comparable with the transfer in the thermally active regions of the Earth. In this respect the Khangai neotectonic uplift, in which all the major hot springs are confined to the watershed regions and slopes of the mountain ranges (which are higher than the drainage base of local streams) and which have altitudes of 1600–3000 m, is especially significant.

Table 7.2. *Heat output from groundwater for some regions of the USSR, Mongolian People's Republic, Iceland, and New Zealand*

Region	Groundwater output (l/s)	Temperature above ambient (°C)	Specific convective heat loss (μcal/cm^2.s)
Russian platform	no data	no data	0.25
Siberian platform	no data	no data	0.12
Kishinev region	no data	no data	0.28
Aral region	$\approx$ 4600	10	0.04
Aral region (local discharge points)	$\approx$ 3460	10	1.20
Caucasus region (thermal springs)	no data	no data	0.10
Georgia, USSR	no data	>5	0.06
Kamchatka (eastern volcanic zone)	no data	no data	0.55
Sochi–Matsestinskii artesian basin (local discharge points)	$\approx$ 12	1	3.50
Western Turkmen artesian basin	$\approx$ 130	34	0.09
Basin of the River Don	no data	1	0.17
Crimea	no data	no data	0.60
Average for the USSR	3445 $\times$ 10^3	1	0.16
MPR (Khangai mountain system – hot springs)	$\approx$ 120	59	0.25
MPR Khangai (local discharge centres of hot springs – Shargalzhut)	51	68	1.30
Iceland (hot spring region)	no data	150	1.00
Iceland (Torfajökull hydrothermal systems)	no data	no data	3.70
New Zealand (North Island region)	no data	no data	1.10

Data taken from: Anon., 1970; Anon., 1974; Frolov, 1976; Anon., 1977; Pisarskii & Shpeizer, 1979.

If the total area of recharge is taken into consideration, then the magnitude of the transfer of heat by groundwater in the Khangai hot spring system is 0.25 μcal/cm^2.s. The most active discharge centre of this system is the Shargalzhut hot spring group which is situated in the south at an altitude of 2100 m. Here, in an area of 0.25 km^2 there are more than 80 springs with temperatures of more than 40 °C (Fig. 7.3). Some are of the geyser type of outflow and have a temperature of 88–92 °C. With a total flow of 51 l/s and an average excess temperature of 68 °C (obviously understated) the total convective heat loss amounts to about 3600 kcal/s, and the specific transfer of heat by hot water, referred to the area of recharge (provisionally taken as the boundary of the watershed), is calculated to be 1.3 μcal/cm^2.s.

All the data on the magnitude of the convective heat transfer still suffer from the common defect that there are insufficiently accurate methods for the determination of the areas of 'heat output'. In many regions of the world researchers confine themselves to an estimate of the magnitude of the total convective heat transfer, not citing the specific value for the different areas and relying on high average values.

The authors of calculations on the quantity of the convective heat transfer proceed in roughly the same way in order to ensure the reliability of the results obtained. This naturally makes it difficult to determine the boundaries of the actual thermally active areas and to compare different regions and local discharge centres one with another. Therefore the working out of methods for the determination of the areas over which heat is transferred by groundwater is one of the most important problems of hydrogeothermics.

In evaluating the transfer of heat by groundwater note must also be taken of forced convection, i.e. the additional entry of heat when deep thermal water is met in boreholes. In the total heat flow from the interior forced convection plays a very insignificant role, but in local discharge centres the heat loss because of this can exceed the natural heat loss by a factor of two or more.

The steam springs of Kamchatka may be quoted as an example (Anon., 1972). Thus in the Bol'shoe Bannoe springs the total heat transfer out of experimental boreholes is more than 2.5 times greater than the natural convective heat transfer. In the Pauzhetka springs this ratio is 2.2.

The redistribution of heat in the crust by groundwater. As has already been noted, the role of groundwater in the transport of heat within the crust is determined by the mobility and the physical state of the water. These two basic parameters of the convective heat transfer in the crust vary with depth.

In the vertical section of the subsurface hydrosphere there are thermodynamic boundaries which divide it into three zones: 'solid' water, saturated ('liquid' water), and water in the supercritical state (see Chapters 2 and 3).

In the first zone, where voids are mainly filled with ice, the movement of water is impeded. And although in frozen rocks there is water in the form of pellicular water, and liquid which moves under gravity (Tyutyunov, 1961), its role in the transport of heat is extremely insignificant. Supercooled water, held in the liquid state between frozen layers, because of its high degree of mineralization and lowered freezing point, plays a much greater role in the convective transfer of heat.

Figure 7.3. Diagram of temperature distribution and discharge centres of thermal waters of the deposits in the Shargalzhut area, Mongolian People's Republic (after B.I. Pisarskii).
Areas of water temperature of springs (°C): I, 89–92; 2, 81–88; 3, 71–80; 4, 61–70; 5, 41–60; 6, lines of tectonic disturbance which conduct hot water to the surface; 7, major fault; 8, locations of hot springs.

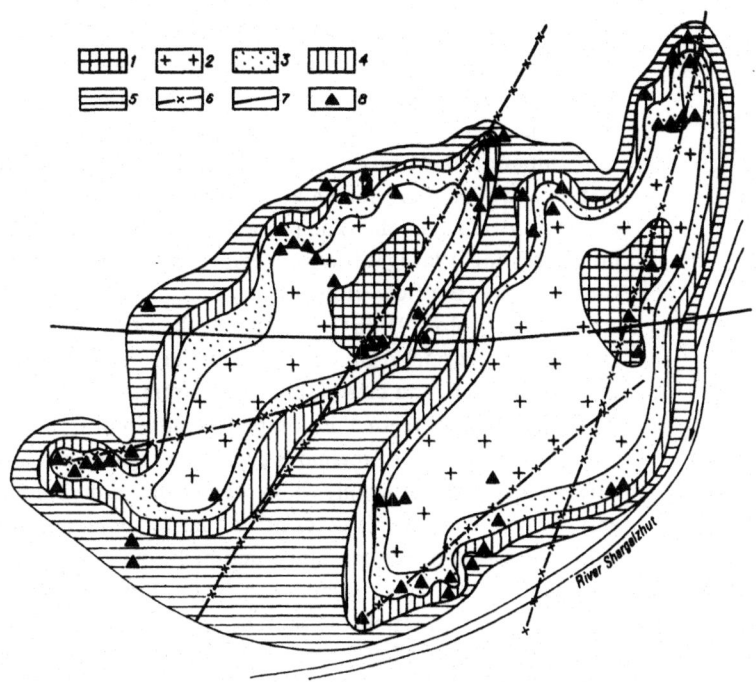

In the zone of saturation the redistribution of heat by moving groundwater plays a decisive role in the formation of its geothermal regime. In the upper parts of the zone the main parameters are the quantity of moving water and its velocity. This is shown to good effect in the vertical section of the Nal'chikskii artesian basin (Fig. 7.4) where the geothermal configuration changes with the water velocity. In the deeper layers of the crust which are characterized by an impeded or very impeded water exchange, the role of convective heat transfer falls, but because of the high excess temperatures, remains quite significant.

Regions of active neotectonic activity and volcanically active regions, where the great velocity of infiltration is combined with high excess temperatures, are of especial interest. Here conditions are often created in which steam and water mixtures can exist and which because of their physical composition must be assigned to a third zone.

Water in the supercritical state is characterized by a temperature that is greater than the critical temperature for the given thermodynamic conditions (for pure water, more than 374 °C). This zone occurs at a considerable depth where in conditions of high pressure the boiling point (the transition to the vapour state) of mineral aqueous solutions can vary considerably and reach 450 °C. Here the convective and diffusive heat exchange, the quantitative relationships of which exist as yet only as hypotheses, are combined. It is possible only to assume that convective heat transfer in the third zone plays a significant role.

The heliothermozone and the geothermozone, with regard to the origin of heat in the crust, differ in depth. The heliothermozone is the upper layer of the crust in which the hydrogeothermal regime is determined by the action of insolation. The geothermozone includes the lower layers of the crust and the upper mantle where the temperature regime of the subsurface hydrosphere depends on internal sources of heat.

Figure 7.4. Hydrothermal cross-section of the southern part of the Nal:chikskii artesian basin (Ovchinnikova, 1964).

Water-bearing deposits (succession of limestones, sandstones, and marls): 1, Upper Chalk; 2, Albian; 3, Aptian; 4, Barremian–Hauterivian; 5, cover of semi-permeable Paleocene–Neocene rocks (aleurites, clays, conglomerates).
Underlying formation of semi-permeable rocks (marls and aleurites): 6, Valanginian; 7, Jurassic.
Other symbols: 8, direction of movement of thermal pressure water; 9, isotherms (°C).

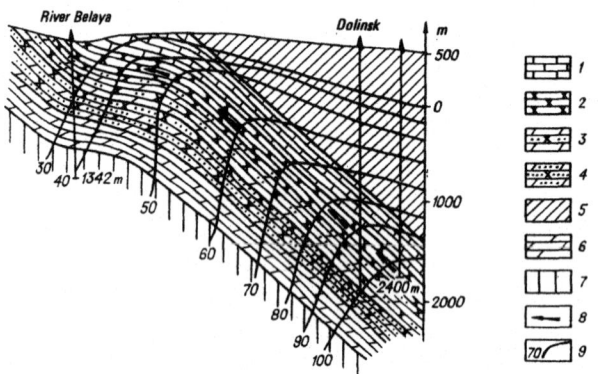

Many researchers (A.M. Ovchinnikov, P.P. Klimentov, and others) take the neutral layer as the lower boundary of the heliothermozone. N.M. Frolov has another point of view: in his opinion the sun's influence is felt throughout the whole crust. He assigns the lower shells of the Earth, beginning with the mantle, to the geothermozone. It is difficult to agree with this, the more so because data on the dependence of the temperature on the climatic zonation regionally are only available down to a depth of 3 km (Frolov, 1976).

It is obviously more correct to divide the section into three zones vertically: the *heliothermozone*, the *heliogeothermozone*, and the *geothermozone*. The depth of the first zone is determined by the influence of the insolation and is bounded by the neutral layer. The heliogeothermozone is situated between the neutral layer and the boundary of the zone in which the temperature varies over a period of centuries under the influence of the climate, and reaches a depth of 1–3 km. Its hydrogeothermal regime is influenced by both the sun's heat and that of the Earth's interior. The geothermozone is located lower than this and here the influence of the interior heat predominates.

Types of hydrogeothermal regime of the subsurface hydrosphere. A.M. Ovchinnikov (1946) considered that in that part of the crust which is accessible to study in the present era, it is more correct to talk of its hydrothermal or hydrogeothermal regime and not the geothermal regime. Emphasizing the leading role of groundwater in determining the Earth's heat field, he proposed that the hydrogeothermal regime be divided into three: the *normal*, the *magmatogenic*, and the *cryogenic*. A normal regime is characteristic of territories where the heat balance of the subsurface hydrosphere is disturbed neither by deep layers of permafrost, nor by the intrusion of magmatic materials into the upper layers of the crust. The Russian platform is an example of a normal regime.

A magmatogenic regime is formed in areas of active or recently active volcanoes close to the surface of the Earth. In such places the heat current forms active thermal anomalies.

The vast areas of permafrost are assigned to the cryogenic regime. Here in the upper layers of the geological section the temperature is below zero to depths of several hundred metres and even more than a kilometre, and the temperature of the neutral zone falls. The presence of a cold zone also influences the hydrogeothermal regime of the deep layers of the crust, and the flow of heat.

N.M. Frolov (1976) also distinguished three main types of hydrogeothermal regime in the upper layers of the Earth: open, combined, and closed, but his interpretation of the term 'hydrogeothermal regime' is completely different.

An open type of regime is typical of regions where at the surface there are permeable rocks which form a draining layer, and the heat exchange is determined mainly by convection (the infiltration of water). The combined type of regime can be found in those places which consist of alternating layers of permeable and impermeable rock beginning with an impermeable layer more than 50 m in thickness. Here heat transfer takes place both as result of molecular heat exchange, and by convection. And, finally, the closed type of regime is found in regions in which thick layers of impermeable rocks occur and in which heat transfer takes place by conduction. Permafrost regions are also of this type.

Such a classification scheme is, no doubt, of considerable practical significance for the upper parts of the subsurface hydrosphere itself in which intensive water exchange takes place, but does not characterize the hydrogeothermal regime of its deeper parts.

7.2   The laws governing the distribution of hot springs

It has already been pointed out that the main agents of heat transfer in the subsurface hydrosphere are water and steam, the temperature of which varies widely (up to 450 °C for the liquid phase and 700 °C for the gas phase).

Strictly speaking, subsurface water and vapour which are heated by the internal energy of the Earth should be classified as hot springs, i.e. those with temperatures greater than that of the neutral layer (Makarenko, 1961), or the average temperature of the air for the given locality (Ovchinnikov, 1963). However, in a large number of classifications, and in practice, a temperature of 20 °C is taken as the boundary between cold water and hot springs. B.F. Mavritskii (1971) bases this boundary not only on external factors (the perception of heat by the human body), but also on the properties of water itself, the most important of which is the viscosity, the unit of which (the centipoise) is determined at 20 °C exactly.

Depending on the physical composition under different combinations of temperature and pressure hot springs may be divided into: (1) thermal waters (the liquid phase is totally dominant); (2) steam springs (steam and water mixtures); (3) hot steam.

Hot springs appear on the surface as hot and 'boiling' springs and are also found at different depths (down to 10 km) in boreholes, wells, and mineshafts. They form a compact layer in the subsurface hydrosphere.

According to present day concepts, the bulk of hot water and water–steam springs are formed at comparatively great depths with the participation of meteoric water (see Section 3.2 and Fig. 6.19), and consequently, as a result of ascending infiltration, they overflow on to the surface or form stratal accumulations in the permeable horizons of sedimentary rocks. The systems of fracture zones in the crystalline rocks are the main channels for the ascending flow to hot springs. The density of such fissures, the depth to which they penetrate into the interior, the thickness of the covering layers of sedimentary rocks, and many other factors predetermine the different conditions of distribution and formation of the composition and resources of hot springs in the subsurface hydrosphere. Hence researchers (Dzens-Litovskii & Tolstikhin, 1940; Makarenko, 1961; Ivanov, 1961; Mavritskii, 1971; Baskov & Surikov, 1975; White, 1968; and others) approach the evaluation of the laws of the regional distribution of hot springs from different points of view (geological–structural, hydrogeochemical, hydrogeodynamic, geothermal, etc.). The most logical is the geological–structural principle which takes into account the intensity of the tectono–magmatic activity,

and to the development of which the ideas of F.A. Makarenko and B.F. Mavritskii have made a great contribution. On this basis are distinguished the folded and the platform regions which differ considerably in the conditions in which hot springs are distributed, are formed, and discharge.

The hot springs of folded regions. In folded regions, hot springs are confined to regions of intensive Cenozoic volcanic activity and current tectonic movement, where a large number of thermal springs and water–steam springs arise on the surface of the Earth, which are associated with zones of large tectonic faults and especially with the junctions of intersecting faults. The distribution of outlets on the surface is either linear (hydrothermal lines according to N.I. Tolstikhin), or areal (outlets are found in all geostructural elements). This is mainly vein water with a degree of mineralization less than 1 and rarely as much as 20 g/l (Chukotka). In some circumstances the mineralization reaches 40–50 g/l and even 100 g/l; such springs are found in New Zealand (Wilson, 1953).

In very active Cenozoic volcanic regions hydrothermal activity is intense. Hot springs and water–steam springs, the temperature of which in natural outlets exceeds 100 °C, and in boreholes of depths of 2–3 km reaches 270–390 °C (California, Mexico, New Zealand), are found in formations of strongly dislocated volcanogenic and volcanic-sedimentary rocks near active volcanoes and form large numbers of water–steam spring systems. The chemical composition of the water is extremely varied and shows increased concentrations of the most varied microcomponents. The gaseous composition is also very varied, but is mostly hydrogen sulphide – carbon dioxide, carbon dioxide, more rarely nitrogen-carbon dioxide. A more detailed description of hot springs in the volcanically active regions was given in Section 6.8.

In regions of intense current tectonic movements, which are of wide occurrence in folded regions, hot springs occur in the mountain massifs which are mainly composed of crystalline rocks, and in intermontane valleys which are filled mainly with Cenozoic deposits. Here also centres of recent volcanic activity are to be found. Intense hydrothermal activity is usually observed in places with a high degree of seismicity.

In fold mountain structures, which form fissure-water massifs, hot springs appear on the surface along zones of seismically active and exposed faults at the most varied altitudes, and are mainly of the fissure-vein type. They occur mainly in intrusive granitic massifs. The occurrence of hot springs has a mainly areal character. This distribution is most clearly seen in the Mongolian People's Republic where hot spring outlets are clearly confined to the limits of the Mongolo–Altai and Khentei–Daursk neotectonic uplifts (these are within the USSR) (Pisarskii, 1977).

The hot springs of mountain structures are nitrogen, hydrogen sulphide-nitrogen, and carbon dioxide types. The temperatures of such springs usually vary from 20 to 90 °C and at high altitudes are close to the boiling point. There are considerable differences in the degree of mineralization and composition. Hot springs with a low degree of mineralization (up to 1 g/l) are usually carbonate-hydrocarbonate and sulphate-sodium; with a degree of mineralization of 1–5 g/l they are more often sulphate- and chloride-sulphate-sodium, and at high values (15–27 g/l) they are chloride-sodium and calcium-sodium.

In the intermontane valleys of fold mountain regions thermal water enters the sedimentary layers along faults in the basement. When there are layers of permeable sedimentary rocks they form aquifers and complexes containing poro-stratal and fissure-stratal water; drilling has met these at considerable depths (1000–4000 m) and their temperatures are high (in places up to 100 °C and more). Water in deeper layers contains methane, and is mainly chloride-sodium, more rarely calcium-sodium. Usually the water is slightly saline, saline, or consists of brines with a degree of mineralization from 2–3 (in the Lake Baikal depression) up to 100 g/l. In the Western Turkmen (Trans-Caspian) depression there are thermal waters with a degree of mineralization of up to 320 g/l (Mavritskii, 1971).

In many intermontane depressions the sedimentary formations consist of semi-permeable rocks, and there are large fault zones in the margins where the sedimentary or sedimentary-volcanic deposits are in contact with the crystalline rocks surrounding the depressions. The exits of hot water in the margins of the depressions form linear patterns on the ground. Such hydrothermal lines are found in the Baikal rift zone, over the whole of the belt of Alpine folding (Barabanov & Disler, 1968), and in other regions. These hydrotherms are characteristically only mineralized to a small degree (up to 1 g/l, rarely more), are hydrocarbonate or sulphate-sodium in composition, and the predominant gas is nitrogen.

Hot springs of the platform regions. In the platform regions which occupy a great area of the Earth's surface, and which are characterized by a considerable depth of crystalline basement (the thickness of the sedimentary cover exceeds 5 km), considerable resources of hot water are concentrated, of which the temperature, composition, and degree of mineralization vary greatly.

Platform regions are usually subdivided according to the conditions under which hot springs occur (Mavritskii & Antonenko, 1967; and others) into: (1) epi-Palaeozoic plates and local downwarpings, and (2) ancient (Palaeozoic) platforms.

In epi-Palaeozoic plates basins and local downwarpings there are considerable resources of hot water which are heated up to 100–150 °C in the deep formations. Hot springs with temperatures of 70–100 °C are found in a large number of boreholes at depths of 1500–2500 m and, as a rule, water gushes from them. The flow from boreholes is quite high (up to 8000 m^3/day). The water is fissure-stratal or poro-stratal according to the conditions in which it has accumulated. In these basins there are very often aquifers of hot water whose temperatures increase with depth. The degree of mineralization and composition of the hot water vary both with depth and with the area of the basin. In some localized areas the water is fresh or slightly saline with a degree of mineralization of 2–3 g/l. It is hydrocarbonate-sodium in composition, more

rarely sulphate-sodium, and contains nitrogen. In central regions of the basin saline water and brines with a degree of mineralization of 100–320 g/l are found at considerable depths; they are chloride-sodium and calcium-sodium in composition, and the predominant gas is methane.

Hot springs are best represented in the Western Siberian basin, where they are found in the Jurassic and Cretaceous formations. The water is sodium, with a high concentration of iodine and bromine. The temperature varies with depth from 80–96 °C (at a depth of 1600–2800 m) to 30–40 °C (at 800–1000 m), the degree of mineralization is 45 and 13–16 g/l respectively, and the composition of the gases varies between these depths from methane to nitrogen respectively.

Ancient platforms (the Russian, Siberian, North American, etc.) are heated to a considerably less extent than the epi-Palaeozoic. In these regions brines with temperatures of 40–70 °C are of widespread occurrence and are found at considerable depths by drilling. There is some flow from these wells but it is small. The waters are usually sodium-chloride and sodium–calcium brines with a degree of mineralization of 200–400 g/l (the Russian and North American platforms) and up to 620 g/l (the Western Siberian platform).

References

Afanas'ev, A.N. (1967). *Variations of the hydrometeorological regime of the USSR.* Moscow: Nauka. 231 pp.

Anon. (1970). *The thermal regime of the USSR.* Moscow: Nauka. 171 pp.

Anon. (1972). *Hydrogeology of the USSR. Vol. 26, Kamchatka, the Kuriles, and the Komandorskie Islands.* Moscow: Nedra. 364 pp.

Anon. (1974). *Deep heat flow of the European part of the USSR.* Kiev: Naukova dumka. 191 pp.

Anon. (1977). *Iceland and the mid-oceanic ridge: deep structure, seismicity, and geothermics.* Moscow: Nauka. 195 pp.

Barabanov, L.N. & V.N. Disler (1968). *Nitrogen hot springs of the USSR.* Moscow. 120 pp.

Baskov, E.A. & S.N. Surikov (1975). *The hot springs of the Pacific area.* Moscow: Nedra. 172 pp.

Budyko, M.I. (1964). Heat balance of the Earth. *Geophysical Bulletin.* No. 14, 39–44.

Bullard, E.C. (1954). The flow of heat through the floor of the Atlantic Ocean. *Proc. Roy. Soc., London. A.*, 222 (1150), 408–29.

Cheremenskii, G.A. (1972). *Geothermics.* Moscow: Nedra. 272 pp.

D'yakonov, D.I. (1958). *Geothermics in petroleum geology.* Moscow: Gostoptekhizdat. 277 pp.

Dzens-Litovskii, A.I. & N.I. Tolstikhin (1940). Mineral waters of Northern Asia in connection with the geological structure. *Proceedings 17th International Geological Congress,* 5, 163–81.

Frolov, N.M. (1976). *The heliothermozone.* Moscow: Nedra. 156 pp.

Frolov, N.M. (1968). *Hydrogeothermics,* 1st edn. Moscow: Nedra. 314 pp.

Goguel, J. (1976). *Geothermics.* New York: McGraw-Hill. 200 pp.

Ivanov, V.V. (1961). Steam and water springs of the Kurile–Kamchatka volcanic zone. In *Problems of geothermics and exploitation of the Earth's heat,* Vol. 2, pp. 43–65. Moscow: Izd-vo AN SSSR.

Kesarev, V.V. (1967). *The moving forces in the development of the Earth and planets.* Leningrad: Nedra. 152 pp.

Kropotkin, P.N. (1948). Fundamental problems of the energetics of tectonic processes. *Izv. AN SSSR, Ser. geol,* No. 5, 89–104.

Kruglikov, N.M. (1963). The problem of the geothermal role of ground-water movement. *Trudy VNIGRI,* No. 220, 260–72.

Kutas, R.I. (1978). *The field of heat currents and a thermal model of the Earth's crust.* Kiev: Naukova dumka. 145 pp.

Lysak, S.V. (1968). *Geothermal conditions and the hot springs of the south of Eastern Siberia.* Moscow: Nauka. 119 pp.

Lyubimova, E.A. (1966). The sources of the deep heat of the Earth and the thermal properties of an Earth-type planet. In *Problems of deep heat flow*, pp. 3–30. Moscow: Nauka.

Lyustikh, E.N. (1959). The role of volcanoes and hot springs in the transfer of heat from the Earth's interior. In *Problems of geothermics and the exploitation of the Earth's heat*, Vol. 1, pp. 31–6. Moscow: Izd-vo AN SSSR.

Makarenko, F.A. (1948). The geothermal conditions of the Caucasian mineralized water region. *Trudy Laboratorii gidrogeol. problem AN SSSR, Moscow*, 1, 171–211.

Makarenko, F.A. (1961). General features of the formation of hot springs and their occurrence in the USSR. In *Problems of geothermics and the exploitation of the Earth's heat*, Vol. 2, pp. 3–20. Moscow: Izd-vo AN SSSR.

Mavritskii, B.F. (1971). *The hot springs of the folded and platform regions of the USSR*. Moscow: Nauka. 242 pp.

Mavritskii, B.F. & G.K. Antonenko (1967). *The prospecting for, investigation of, and practical application of, hot springs in the USSR and abroad*. Moscow: Nedra. 176 pp.

Ogil'vi, N.A. (1956). The geothermal field as a factor in the formation of groundwater. In *Proceedings of First All-Union Congress on Geothermal Research in the USSR*, pp. 19–23. Moscow.

Ogil'vi, N.A. (1959). Geothermal field theory as a supplement to geothermal methods of prospecting for groundwater. In *Problems of geothermics and the exploitation of the Earth's heat*, pp. 53–85. Moscow.

Ovchinnikov, A.M. (1946). The geothermal regime of the Earth's crust. *Dokl. AN SSSR, Nov. seriya*, 33 (7), 649–52.

Ovchinnikov, A.M. (1963). *Mineral waters*, 2nd edn. Moscow: Gosgeoltekhizdat. 376 pp.

Ovchinnikova, L.K. (1964). Nal'chikskii artesian basin and the prospects of using its hot springs. In *The hydrogeothermal conditions of the upper parts of the Earth's crust*, pp. 138–41. Moscow.

Pisarskii, B.I. (1977). The geological basis for the formation of minerals in the Mongolian People's Republic. In *Natural conditions and resources in some regions of the MPR*, pp. 48–9. Irkutsk.

Pisarskii, B.I. & Shpeizer, G.M. (1979). Hot springs of the MPR. In *Proceedings of the All-Union Conference on the Waters of the Eastern Regions of the USSR, Irkutsk–Petropavlovsk–Kamchatskii*. 151 pp.

Polyak, B.G. (1966). *The geothermal features of active volcanic regions*. Moscow: Nauka. 180 pp.

Shvetsov, P.F. (1974). *Geothermal conditions of Mesozoic–Cenozoic oil-bearing basins*. Moscow: Nauka. 131 pp.

Smirnov, Ya.B. (1966). Submarine heat flow. *Dokl. AN SSSR*, 168 (2), 428–31.

Tyutyunov, I.A. (1961). The physical and chemical changes of hard rock in the extreme North. In *Physical and chemical processes in frozen rocks*, pp. 7–27. Moscow: Izd-vo AN SSSR.

White, D.E. (1968). Thermal and mineral waters of the United States. Brief review of possible origins. *International Geological Congress, Report of Twenty-third Session, Prague*, Vol. 19, pp. 269–86.

Wilson, S.H. (1953). The chemical investigation of the hot springs of New Zealand thermal region. *Seventh Pacific Sci. Cong., New Zealand, 1949*, Vol. 2, Geology, 449 pp.

8

Regional hydrogeological laws

The task of regional hydrogeology is to study the groundwater of some definite region. Regional hydrogeological research is concerned primarily with establishing the laws governing the extent of the groundwater both vertically and horizontally. These laws describe the hydrogeological conditions of the territory under study, and hence are the basis of regional descriptions. From their great variety we will limit ourselves to the most important — to information on hydrogeological zonation and a short description or characterization of the distribution of the main groundwater reservoirs of the Earth.

8.1 Hydrogeological zonation

In any system of zonation territorial units are established which in some way or other differ from their neighbours. Zonation (the division of the whole into parts) is a research method which is used in all geological–geographical sciences. The process consists of: (1) the choice of the parameters by which the territory is to be divided, and (2) the mapping of the boundaries of these parameters.

In so far as hydrogeology is concerned basically with groundwater, the task of hydrogeological zonation consists of the systematization of the laws governing the spatial distribution and formation of groundwater. This is achieved by summarizing the regional data on groundwater, putting it in order, and mapping the hydrogeologically different regions.

A region is a name which summarizes the taxonomic units of zonation irrespective of rank. It is borrowed from physical geography (Mil'kov, 1970) and is equally applicable to various areas of territory from large to small. That part of the Earth's surface and interior which is distinguished by a unity of the parameters of occurrence, formation, and in the final analysis the exploitation of the groundwater, is necessarily described as a *hydrogeological region*.

8.1.1 The principles and types of hydrogeological zonation

The basis of the zonation of a territory is its natural features which are examined from the position of the historical development of, and the action on, the groundwater environment. The purpose of hydrogeological zonation is the isolating of territorial units which differ from each other in hydrogeological features and parameters. The natural-historical approach must be regarded as an important principle of hydrogeological zonation and a guarantee of objectivity.

A whole series of natural factors are used as classification parameters in hydrogeological zonation, the most important of which are: (1) the physical-geographical features (climate, relief, hydrology); (2) the geological aspects (lithology, structure, volcanic activity, etc.); (3) the hydrogeological conditions. The concept of 'hydrogeological conditions' serves to describe the diverse aspects of the life of groundwater beginning with its appearance in the crust and ending with its use in the various branches of the economy.

The laying down of the principles of hydrogeological zonation has received attention from many of the leading hydrogeologists of the USSR (V.S. Il'in, M.M. Vasil'evskii, B.K. Terletskii, B.L. Lichkov, F.P. Savarenskii, G.N. Kamenskii, N.I. Tolstikhin, O.K. Lange, I.K. Zaitsev, A.M. Ovchinnikov, F.A. Makarenko, B.I. Kudelin, and others). Particularly in the problem of the procedures of zonation there are, together with considerable achievements, some controversial viewpoints.

Groundwater or territory can be divided into zones according to the most varied features and combinations of features. The use of one or other set of parameters for the division of a territory into hydrogeological regions depends upon the scale of the map and the purpose for which the zonation is being carried out, and also on the methodology which the tasks of hydrogeological research require.

Zonation is not an end in itself. The need for hydrogeological zonation arises as a result of both scientific and practical demands. Hydrogeological zonation is stimulated on the one hand by a comprehensive knowledge of the hydrogeological laws of the territory, and on the other hand by the variety of the problems which arise from the exploitation of the groundwater.

There is a marked tendency to distinguish several types of hydrogeological zonation. The scale of the zonation is determined by the scale of the hydrogeological map. It is true that these conditions are not always observed. In the general case in regional hydrogeological research two types of zonation are distinguished – *review* and *areal*. Such a division is provisional, but simplifies the solution of the scientific and practical tasks.

Review zonation (small scale 1 : 1 000 000) is carried out on the most general aspects of classification, distinguishing large territorial units and reflecting only the essentials of the problem. It is shown on review hydrogeological maps. Detailed and more complete examinations of small territories are carried out using areal zonation (1 : 1 000 000 and larger), when hydrogeological surveys are made. The work is based not only on qualitative but also on quantitative groundwater parameters. The detail of the zonation and the nature of the recommendations for the use of the groundwater increase with increase in the scale used. The extent to which the groundwater has been studied is of no small significance. This to a certain extent determines the scale of the zonation: the less it has been studied, the smaller the scale, and vice versa. However, in the zonation of any particular territory the degree to which it has already been studied from the hydrogeological point of view influences the choice of classification parameters.

If the area has been poorly studied hydrogeologically then zonation must be based mainly on indirect parameters (physical-geographical and geological). Parameters which characterize the distribution and formation of groundwater directly are of little use in such a situation. As knowledge of the groundwater increases, or when a zonation exercise has been carried out on a territory which has been relatively well studied, direct parameters and factors which come into the concept of the 'hydrogeological conditions', play an ever increasing role. The hydrogeological regions thus distinguished receive not only a qualitative but a quantitative appraisal and it becomes possible to study the features and properties of the water more fully.

The methods of zonation by using direct and indirect parameters supplement one another. Even in comparatively well studied areas indirect parameters (for example geological–structural, lithological, permafrost, geomorphological, and other factors) are nevertheless taken into consideration, although they do not have such great importance. They are also indispensable in a detailed zonation.

An examination of the various schemes of hydrogeological zonation of the USSR illustrates well the growing role of direct parameters in comparison with indirect parameters as the amount of hydrogeological study carried out increases. The first hydrogeological zonation of the USSR which was carried out under the direction of M.M. Vasil'evskii reflected the state of knowledge at the end of the 1930s. The geological–structural factor was mainly used. Considerable progress in respect of the bulk of the calculation of the direct parameters can be seen in the scheme of G.N. Kamenskii *et al.* (1959), and also in the hydrogeological maps of the USSR to scale 1 : 2 500 000, which were published under the editorships of I.K. Zaitsev in 1959 and N.A. Marinov in 1964. When the degree to which the groundwater had been studied had increased, the multivolume publication *The hydrogeology of the USSR* was published at the end of the 1960s and at the beginning of the 1970s. The volumes and maps contain appendices of schemes of hydrogeological zonation which take account of the direct parameters.

There are two approaches to hydrogeological zonation which differ in principle. B.L. Lichkov, A.N. Semikhatov, O.K. Lange, and others have produced zonation schemes for subsurface and artesian water separately. 'If in the life of subsurface water', wrote O.K. Lange (1960, p. 15), 'a leading role is played by such factors as climate, relief, and lithological composition of the water-containing hard rocks, then for artesian water the influence of climate and geomorphological conditions is reduced, but then geostructural features become of great significance in its distribution in the crust.' We see another different approach by the geologists M.M. Vasil'evskii, N.K. Ignatovich, N.I. Tolstikhin, A.M. Ovchinnikov, I.K. Zaitsev, N.A. Marinov, and others, who think it better to consider subsurface and artesian water together, basing their zonation on the *structural–hydrogeological principle*. As N.I. Tolstikhin (in Mitgarts & Tolstikhin, 1961) points out, in this case it is not subsurface water which is treated but the natural aquifers. The treatment of subsurface water and artesian water together is in keeping with the concept of the unity of the water of the Earth. This is why it ought to be accorded preference.

The above methods of zonation are not mutually exclusive. Each is used according to the aims of, and the problems met by, any particular piece of hydrogeological research. But in discussing the methodology of zonation it is necessary to define clearly the purpose for which it is to be carried out.

In order to define the aims of the research it is recommended that two types of hydrogeological zonation be distinguished: *specialized*, and *general*. The problems posed by the above types of zonation differ considerably.

A specialized zonation is based on the solution of comparatively narrow, but completely real sets of applied or scientific problems. Usually this is the problem of the use of groundwater to satisfy the needs of some branch of the economy (water supply, irrigation, balneology). The zonation of a territory by properties, composition, or quantitative parameters which determine possible use of the groundwater also comes into this category of zonation (there are zonation maps of fresh, medicinal, industrial, or thermal water, and also hydrogeochemical, hydrogeodynamic, and other maps). M.R. Nikitin (1969) calls such a zonation in which some single aspect of hydrogeology is considered *particular*, in contrast to *subject* zonation which is carried out to satisfy the requirements of some branch of the economy.

A general zonation is carried out with the aim of distinguishing natural hydrogeological regions which differ in the totality of parameters of the distribution and formation of groundwater and does not concern itself with the solution of any problems which have been stipulated previously by economic considerations. This of course does not mean that it is completely divorced from practice. To present day natural scientists such a zonation serves to solve a varied complex of problems in the exploitation of groundwater. Hence the results of a scientifically based general zonation always have great practical significance too.

The regions distinguished by a general zonation are properly hydrogeological in contrast to those distinguished by a specialized zonation and which are known by various names depending upon the aim of the exercise (water supply, hydrogeological–irrigational, thermal waters, etc.).

In the hydrogeological zonation scientific interests are closely combined with practical recommendations. But economic considerations must not determine the principles of zonation. For the basis of both general and special purpose zonations there must be natural-historical parameters.

Any hydrogeological zonation, whether it be separate or combined (subsurface and artesian water), specialized or general, must be complex. Apart from natural-historical parameters, it is also necessary to take into consideration a number of artificial (technogenic) factors. Even a specialized zonation which takes into consideration only one, even though it may be quite a fundamental factor, must rely upon a complex of natural factors. If it were not so it would be artificially adapted to the demands of the different branches of the economy and be deprived of its scientific basis.

Thus the practice of hydrogeological research has devised various methods and procedures for zonation. The zonation of the groundwater or the hydrogeological reservoirs is carried out according to the method used; thus various maps are made (Mitgarts & Tolstikhin, 1961):

(1) the zonation of groundwater *per se* (subsurface or artesian, fresh or saline, hot springs or otherwise), which differs as to properties, composition, or qualitative parameters;

(2) the zonation of maps of natural groundwater reservoirs, the bulk of which are special purpose, but which sometimes (for example maps of groundwater or artesian water) are subordinated, as are maps of hydrogeological reservoirs, to the general problems of discovering the laws governing the distribution and formation of groundwater.

Accordingly two methods of zonation are also distinguished.

The first method provides for the classification of groundwater into types according to some defined parameters; hence such a zonation must be called a *typological* hydrogeological zonation. In a typological zonation the taxonomic units are mapped according to the most important factors, which also determine the contents of the maps. The well known scheme of provinces and zones of the groundwater of the USSR (Lange, 1969) is an example of a typological zonation map. The zonation maps constructed for the purpose of the exploitation of groundwater (water supply, balneology, etc.) come into this category.

The second method is based on the structural–hydrogeological principle and also consists of the distinguishing of hydrogeological regions, which have a unity of the whole totality of hydrogeological conditions. Such a zonation may be called a *structural* hydrogeological zonation. Structural zonation reveals the hydrogeological laws of the territory studied (Fig. 8.1).

8.1.2 The methods of distinguishing hydrogeological regions

In hydrogeological zonation regions of different taxonomic order are distinguished and mapped as a basis for the evaluation of the groundwater. Concerning the methods used for mapping hydrogeological regions, it is necessary to have a clear conception of the objectives of zonation and their relative importance in the order of things. It is concerned in the first place with structural–hydrogeological zonation whose methodology we will discuss.

There is so far no unanimity concerning the objectives of hydrogeological zonation. A uniform nomenclature for the naming of hydrogeological regions has not so far been devised. In practice hydrogeological researchers concerned with isolating regions more often than not start with 'basic element' or the 'basic unit' of zonation.

Since the first All-Union Hydrogeological Conference (1931) the basic element of zonation has been considered to be the hydrogeological region; this is a part of the Earth's crust which is distinguished by a dominant type of groundwater and a unity of historical development. According to the resolution of the Conference these are artesian basins, basins of underground runoff isolated by water divides, and aquifers. The vagueness of such a formulation is obvious.

Research in recent years has made the content of the basic element of hydrogeological zonation more precise. In the process of summarizing a great amount of factual material, and in the drawing of regional hydrogeological maps, the basic taxonomic unit of hydrogeological zonation, which is called a region of the first order, is taken as some definite type of groundwater basin. It is distinguished by the structural–

hydrogeological principle, which ensures that the maximum number of various factors is taken into account (physical-geographical, geological, and hydrogeological).

Nevertheless various opinions exist as to the treatment of the contents of a hydrogeological region and its taxonomic order. A.M. Ovchinnikov (1961) included the pressure water system, i.e. the artesian basin, the fissure-water basin, etc. (see Table 5.2). He took the pressure water system as the hydrogeological region of the first order and within this distinguished second and third orders; several large pressure water systems form a hydrogeological region which is the largest taxonomic unit of zonation. For G.N. Kamenskii *et al.* (1959) this is a geostructural subdivision in which groundwater is formed according to the conditions associated with a single system and governed by laws of distribution.

B.I. Kudelin & I.F. Fideli (1970) call the groundwater runoff basin the elementary unit of zonation, i.e. the territory with a common water budget. A principle for distinguishing hydrogeological regions of higher order has been put forward by N.A. Marinov. In his opinion 'zonation must be based on the distinguishing of large regions which consist of one, two, or a whole system of artesian basins together with areas of internal and external recharge, accumulation, and discharge of

Figure 8.1. Map of the Baikal Amur Magistral.*
Genetic types of groundwater basins and massifs: 1, platform, central, and marginal stratal water basins; 2, intermontane and superimposed stratal water basins; 3, basement fissure water massifs; 4, orogenic fissure water massifs.
Boundaries: 5, of the hydrogeological provinces (kratogens and orogens); 6, hydrogeological sub provinces (systems of basins and massifs); 7, hydrogeological regions (basin complexes or massif complexes); 8, mainly continuous permafrost.

Hydrogeological divisions: *Provinces*: A, Eastern Siberian hydrogeological kratogen; B, Sayan–Altai hydrogeological orogen; C, Mongol–Okhotsk hydrogeological orogen; D, Amur–Sikhote Alin hydrogeological orogen. *Subprovinces*: I, Eastern Siberian stratal water basin system; systems of fissure water massifs; II, Vitim–Potomsk; III, Aldansk, with superimposed basins; systems of groundwater massifs and basins; IV, Eastern Sayan; V, Baikal; VI, Daursk (Trans-Baikal); VII, Okhotsk–Stanovoye; VIII, Amur–Okhotsk (Khingan–Amgurskoye); IX, Sikhote Alin. *Regions*: stratal water basin complexes: I_1, Angara–Lena; I_2, Tungunska; I_3, Yakutsk; III_1, Chul'man; III_2, Gonam–Tokko group; V_1, Baikal region; V_2, Tunkinskiy; V_3, Bargazinskiy group; V_4, Upper Angara; V_5, Muisk–Kuandinsk group; VI_1, Dzhidino–Udinsk group; VI_2, Eravninsk group; VI_3, Chikoy–Khilok group; VI_4, Chita–Ingoda group; VI_5, Nercha–Shilka group; VI_6, Onon–Borzya group; VI_7, Upper Onon; VI_8, Argun region group; VI_9, Baley; VI_{10}, Upper Amur group; VII_1, Upper Zeya; VII_2, Uda; VII_3, Torom; VIII_1, Amur–Zeya; VIII_2, Bureya; VIII_3, Middle Amur; VIII_4, Lower Amur group; VIII_5, Tugur–Amgun group.

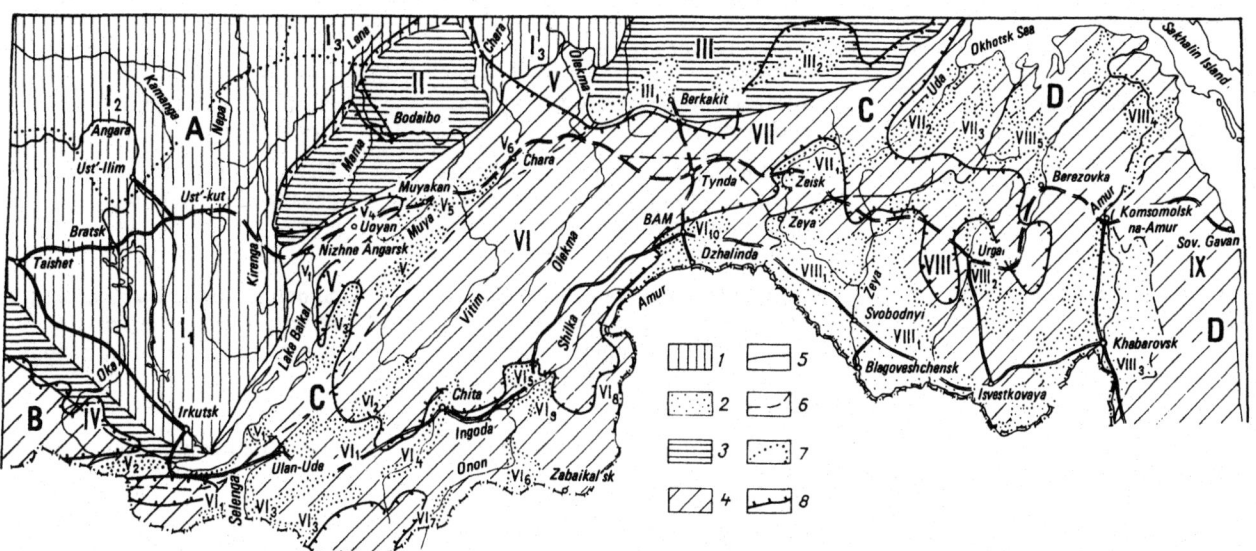

*Regions of fissure water massif complexes have been omitted to avoid overcrowding the map.

groundwater' (Marinov, 1971, p. 40). Hydrogeological regions comprise hydrogeological districts which are united by a unity of underground and surface runoff. On the Asiatic mainland N.A. Marinov distinguished the hydrogeological regions of the Arctic, Indian, and Pacific Oceans and the enclosed hydrogeological regions of the Mediterranean and the Black Sea.

I.K. Zaitsev & N.I. Tolstikhin (1963) take as the basic unit, i.e. the hydrogeological region of the first order, the natural reservoir full of water within which the processes of recharge, accumulation, and discharge of groundwater occur. They distinguish two types of hydrogeological region, simple in the taxonomic sense: the artesian basin and the hydrogeological massif. Areas are combined into hydrogeological regions, which are artesian or folded. To the first belong those regions which contain mainly stratal water and to the second those which contain fissure and vein-fissure water. A province of artesian or folded regions is considered to be the largest taxonomic unit.

According to I.K. Zaitsev and N.I. Tolstikhin, artesian basins consist of hydrogeological regions of the second or third order. In order to establish their boundaries it is recommended that one should be guided by the position of the groundwater watersheds, which determine the direction of the runoff within two upper hydrogeodynamic zones. In the detailed zonation given by these authors subregions are outlined which differ according to the predominant accumulation of groundwater and usually coincide with the aquifers, complex zones, etc.

N.V. Rogovskaya and V.N. Kunin, touching on the subject of future international hydrogeological maps, consider the main objective of mapping to be the hydrogeodynamic and physical—chemical parameters of groundwater. 'On such maps', write the authors (Rogovskaya & Kunin, 1969, p. 143), 'the natural reservoirs of groundwater must be mapped (in several stages) together with the rates and volumes of the

natural water exchange. The hydrogeological map will then reflect the "limiting" boundary and conditions between the different natural reservoirs of groundwater.'

In the opinions stated above, in spite of the different concepts of the basic unit of regional hydrogeological zonation, there can be seen a desire to endow the term hydrogeological region with a spatial concept. It is very likely that it is the reservoir itself and not some other which ought to be regarded as the object of zonation.

What then is to be taken as the basic element of regional hydrogeological zonation? Apparently, just as in physical-geographical zonation, it is necessary to base regional hydrogeological zonation not on a 'basic unit', but on an elegant and properly ordered classification of the regions. We find elements of such a classification in I.K. Zaitsev & N.I. Tolstikhin (1963).

As a basis for devising the gradations of hydrogeological regions E.V. Pinneker (1977) makes use of a fractional systematization of groundwater systems which has been shown above in Fig. 5.3. Such a nomenclature of hydrogeological regions is shown in Table 8.1. Each region corresponds to defined groundwater reservoirs. From the largest to the smallest the regions are placed in the following taxonomic order: province, subprovince, region, area, district.

Thus a hydrogeological region is a projection of the capacity of the groundwater on the surface of the Earth. Because grading of regions is taken to give numerical indices from the largest to the smallest, we will begin the examination of the grading with the region of the first order.

A hydrogeological *province* is taken to be a region of the first order. It corresponds to the largest reservoirs of groundwater, to the hydrogeological kratogen or the hydrogeological orogen. A hydrogeological *subprovince* is a region of the second order and corresponds to a system of stratal water basins (artesian region), a system of massifs and groundwater

Table 8.1. *Nomenclature of hydrogeological regions*

Order (rank)	Hydrogeological region	Groundwater capacity	Geological body (structural element)
I	Province	Hydrogeological kratogen, hydrogeological orogen	Platform, part of the geosynclinal (folded) belt
II	Subprovince	Stratal water basin system (artesian region), system of groundwater massifs and basins (folded region), fissure water massif system	Plate, medium size massif, shield, geosynclinal system
III	Region	Complex stratal water basin (artesian basin), complex fissure water massif (complex hydrogeological massif or complex fissure water basin)	Syncline, local downwarping, folded structure, major outcrop of the basement, group of intrusions
IV	Area	Simple stratal water basin (artesian basin), simple fissure water massif (hydrogeological massif, or fissure water basin)	Bowl, depression, folded uplift, lava cover
V	District	Aquifer, water-bearing complex, water-bearing 'zone' in fissured or broken rock	Stratum, suite of strata, fissured formation of crystalline rock, fractured rock and faults

basins (folded region), or a system of fissure water massifs. In order to avoid terminological confusion in distinguishing some orders of hydrogeological regions, a complex basin of stratal water and a complex massif of fissure water are assigned to an independent grade called a hydrogeological *region* (*sensu stricto*). Then follow the regions of the fourth order, the hydrogeological *areas*. It is not proposed that every basin and massif should be called a hydrogeological area, but only simple stratal water basins, in other words, simple artesian basins and simple fissure water massifs, i.e. a hydrogeological massif (according to I.K. Zaitsev and N.I. Tolstikhin) or (in the terminology of A.M. Ovchinnikov, A.A. Kartsev, and others), a fissure water basin. The concrete definition of the content of this taxonomic unit is a great convenience when distinguishing hydrogeological regions in practice, and simplifies their mapping (see Table 8.1).

A fifth order region of the Earth's surface represents the district of distribution of the main aquifers or complexes, water-bearing zones of fissuring or faulting (water-bearing veins), therefore it is proposed that it be called a hydrogeo-logical *district*. The name is provisional but its meaning is fully defined. It is a reservoir collector. N.V. Rogovskaya & V.N. Kunin (1969) propose to include just such natural reservoirs of groundwater on hydrogeological maps.

The result of the distinguishing of hydrogeological zones is that the nomenclature of hydrogeological regions (see Table 8.1) now becomes applicable not only to small-scale but also to large-scale hydrogeological zonation. On the basis of such a wide gradation a territory of any size lends itself to zonation in descending order of size (from the largest units to the smallest), or upwards (by combining units of lower order into units of higher order). If necessary more detailed subdivisions may be made for any region. In the first place this concerns areas and regions within which it is expedient to differentiate

subareas and subregions respectively. For a fissure water massif (hydrogeological massif) the subarea will be a reservoir with a single water budget.

Sometimes superorder, suborder, and extraorder sub-divisions of the hydrogeological zonation are spoken of.

For example global *belts* of artesian basins, as proposed by N.I. Tolstikhin (1971), are classed as superorder units. There are seven global belts separated by critical parallels which consist of individual artesian basins or artesian areas, which also include marine basins: (1) Arctic, (2) Boreal, (3) Mediterranean, (4) Equatorial, (5) Southern, (6) Peri-Antarctic, (7) Antarctic basins, (Fig. 8.2). Furthermore, around the Pacific Ocean along its shores there is situated the Pacific belt of artesian basins.

The term *unit* is recommended as the suborder division dividing the interior of a hydrogeological district according to some or other particular parameter (the degree of water abundance in the rocks, the composition of the groundwater, etc.). Incidentally, in physical geography (Armand, 1952), engineering geology (Pal'shin, 1959), and irrigation (Tkachuk & Molodykh, 1972) a unit is the smallest subdivision of zonation.

A groundwater *deposit* is the extraorder unit. It forms a complex dynamic system the contours of which vary with time and which occupy partly or completely some or other reservoir of groundwater. A.M. Ovchinnikov (1963, p. 124) regards a mineral water deposit as a 'spatially bounded accumulation of water of definite composition (which corre-sponds to established conditions) in quantities which are sufficient for their economic use'. Ovchinnikov's definition applies to practically all deposits of groundwater including fresh water. N.I. Plotnikov (1965, p. 32) assigns such reservoirs or parts of them within whose limits 'the accumulation of groundwater accumulates under natural conditions in

Figure 8.2. Artesian basin belts of the Earth (Tolstikhin, 1971).
Ar, Arctic; B, Boreal; M, Mediterranean; E, Equatorial; S, Southern; P-A, Peri-Antarctic; An, Antarctic.

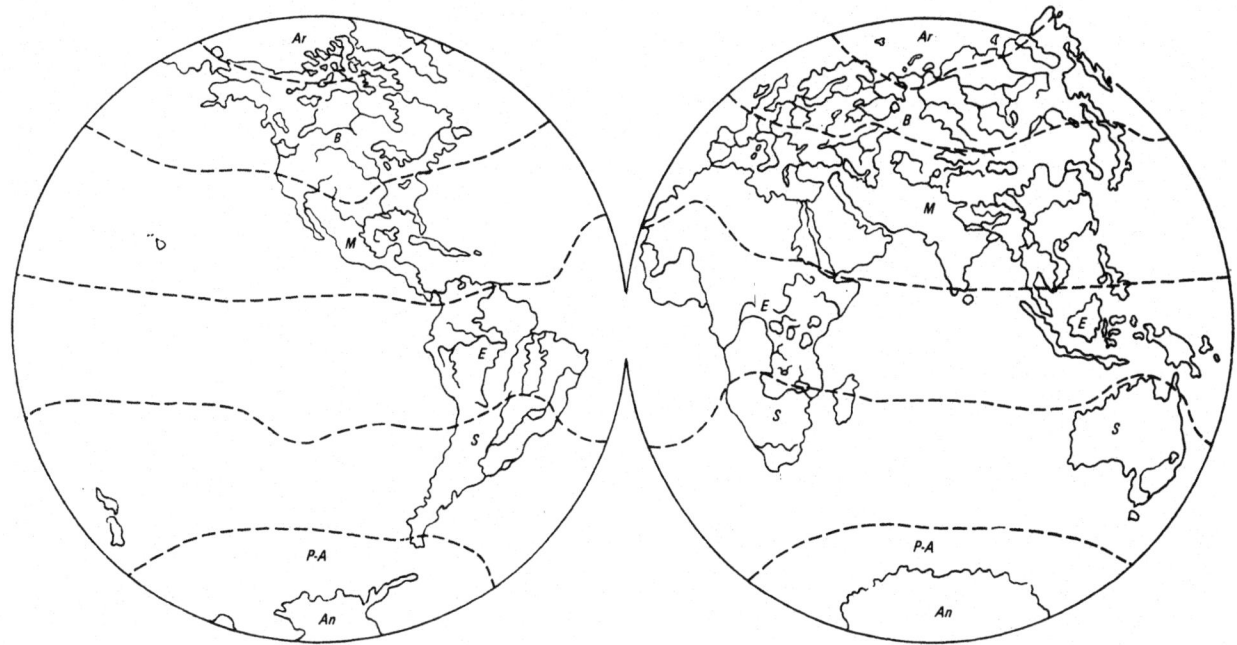

sufficient quantities to be economically useful (water supply, irrigation, etc.)' to the latter.

Hydrogeological belts and units are not related to regions of some or other order: the first is the totality of the largest reservoirs; the second is a part of the smallest reservoir. Therefore it is impossible to assign them to a category of ordered taxonomic units. A groundwater deposit corresponds to a hydrogeological unit of a hydrogeological region of a different order (district, area, region), and sometimes combines regions of the same order, i.e. is comparable with the various taxonomic units, being an extraorder subdivision.

Consequently, it is by no means obligatory to carry out a hydrogeological zonation from the 'basic unit' with subsequent increase or division into regions of different orders. Such a method is justified only for defined stages of zonation in which the distinguishing of a limited number of taxonomic units is necessary.

Both structural and typological zonation is made easier if it proceeds from the largest units to the smallest via a multi-stage scheme which reflects the parameters of the territory, which are introduced in sequence for each stage of the zonation. The process of distinguishing hydrogeological regions must begin with the establishing of a number of stages of zonation and classifying those parameters which are necessary for the determination of the taxonomic units. Each stage reflects the degree of the differentiation of the territory into its component parts. The higher stages of zonation correspond to the large regions which are defined according to the most general parameters, which are themselves based on the structural–hydrogeological principle. In proceeding to the lower stages of zonation and the stage of small regions, it is mainly those particular factors relating to the 'leading' and direct parameters which are taken into account. The rank of taxonomic units is established by means of dividing large regions into smaller ones or combinations of small regions.

Hydrogeological regions are drawn directly on the hydrogeological map, or are represented in the form of specially constructed schemes of hydrogeological zonation. The boundaries of regions are shown on the main map when there is space to do so without sacrificing clarity. More often a separate scheme of hydrogeological zonation is constructed as a small-scale appendix to the main hydrogeological map.

The exposition of the methods of distinguishing hydrogeological regions is useful both in small-scale and large-scale zonation. The hydrogeological zonation carried out along the route of the Baikal Amur Magistral (Fig. 8.1) is an example of zonation carried out by distinguishing the highest divisions of Table 8.1 (province, subprovince, and region) and an example of the above scheme. A lower stage of zonation corresponding to the stage of hydrogeological areas and districts may be found, for instance, on the hydrogeological maps constructed during hydrogeological surveys (Fig. 8.3).

8.2 World distribution of aquifers

The laws governing the occurrence of groundwater differ for the continents and the ocean floor. On the accompanying hydrogeological scheme of the Earth (Fig. 8.4) are shown the continental (subaerial) and oceanic (submarine) groundwater reservoirs. The transitional reservoirs are also shown. Regional monographs and compendiums, containing descriptions of groundwater, were used in its compilation: Soviet – (Anon., 1967–1978; Anon., 1974a; Anon., 1974b; Anon., 1978; Balaev, 1976; Bogolepov & Chikov, 1976; Grigorkina, 1977; Dzhamalov *et al.*, 1977); foreign – (Furon, 1955; Anon., 1959; and others). Furthermore, a multitude of maps from various countries, reports on water resources from the Geological Survey of the USA (Geological Survey Water Supply Papers, the total number of which at the present time exceeds 2000), and hydrogeological articles in periodicals have served as sources of information.

The depth at which groundwater occurs on the continents varies. In tectonic fault zones groundwater is found at great depths. The granite and basalt layers of the continents have been poorly studied in general and in respect of their water resources it is possible only to make hypotheses.

Under the ocean floors the thickness of the sedimentary deposits varies from 0.2 to 2 km. Having a porosity of 15–35%, these rocks contain stratal water. Below these there follow sedimentary-volcanic and basalt layers of thickness 1–3 and up to 5 km respectively, in which there are mainly aquifers of the fissure-vein type.

We will examine in the most general way the distribution of aquifers within continents and oceans.

Figure 8.3. Hydrogeological zonation map taken from a medium-scale hydrogeological survey.
1, boundary and number of hydrogeological areas (I, birch forest artesian basin; II, pine forest hydrogeological massif; III, larch forest hydrogeological massif).
2, boundary and number of hydrogeological districts (I_1, poro-stratal water of sandy–pebbly alluvial deposits; I_2, fissure-stratal water of sandy clay deposits; I_3, fissure-stratal water of well-sorted sandstones; II_1, fissure water of karst marbles; II_2, fissure water of the gneiss-schist layer; II_3, fissure-vein water of zone of faulting in deep-lying deposits which give rise to carbonated thermal water springs; III_1, fissure water of kataclastic granites; III_2, fissure water of dense granites; III_3, fissure water of the near-surface fault zone, which contains fresh water).

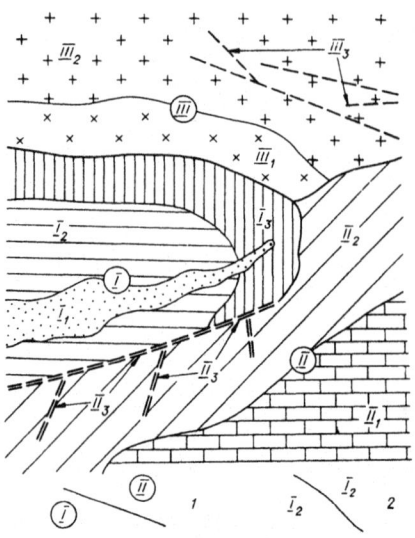

Europe, occupying an area of 10.5 million km², is represented by systems of stratal water basins (artesian regions) and fissure water massifs (folded hydrogeological regions).

The largest region of stratal water is the Eastern European and is confined to the shield of the Russian platform, which consists mainly of terrigenous, Cretaceous, and sometimes halogenic Palaeozoic formations of a general thickness of 4 km (in some downwarps 8–12 km). It is divided according to the scheme of I.K. Zaitsev & N.I. Tolstikhin (1972) into Baltic–Ural, Northern Caspian, Donets, Dnieper, Pripet, and L'vov–Volynskii complex artesian basins. The central part of the first of these is occupied by the Moscow artesian basin, where the thickness of the sedimentary cover reaches 5 km. The zone of fresh water here is several hundred metres thick. The high flow rate is in layers of karst rocks. At a depth of 200–300 m brackish and saline water is found and below that brines occur; the thickness of the strong brine zone (more than 150 g/l) reaches 2000 m. Brines with a degree of mineralization of up to 300 g/l are found in some boreholes in the deepest part of the basin.

In the south of the eastern part of Europe is the Caspian–Black Sea artesian region which includes the Black Sea basin, the Azov–Kuban, and the Tersko–Caspian artesian basins. The degree of mineralization of the groundwater reaches 100 g/l.

The Timan and Ural fissure water massif systems form the north-eastern boundary of the Eastern European artesian region. Between them is the complex Pechora artesian basin in which the sedimentary cover is up to 8 km in thickness and the hydrogeological profile goes from fresh hydrocarbonate water to chloride brines with a degree of mineralization of 200–300 g/l.

The basin of stratal water in Central and Western Europe is also of considerable size. Here the basement consists of Caledonian (in Central Europe) or Hercynian (in Western Europe) folded formations. According to the latest scheme of geotectonic zonation (Khain, 1977), the Central European and Western European artesian regions are differentiated as separate entities here.

The largest aquifer in Central Europe is the Polish–German complex artesian basin. In its sedimentary profile, which is up to 8 km in thickness, Devonian molasse rocks are found, and Cretaceous, terrigenous deposits are of widespread occurrence. The central part of the sequence is occupied by the Permo-Triassic evaporite formation which is more than 1000 m thick. Associated with this are the phenomena of salt tectonics. Such structural features of the basin have led to complex hydrochemical zonation, which is characterized by the occurrence of fresh, brackish, and saline water, and brines (with a degree of mineralization of up to 500 g/l).

The Iberian, Aquitaine, Paris, Southern Germany, and other artesian basins form a second region. In the cover of the basins terrigenous carbonate Mesozoic and Cenozoic deposits occur with a general thickness of 3–5 km. In the Aquitaine basin Upper Triassic evaporites are found. Just as in the above-mentioned basins, there is here a variegated hydrogeochemical sequence and deposits with different water capacities. The natural level of the groundwater in the Paris basin has fallen more than 120 m as a result of centuries of exploitation.

Stratal water basins are bounded by hydrogeological folded regions: in the north of the Baltic and Scandinavian regions they are confined to the shields of the same names, in the east by the Urals which we have already mentioned; in the south by the Pyrenean, Apennine, Alpine, Balkan, Carpathian, and other ranges of folded rock formations. Intermontane artesian basins are often found in hydrogeological folded regions (Transylvanian, Hungarian, Po, etc.). The bulk of the cover in these basins is up to 5 km thick and consists of Neocene molasse deposits, but there are also more ancient terrigenous carbonate sediments (Hungarian artesian basin). In these deposits there are significant resources of thermal water. The massifs of the folded regions are usually elevated, very much faulted into blocks, and fissure-karst water is often of widespread occurrence in them. The classical karst regions, the Balkans, the Alps, and the Crimea are remarkable in this respect. Fissure-karst water is discharged in great quantities along the coast of the Mediterranean and other seas.

Asia. Artesian basins also occupy a considerable part of Asia (total area 43.5 million km²). They are confined both to the ancient and epi-Palaeozoic platforms. The greatest systems of stratal water basins are the Eastern Siberian, Chinese, Arabian, and Hindustan systems. The tectonic activity of the majority of the ancient platforms of Asia distinguishes them quite well from the kratons of Europe, and is responsible for the formation of the sedimentary-volcanic cover and the block structure.

The Eastern Siberian stratal water basin system includes the Northern Siberian, the Tungus, Yakutia, and the Angara–Lena complex artesian basins (Pinneker, 1977). The hydrogeochemical and hydrogeodynamic zonation of these basins is determined by the presence of interleaving terrigenous sedimentary Cretaceous and Cenozoic deposits with tufa, by permafrost, and by the existence of 'open' fault zones, which penetrate right into the crystalline basement. The saline layer has the greatest influence in determining the composition of the groundwater. In the Tungus artesian basin below the permafrost layer there are only brines. Chloride-calcium brines occur widely in the Lower Cambrian deposits of the Angara–Lena artesian basin (the mineralization reaches 500–620 g/l). In the Mesozoic terrigenous deposits of the Yakutia artesian basin the degree of mineralization of groundwater rarely exceeds 50 g/l.

The Chinese artesian region, consisting of a cellular honeycomb mosaic of stratal and fissure-stratal water basins, is distinguished by complex hydrogeological conditions which are the result of the block structure and strongly dissected relief. The zone of fresh water usually reaches a depth of 200–400 m, but in the western parts of the region, where arid conditions predominate, saline water can be either at or near the surface. The brines of the Szechwan artesian basin (degree of mineralization from 150 to 250 g/l) are well known.

The Arabian stratal water basin system occupies the

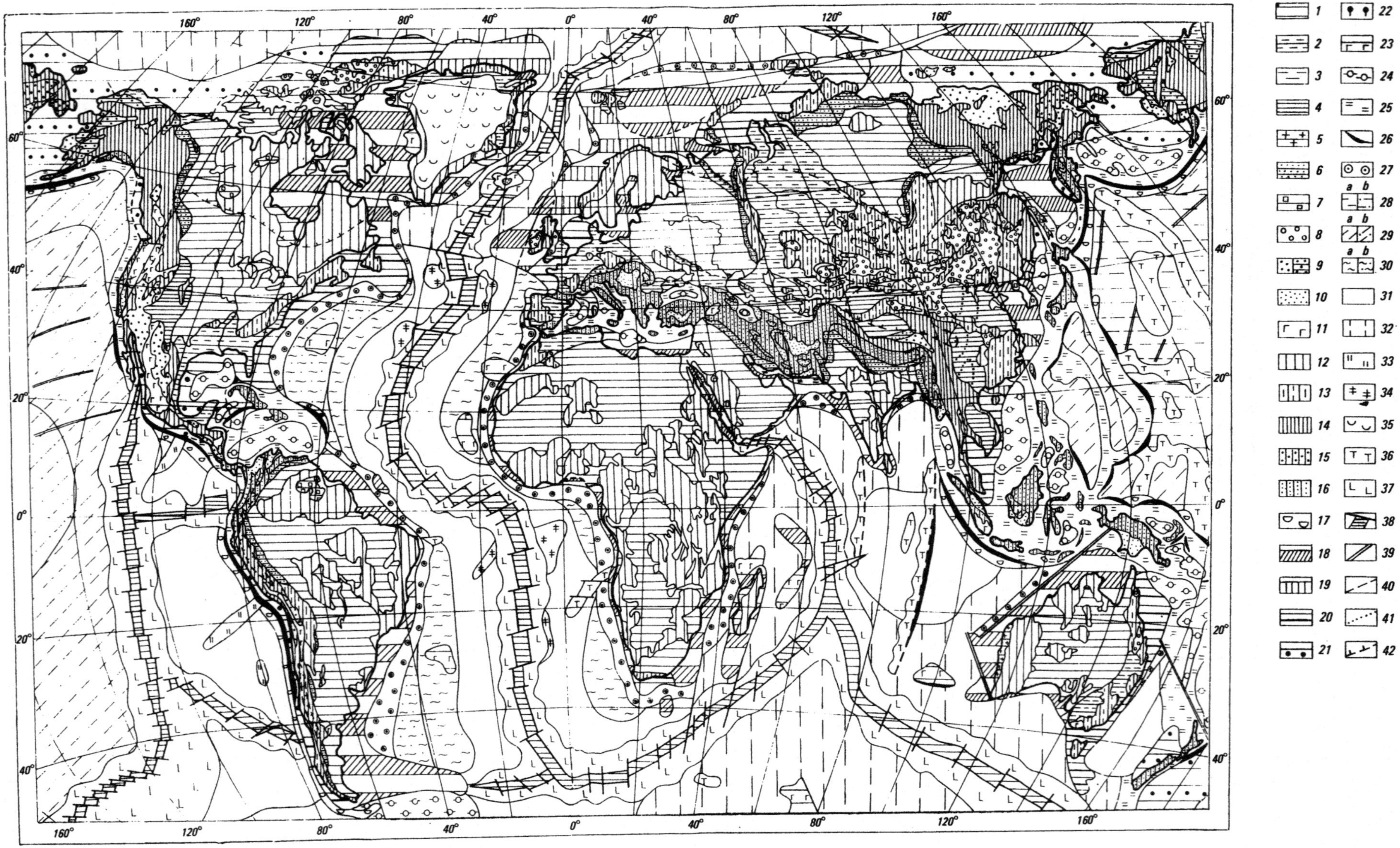

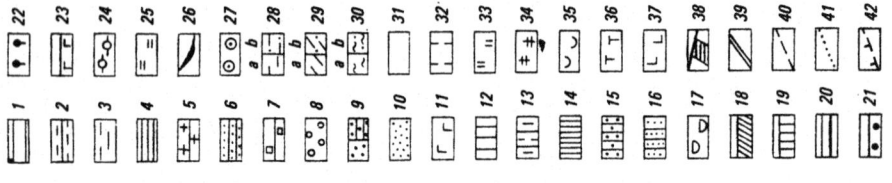

Figure 8.4. World distribution of aquifers (after K.P. Koranov).
(See opposite for explanation of key.)

Figure 8.4. World distribution of aquifers (after K.P. Karavanov).

Continental (subaerial) aquifers.

Platform and localized (artesian basins): 1, ancient platform (sometimes local downwarps); 2, epi-Caledonian and epi-Hercynian platform; 3, epi-Mesozoic platform (continental part of the Beaufort Sea platform); 4, depressions of the Chinese platform, Indo–Sinian plateau and the epi-platform orogenic systems (combination of stratal water basins and stratal-fissure water); 5, allochthogenes (mainly basins of stratal fissure water; 6, local downwarps.

Within the intermontane and superimposed stratal water basins (artesian basins): 7, crystalline shields of ancient platforms and crystalline massifs of the China platform; 8, Baikalian, Caledonian, and Hercynian folded regions; 9, Mesozoic folded regions (a, poro-stratal water, b, stratal-fissure water); 10, Cenozoic folded regions; 11, superimposed volcanogenic basins.

Orogenic and plate massifs of fissure water (hydrogeological massifs), sometimes with intermontane or superimposed stratal water basins: 12, shields, crystalline, and central massifs; 13, Baikalian, Caledonian, and Hercynian folded regions; 14, Mesozoic folded regions; 15, Mesozoic–Cenozoic volcanic belts (often with stratal fissure water); 16, Cenozoic folded regions; 17, aquifers covered with thick layers of ice.

Oceanic (submarine) aquifers.

Platform (shelf) stratal water basins: 18, ancient platforms; 19, epi-Caledonian (occasionally epi-Baikalian) platforms and local downwarps; 20, epi-Hercynian platforms and local downwarps on the Caledonian–early Hercynian basements (continuation into the massive areas of the Innuitian orogenic belt in North America); 21, epi-Mesozoic platforms and local downwarps or structures of pre-Cenozoic consolidation; 22, Cenozoic local downwarps; 23, superimposed volcanogenic basins on continental margins.

Stratal water basins of recent geosynclines: 24, deep trenches of the oceanic margins; 25, recent geosynclinal deeps within the western Pacific sector, Cenozoic incompletely folded regions and Cenozoic folded regions; 26, localized trenches of the ocean and oceanic margins; 27, assumed geosynclinal downwarps along the continental slopes with a sedimentary cover of 2–4 km, sometimes up to 10 km, thickness.

Two-terraced (upper terrace is of poorly lithified sediments and the lower terrace consists of volcanogenic sedimentary rocks) stratal water basins of oceanic plates or thalassokratons: 28, pre-Palaeozoic oceanic plate with Mesozoic–Cenozoic sedimentary cover of thickness: (a) up to 300 m, (b) up to 1000 m; 29, as 28, with Cenozoic sedimentary cover of thickness: (a) up to 100 m, (b) up to 500 m; 30, Palaeozoic and early Mesozoic oceanic plate with Mesozoic–Cenozoic sedimentary cover of thickness: (a) up to 500 m, (b) up to 1000 m; 31, as 30, with Cenozoic sedimentary cover of thickness up to 300 m; 32, early Mesozoic and younger oceanic plates with Mesozoic–Cenozoic or only Cenozoic sedimentary cover of thickness up to 500 m.

Fissure water massifs with intercalated stratal water basins which are characterized by a thickness of poorly lithified sediments up to 100 m, or which are covered with argillaceous sediments of thickness up to 100 m: 33, zone of block folding within thalassokratons; 34, ancient mountain chains on the flanks of the Mid-Atlantic ridge; 35, localized swells; 36, Mesozoic(?) and Cenozoic volcanic belts mainly in the form of raised swells (thalasso-anticlises) and isolated volcanogenic or sedimentogenic stratal fissure water basins; 37, Cenozoic mix-oceanic (mainly volcanic) belts with sedimentary cover in some places.

Vein water reservoirs: 38, central parts (rifts) of recent mid-oceanic ridges; 39, transcurrent faults of the Mendocino zone or transcurrent faults of the mid-oceanic belts.

Boundary symbols: 40, boundary of the fissure water massifs of folded regions of various ages within continents and of stratal water basins within oceanic regions; 41, boundary between basins of stratal and stratal-fissure water within the synclises of the Chinese platform; 42, boundary of the distribution of the permafrost (ticks point towards the permafrost).

south-west of Asia, including the Arabian and Mesopotamian artesian basins. In the Arabian basin the cover, which is up to 5 km thick, consists of rocks of almost all systems with the exception of the Carboniferous (Anon., 1974a). Among the Palaeozoic terrigenous limestone deposits there are also gypsiferous beds (Permian), and these are also well known in Triassic, Jurassic, Palaeogene, and Neocene formations. The Cretaceous system includes volcanic rocks — diabases and spilites. The aridity of the climate and the presence of salt-bearing facies in the sequence lead to the widespread occurrence of saline waters and brines, although fresh water is no rarity here (near the external recharge regions). In the Upper Cretaceous deposits brines with a degree of mineralization of 200 g/l are found at a depth greater than 1000 m.

The Hindustan system of stratal water basins occupies the southern part of Asia within which basins can be distinguished which coincide with the boundaries of the Hindustan platform (among these are the well known Deccan, Thar, and Lower Ganges), and basins that are confined to the Cenozoic local downwarps which surround the Himalayan folded region. Among these can be distinguished, from west to east: the Indus, Ganges (Siwalik), and the Arakan regions. The external recharge regions, which reach altitudes of 8000 m, have a marked influence on the formation of groundwater in the latter group of basins. Just as on the Siberian platform, intrusive bodies of basic igneous rocks are of widespread occurrence and form impermeable layers and barriers to groundwater.

In the hydrogeological folded regions of Asia, in the Sayan–Altai, Baikal–Okhotsk, Central Asiatic, Pamir–Himalayan, and other hydrogeological orogens, there are concentrated a large number of intermontane stratal water basins. About 90% of the basins are characterized by the occurrence in them of Mesozoic or Cenozoic cover consisting of terrigenous, terrigeno-volcanic, or terrigeno-carbonate deposits. They contain considerable resources of fresh groundwater. Saline water and brines are found where there are evaporites in the sequence.

The different character of the water-bearing rocks and the positions of the basins in zones with frigid, humid, or arid climates leads to the intermontane artesian basins of northeast Asia being characterized by the occurrence of zones (up to 1 km thick) of fresh hydrocarbonate water and brackish water at depths of 2–3 km, and within central and western Asia at insignificant depths far from the external recharge regions there are saline water and brines (Anon., 1974a; Karavanov, 1977).

In north and north-east Asia rocks at depths of 0.5–1 km and more are frozen. On the Anabar shield for example, groundwater is found only in the active layer; in other aquifers it is usually confined to taliks along fault zones.

The western part of Asia is the territory of large artesian regions, which coincide territorially with the epi-Palaeozoic plates, the Western Siberian and the Turansk. Only the terrigenous deposits (those of the Western Siberian region) in which the mineralization of the groundwater does not exceed 80 g/l, or the complex set of formations (terrigenous, carbonate, halogenic rocks with saline water and brines, up to 400 g/l) which are typical for the Turanian stratal water basin systems are water-bearing rocks.

Africa. The greatest part of Africa (total area 30.3 million km^2) is occupied by the artesian regions in the ancient platform. Aquifer rocks are Phanerozoic marine and continental deposits. The most interesting of these in the northern part of Africa are the continental deposits of the Nubian Series, which were formed in the Carboniferous to Cretaceous periods, and which are characterized by the occurrence in them of fresh groundwater. Among the artesian basins of the African continent there are (from north to south): Tindouf, Western Saharan, Eastern Saharan, Eastern Libyan, Nile, Senegal, Taoudeni, Mali–Nigerian, Chad, Upper Nile, Ethiopian–Somalian (Eastern Africa), Rudolf, Volta, Benue and Lower Niger, Okovango, Kalahari, South-east African, Madagascar, and Karroo Basins, which are named after the geological structures (Khain, 1971).

The laws governing the formation and occurrence of groundwater resources which have been elucidated to date (Anon., 1978), reflect both the complex geological–structural features of the territory and the distribution of the various aquifer rock formations, and also the continental climate and nature of the external groundwater recharge regions.

As a consequence of the strong evaporation, over a considerable territory (Anon., 1959), the water of the upper aquifers is chloride-sodium with a degree of mineralization of 5–10 g/l or sulphate-chloride (or chloride-sulphate) of mixed cation composition with a degree of mineralization of 1–5 g/l. The subhumid and humid climatic zones are areas over which hydrocarbonate and silicate-hydrocarbonate groundwater with a degree of mineralization of up to 1 g/l and a mixed cation composition occur. In the artesian basins two types of chemical zonation of groundwater are found: (1) direct or humid (in which the degree of mineralization increases from 1 to 100 g/l and the water changes from hydrocarbonate to sulphate and then to chloride), (2) inversion or arid (here the degree of mineralization falls from the surface from 5–35 to 1 g/l, and then increases to 200 g/l or more). The areas in which the artesian basins of these types of hydrochemical zonation are found coincide spatially with the corresponding climatic zones.

The water supply of innumerable oases of the Sahara is obtained from groundwater. In the aquifers of the Nubian Series (north-east Africa) a well-defined latitudinal hydrogeochemical zonation of groundwater is observed from south to north. In this direction the degree of mineralization increases from 0.1–0.7 (depths from 50–700 m) to 200 g/l at a depth of 3150–3500 m (Anon., 1978).

The Atlas zone of Alpine folding and adjoining Hercynian structures of north Africa, the Cape Hercynian region in the south of the continent, and also the outcrops of the Precambrian basement in various parts of the ancient platform make up a mosaic network of hydrogeological folded regions with the occurrence of fresh groundwater in the fissured zones. There are many hot springs associated with the local faults of the East African rift.

Australia. A considerable part of this continent (area 1.6 million km^2) is occupied by artesian basins, which are confined to the cover on the Australian platform (Grigorkina, 1977; Dorokhov, 1973): Great Artesian, Canning, Murray, Georgina (Barkly and Daly), Eucla, Carnarvon, Ord–Victoria, etc. The aquifers are Palaeozoic, Mesozoic, and Cenozoic terrigenous and carbonate deposits of thickness up to 2000 m. The groundwater has been studied to about the same depth. The degree of mineralization varies from 0.3 to 37 g/l (Grigorkina, 1977). Artesian basins (of the intermontane type) are known in the Australian mountain ranges: these are the Bowen, Clarence, Sydney, East Gippsland, etc. Fissure water massifs are represented by outcrops of the crystalline basement of the Australian platform (Western Australian, Kimberley Plateau, Western Carpentaria, Macdonnell Ranges, etc.), or folded structures of the Australian cordilleras. Groundwater in the hydrogeological massifs is confined to the fissured zone of deep-lying metamorphic Archaean and Proterozoic formations (massifs of outcrops of the Australian platform), or Palaeozoic sediments and sedimentary-volcanic rocks.

North America (area 24.2 million km^2). Central and northern plains are occupied by the North American stratal water basin systems which correspond territorially to the ancient platform of the same name. The largest artesian basins are the Northern Canadian, Western Canadian, Williston, Denver, and Anadarko (western interior). The southern part of North America is occupied by the Mexican stratal water basin system, which consists of Mesozoic and Cenozoic terrigenous-carbonate cover. It is underlain over a considerable area by the buried structures of Alpine folding and the Mesozoic of the Cordillera.

The Canadian shield with groundwater in the upper fissured zone of metamorphic rocks or in the Quaternary glacial deposits, occupies a vast expanse of the north-east of the land mass. The hydrogeological orogen of the Cordillera stretches along a comparatively wide belt for several thousand kilometres, and in it, among the Mesozoic terrigenous-effusive rocks, volcanic Cenozoic formations are of widespread occurrence. The saturation of these rocks is high, especially along the tectonic fault zones. Here are often found Cenozoic intermontane artesian basins whose sedimentary formations are up to 5000 m in thickness.

The Southern Appalachians, Northern Appalachians, and the Newfoundland hydrogeological folded regions, which are characterized by the occurrence of groundwater in the fissured zone of the Palaeozoic rocks, stretch from the south-west to the north-east.

The aquifer systems in Alaska, in the north of Canada, and in Greenland are deeply frozen. The permafrost is often intermittent in character.

South America (area 17.8 million km^2). About 80% of the territory is occupied by the South American ancient platform, which reaches to the Andes Cordillera and which includes the Orinoco, Amazon, Maranhão, San Francisco, Central Brazilian, Upper Paraguayan, Parana, Cordoba, Chaco–Pampas, Neuquén, Chubut–Santa Huancache, Patagonian, and other artesian basins. The cover of these is formed by terrigenous, carbonate, and often volcanic formations too. The territory of Patagonia and Tierra del Fuego is located on a part of the epi-Palaeozoic platform.

The shields of the South American platform (Guianan, Central Brazilian, and Eastern Brazilian) and the hydrogeological orogen of the Andes form several hydrogeological folded regions, often with compressed intermontane artesian basins. Among the last the following can be distinguished in the Andes: Colombian (Lower Magdalena), Northern Venezuelan (Maracaibo), the Pacific, Altiplano, and Chilean. In these the aquifers are sandy Cenozoic formations. Hot springs are associated with the active faults of the Andes.

Antarctica (total area 14.1 million km^3). The hydrogeology of this continent has been studied least of all. A significant part of it is an ancient platform, covered with a layer of ice more than 1000 m thick. Ya.V. Neizvestnov & N.I. Tolstikhin (1976), studying the physical and thermal features of ice at such depths, find in similar circumstances subglacial groundwater basins, i.e. a saturated zone of fissured rocks below the ice cap. Outcrops of the basement of the platform and folded structures on the South Shetland Islands are deeply frozen fissure water massifs analogous to the Anabar shield in the USSR.

In conclusion it must be noted that the climatic features of the Earth in Cenozoic times led to the differentiation of the continents into two zones: the first (largest) in which groundwater occurred as a liquid, and the second with frozen rocks 100–500 m thick, less often more than 1 km, where underground water-bearing systems are found only below this depth.

Land occupies one-third of the Earth's surface; about two-thirds is covered by the seas and oceans. The shelf area bounded by the 200 m isobath amounts to 27 million km^2 and that bounded by the 1000 m isobath is 41 million km^2. The world ocean has an area of 361 million km^2 (excluding the shelf; below the 1000 m isobath – 320 million km^2). Below its floor there is groundwater which was discussed in Section 6.9.

The Pacific Ocean. Out of a total area of 179.7 million km^2 (from here the area of the oceans is given including the shelf zone) 43% is occupied by submarine stratal water basins. The water-bearing structures here are poorly lithified Mesozoic–Cenozoic sedimentary layers, rarely more than 300 m thick (North-west Pacific, East Mariana, Melanesia, and Central stratal water basins), only occasionally reaching 1000 m (the North-east artesian basin). In some of the basins

the Cenozoic sedimentary cover is typically less than 100 m thick (West Caroline, East Caroline, Southern, Philippines, Nampo, West Mariana Basins).

Drilling by the American research vessel *Glomar Challenger* revealed stratal collectors off the American coast and to the south-west of the Hawaiian Islands of thickness from a few tens of metres to 200 m and more (boreholes Nos. 155, 165, 167, 177, 183), consisting of sandstone and limestone. In one borehole (06° 07′ 38″ N, 81° 02′ 62″ W) the degree of mineralization of the pore solutions was generally 30 g/l. Below the sedimentary cover thin layers of basalt were found. On the basis of geophysical data, these basalts probably alternate with lithified sedimentary formations which may contain stratal groundwater.

Submarine crystalline massifs containing fissure, fissure-vein, and stratal-fissure groundwater coincide spatially with the mid-oceanic ridges, volcanic seamounts, rises, and swells. The largest of these are: the South Pacific, Eastern Pacific, Albatross, West Chile, Hawaiian, North-western, and Marcus–Necker.

In the rocks of the floor of the Pacific rift zones are found zones which contain fissure-vein water. The presence of hot springs in the rift of the Galapagos ridge bears witness to this (Anon., 1977).

The Atlantic Ocean. Submarine stratal water basins occupy 38% of the area of this ocean, whose total area is 93.4 million km^2. There are Mesozoic–Cenozoic sedimentary rocks whose total thickness varies from 500 m (the basins of Iceland, Western Europe, the eastern part of the Canary and the Cape Verde Basin, Angolan, Canary, Labrador, the central parts of the Argentine and North American, the eastern part of the Brazilian Basins) up to 1000 m (the eastern parts of the North American and Argentine Basins), or only Cenozoic sedimentary rocks up to 300 m thickness (the western part of the Iberian, Canary, and Cape Verde Basins, and also the slopes of the Brazilian, Argentine, Newfoundland, and Guiana Basins).

Innumerable boreholes, which reveal rocks of the oceanic floor of the Atlantic, indicate the wide distribution of deposits of carbonate, sandy-clayey, clayey-siliceous, and basalt formations. The aquifers are usually represented by sandy deposits.

The northern and southern sectors of the Mid-Atlantic ridge and the Walvis ridge belong to great fissure water massif systems. Morphologically they are mid-oceanic or block structures, the substratum of which consists of basaltic rock. The massifs are associated with longitudinal and transverse faults in which there are deep-lying water outlets.

The Indian Ocean. Its area is 74.9 million km^2, of which 46% consists of submarine stratal water basins with Mesozoic–Cenozoic or only Cenozoic cover up to 500 m in thickness (the Arabian, Somali, Amirante, Mascarene, Madagascar, Crozet, African–Antarctic, Australian–Antarctic, Mozambique, South Australia Basins), and Cenozoic sedimentary cover up to 300 m thick (West Australian).

As a result of deep water drilling by the *Glomar Challanger* in the Indian Ocean collector rocks of thickness 20–200 m have been found; mainly limestones, sandstones, and dolomites (the northern part of the Indian Ocean). Fissure water massifs (Arabian–Indian, Central Indian, Western Indian, Australian–Antarctic, Eastern Indian, Kerguelen, Mozambique, Madagascar, Cocos) occupy a zone of the mid-oceanic or volcanic ridges made of basaltic rocks.

The Arctic Ocean. Within this marine region of total area 13.1 million km^2, about 70% of the surface is occupied by submarine artesian basins. The cover is formed of poorly lithified Mesozoic–Cenozoic sediments up to 500 m in thickness. Here the Norwegian, Lofoten, Greenland, Nansen, Amundsen, Tollya, and Canadian stratal water basins can be distinguished. The submarine fissure water and fissure-vein massifs coincide spatially with the Lomonosov, Mendeleev, Gakkel, Knipovich, and Mohns ridges (see Fig. 8.4).

References

Anon. (1959). *Hydrogeology of the Earth's arid zones*, Vol. 2. Moscow, IL. 432 pp.

Anon. (1967–1978). *The hydrogeology of the USSR*. Vols. 1–45. Moscow: Nedra.

Anon. (1974a). *The hydrogeology of Asia*. Moscow: Nedra. 575 pp.

Anon. (1974b). *Main types of hydrogeological structures of the USSR*. Leningrad. 91 pp. (VSEGEI, nov. ser. Vol. 229.)

Anon. (1977). *Smithsonian Institution Natural Science Event Bulletin*, Vol. 2, p. 15.

Anon. (1978). *The hydrogeology of Africa*. Moscow: Nedra. 372 pp.

Armand, D.L. (1952). Principles of physical-geographical zonation. *Izv. AN SSSR, Ser. geogr.* No. 1, 67–82.

Balaev, L.G. (1976). Hydrogeological zonation of the USA. In *Collection of scientific papers of the VNII on hydrotechnics and irrigation*, No. 4. 71–99.

Bogolepov, K.V. & B.M. Chikov (1976). *Geology of the ocean bed*. Moscow: Nauka. 247 pp.

Dorokhov, L.A. (1973). The water resources of the Australian deserts. *Problemy osvoeniya pustyn'*, No. 1, pp. 23–33.

Dzhamalov, R.G., I.S. Zektser & A.V. Meskheteli (1977). *Underground runoff into the seas and the world ocean*. Moscow: Nauka. 93 pp.

Furon, R. (1955). *Introduction to the geology and hydrogeology of Turkey*. Moscow, IL. 144 pp. (Translated from the French.)

Grigorkina, T.E. (1977). The groundwater of Australia and its exploitation. *Izv. Vses. geogr. ob-va*, 109 (1), 96–100.

Kamenskii, G.N., M.M. Tolstikhina & N.I. Tolstikhin (1959). *Hydrogeology of the USSR*. Moscow: Gosgeoltekhizdat. 366 pp.

Karavanov, K.N. (1977). *Groundwater basins of the fold mountain regions of Eastern Asia. The features of their development and structure, and problems of classification and mapping*. Moscow: Nauka. 142 pp.

Khain, V.E. (1971, 1977). *Regional geotectonics*, 2 vols. Moscow: Nedra. Vol. 1, 1971, 548 pp., Vol. 2, 1977, 360 pp.

Kudelin, B.I. & I.F. Fideli (1970). Contribution to the problem of the hydrogeological zonation of the USSR. In *Fifth Scientific Account of the Geological Faculty of Moscow State University (16–19 March, 1970). Reports*. pp. 270–2. Moscow.

Lange, O.K. (1960). Distribution of groundwater in the Earth's crust. In *Problems of hydrogeology*, pp. 15–25. Moscow: Gosgeoltekhizdat.

Lange, O.K. (1969). *Hydrogeology*. Moscow: Vyshaya shkola. 363 pp.

Marinov, N.A. (1971). Principles of and the scheme of the hydrogeological zonation of Asia. In *Groundwater of Siberia and the Far East*, pp. 33–43. Moscow: Nauka.

Mil'kov, F.N. (1970). *Physical geography dictionary and reference book*, 2nd edn. Moscow: Mysl'. 343 pp.

Mitgarts, B.B. & N.I. Tolstikhin (1961). The hydrogeological zonation of Central Asia. In *Material on regional and prospecting hydrogeology*, pp. 49–76. Leningrad. (Trudy VSEGEI, nov. ser. Vol. 61.)

Neizvestnov, Ya.V. & N.I. Tolstikhin (1976). Subglacial groundwater basins of polar regions. In *Cryology of the Earth*, pp. 174–8. Moscow: Nauka.

Nikitin, M.R. (1969). Fundamental problems of hydrogeological cartography. In *Problems of regional hydrogeology and methodology of hydrogeological mapping*, pp. 187–215. Moscow. (Trudy VSEGINGEO, No. 24.)

Ovchinnikov, A.M. (1961). Pressure water systems of the Earth's crust. *Izv. vuzov. geologiya i razvedka*, No. 8, 85–90.

Ovchinnikov, A.M. (1963). *Mineral waters: studies of mineral water deposits based on hydrogeochemistry and radiohydrogeology*. Moscow: Gosgeoltekhizdat. 375 pp.

Pal'shin, G.B. (1959). The content of engineering geology survey maps. *Papers from the Second Conference on Groundwater and Engineering Geology of Eastern Siberia*, No. 2. pp. 43–9. Irkutsk.

Pinneker, E.V. (1977). *Problems of regional hydrogeology. Laws governing the occurrence and formation of groundwater*. Moscow: Nauka. 196 pp.

Plotnikov, N.I. (1965). Prospecting for fresh groundwater for large-scale water supply. Part 1 of *Fundamentals of hydrogeology*. Moscow: Izd-vo MGU, 243 pp.

Rogovskaya, N.V. & V.N. Kunin (1969). An international hydrogeological map. *Sov. geologiya*, No. 11, 140–3.

Tkachuk, B.G. & I.I. Molodykh (1972). Complex natural zonation for irrigation purposes. *Gidrotekhnika i melioratsiya*, No. 3, 4–11.

Tolstikhin, N.I. (1971). Hydrogeology of the Earth and the cryosphere. In *Groundwater of Siberia and the Far East*, pp. 28–33. Moscow: Nauka.

Zaitsev, I.K. & N.I. Tolstikhin (1963). The basis of the structural–geological zonation of the USSR. In *Material on regional and prospecting hydrogeology*, pp. 5–35. (Trudy VSEGEI, nov. ser. Vol. 101.)

Zaitsev, I.K. & N.I. Tolstikhin (1972). *The laws of the distribution and formation of mineral (industrial and medicinal) groundwater in the USSR*. Moscow: Nedra. 279 pp.

135

Conclusion

This volume is dedicated to the examination of the features of the phenomenon and the general laws governing the occurrence of groundwater in the interior of the Earth. The authors have tried to rely on a unified system of views reflecting the present day state of knowledge. However, in connection with the non-equivalence of the factual basis of various branches of the science of the subsurface hydrosphere one of the positions expressed is based on well tried and tested material and the other is in need of corroboration by additional information or a more convincing theoretical base. The latter concerns mainly the problematical questions of the chapters and sections in which to place the treatment of the material which differs from the traditional, or to make attempts to find the proper approach to the solution of the hydrogeological problems.

New ideas, naturally, do not conquer science. 'Supposition precedes knowledge.' These words of A. Humboldt reflect very accurately the development of the concepts of the subsurface hydrosphere. Nowadays, when new and unique information, a laboratory experiment, and especially a quantitative assessment of processes, permit the use of figures and formulae in the description of hydrogeological laws, many traditional positions lose their strength and they must give way to others which are more fundamental.

Together with the basic aim, that of the discovery of the general laws of hydrogeology, there are the debatable problems which have been mentioned above and which still await solution. Some of these are directly concerned with the general laws of hydrogeology. For the overwhelming majority they are at the stage of hypothesis or proposition. Furthermore there are also concepts being expressed which may already be in practical use in hydrogeological research. Among these are the concepts of the hydrologic cycle in the Earth, the classification of groundwater according to the way in which it is formed, the principle of structural–hydrogeological zonation, etc.

The new definition of hydrogeology as a science deserves special attention. Actually, in order to understand the life of water in the Earth's interior, it is necessary to study all its varieties. An examination of the general laws of hydrogeology shows that this science has gradually changed from a study of groundwater to the science of the whole subsurface hydrosphere, although groundwater up till now remains, and will probably continue to remain, the basic object of its study.

It is difficult to give up a long held position. As yet not everyone shares the notion that groundwater is not the only branch of hydrogeology. In essence this volume is concerned with the subsurface hydrosphere. Hence it is hoped to conclude it with a statement of the laws which result from the dialectical approach to the new subject of hydrogeology.

(1) The natural waters of the world are one, but this unity is completely contradictory. The subsurface hydrosphere, which occupies not only the continental but also the oceanic crust, contains water in different phases (liquid, solid, and gas), and states (free and bound), in the form of solutions of different degrees of mineralization, temperatures, etc. The contradictory nature of the subsurface hydrosphere is thus already plain to see, although all its components are interconnected. Existing as a material form, it is distinguished by the interaction of its opposing origins. The variety of its contradictory branches is integrated in a fundamental contradiction – the opposition of the surface and deep influences, of the spatial and temporal differences. The interaction of exogenic and endogenic processes thus determines the stability and changeability of the subsurface hydrosphere. In the course of geological time the unity and struggle of these contradictions also form the basis of its internal development, and are the sources of its mobility.

(2) The component parts of the subsurface hydrosphere are subject to changes, both insignificant and quantitative and fundamental and qualitative. There occur changes which are only quantitative (the reduction of the velocity of groundwater with depth, the increase of its degree of mineralization with depth), but as a whole uneven transitions of quantitative changes to qualitative are typical of the subsurface hydrosphere. Such transitions take many forms, they may be sudden (the change of water from the liquid phase to the solid and to the gas phases), or gradual (the liberation of bound water from minerals), and sometimes stretched out over long periods of geological time (changes of ion-salt and gaseous composition of groundwater). The varieties of form of the transition of quantitative changes to qualitative reflect the character of the path of development of the subsurface hydrosphere.

(3) The development of the subsurface hydrosphere is cyclic and progressive. The cyclic nature of the processes is seen in the water exchange with the surface hydrosphere, the replacement of magmatogenic and sedimentogenic water by infiltration water, the transition of water from one phase or one state to another and vice versa. Each succeeding cycle terminates the preceding one, although the continuity is maintained. In this there is not a simple repetition, but a qualitatively new and higher stage, which shows a progressive spiral development. The existing reflection of the dialectical law of negation or termination, the cyclic nature, and the progressive character of the processes reflect the historical continuity of the development of the subsurface hydrosphere.

Hydrogeology is a science in which interest is continually growing. As has already been said, the enormous and ever growing demand for water on the one hand, and the very important task of conserving the resources of the subsurface hydrosphere, which arises as a result of the ever increasing load placed on it by industry on the other hand, are the reasons for this interest. The most important part of the environment is the object of study of hydrogeology, i.e. that part which demands systematic, profound, and comprehensive study in order to regulate man's environment. Therefore the interest in the study of the waters of the Earth's interior is not surprising.

Hydrogeology faces complex problems which are very real both in the practical and the scientific senses. We will name the most important of them.

Quantitative–qualitative evaluation of the resources of the subsurface hydrosphere. This problem is probably the most important because every investigation starts with the evaluation of what is considered to be a subject of research. Not for nothing is it said 'To study is to measure.' Furthermore, we still know little about the subsurface hydrosphere: for example, less, no matter how strange it may seem, than the cosmos; even about the resources and the composition of the most important component – the groundwater – our evaluations are extremely approximate. The aqueous solutions of the interior form a dynamic system which contains H_2O in different phases and conditions and is characterized by continually changing composition. Moreover, they are in continuous circulation, being connected with the mantle and the cosmos. And this is one of the basic difficulties of a quantitative–qualitative evaluation of the resources of the subsurface hydrosphere. What are important are not so much the global as the territorial evaluations, primarily the resources and the composition of the groundwater, whether it is suitable for water supply and irrigation, or for medicinal and industrial purposes, and for the extraction of thermal energy.

The final outcome of the quantitative–qualitative evaluations of the resources of the subsurface hydrosphere must be the laws which govern the occurrence of water in the interior of the Earth. In conformity with groundwater they find expression in the hydrogeological zonation, different forms of zonation, etc. The task is to study the laws governing the occurrence of other components of the subsurface hydrosphere more deeply and to understand their interrelationships with the groundwater.

The study of the origin and evolution of the subsurface hydrosphere is the next problem, which may be regarded as the historical basis of hydrogeology. The paths along which water migrates within the Earth, the energetics of the water–rock–gas–living-material, the water exchange and mass transfer in the subsurface hydrosphere – all this in short is the essence of a similar school of research, although frequently hydrogeologists restrict it still further – to the formation of groundwater.

In the study of the formation of groundwater, two sides of this complex matter can be distinguished: firstly the origin of the resources of the groundwater, and secondly, the origin of its composition. Both sides are closely connected and inseparable one from the other. Nevertheless, each has its own value, because the solvent (water) and the dissolved material (ions, salts, gases, organic compounds), over the span of geological history, undergo various changes.

In recent years notable success has been achieved in the diagnosis of the sources of groundwater, which has become possible as a result of the purposeful study, not only of the 'liquid', but also vapour, 'solid', and bound water. Palaeo-hydrogeological reconstructions, analysis using isotopes and thermodynamic methods, hydrogeodynamic, hydrogeo-chemical, and hydrogeothermal data find fruitful application. The study of the origin of dissolved material enables factors (the driving forces) which determine the composition of the groundwater, the processes which create it, the circumstances in which these factors operate, and the processes, to be studied. The processes which form the composition of the groundwater, and the water–rock–gas–living-material system interact closely and it is impossible to treat them in isolation.

With regard to the origin and evolution of the subsurface hydrosphere, including the formation of groundwater, because of lack of sophisticated information there sometimes exist different, often mutually exclusive, points of view in which the truth is concealed. How does water behave in the interior of the Earth – does it move, or is it stationary? A large number of facts, obtained over the last 25 years, demonstrate convincingly that there is no stagnation in the deep horizons. 'Everything is in a state of change' – this thesis of natural scientists in ancient times applies to the whole of the hydro-sphere. Moreover, in the crust there is a ramifying system of deep runoff and drainage. Water in the geological history of the Earth is continually generated at different levels and in different currents. As a consequence of its high dissolving power it has assumed the role of universal carrier of chemical elements, both in carrying them from the mantle and also in other situations, redistributing them within the crust. Water is the most important element of the geological form of the movement of material.

Further research must be directed towards the study of the laws governing the formation of groundwater, taking into account too the indissoluble link between water exchange and mass transfer for the subsurface hydrosphere as a whole. In order to understand the laws and processes it is necessary to translate them into the language of mathematics; this means the use of such parameters as number and magnitude.

The geological activity in the interior of the Earth has long been the subject of study, but it is only in recent times, when geophysical and geochemical techniques came to the aid of hydrogeologists, that it has begun to be seen in a com-pletely new light.

Water is an indispensable component of literally all geo-logical processes. Not only at the surface of the Earth where its work is most noticeable, but also deep in the Earth's interior is this fact seen. If the zone of hypergene processes as a result of geological activity is completely reformed, then in the deep layers it regulates the intensity of these processes (magmatic, seismic, etc.) which take place there, influencing them decisively. Even the activation of tectonic movements, according to new research, is frequently caused by the pro-cesses originating in the subsurface hydrosphere (in particular the dehydration of minerals). Without any exaggeration water is the sculptor of the Earth's crust. As V.I. Vernadskii said 'The map of visible nature is determined by water.' It is important to note that it is not a passive participator, but a geological agent, overflowing with active energy.

Water plays an extremely important role in the for-mation of mineral deposits. The genesis of oil and gas struc-tures, and of sedimentary and metamorphic ores, the mineralizing activity of gas–liquid solutions, all these pro-cesses take place in an aqueous medium. But water is not only the creator – it is also a destroyer: whether a deposit is preserved or not depends on the nature of its action. Hence without knowing the geological role of water, it is impossible to understand the laws of distribution of minerals and prospect for them competently.

As an example of geological activity it is plain that 'liquid' free water must necessarily be studied in conjunction with other components of the subsurface hydrosphere (pore solutions, gas–liquid hydrothermals, water of hydration, etc.). The subsurface hydrosphere as a whole, the different varieties of H_2O, i.e. in all phases and states, must be the subject of research.

Taking part in various processes as it does, water carries information about them. The theme 'As the rocks, so the water' provides a solution to the reverse problem which forms the basis of the hydrochemical method of prospecting for minerals, in which deposits are discovered by finding the composition of the groundwater which lies deep in the interior of the Earth. In precisely the same way observations on the groundwater regime in seismically active regions enables the probability of earthquakes to be forecast.

It is becoming obvious that our concepts about the geo-logical activity of water are inadequate, especially about its role in endogenic processes. On the basis of present day geo-logical concepts, geochemical and geophysical data offer the possibility of interpreting the geological activity of water at different levels (in the crust and the mantle) and in different thermodynamic circumstances. There can scarcely be any doubt as to the fruitful outcome of such an approach.

Man, exploiting the reserves of the subsurface hydrosphere, has an ever greater influence on its components. To halt the influence of industrial activity, or, in modern terminology, the industrial load, is difficult – it is progressive. Therefore life persistently puts forward the problem of the *rational use and preservation of the resources of the subsurface hydro-sphere*. The industrial load is noticeable at various depths;

primarily because of its influence groundwater and soil moisture become exhausted and polluted. Gradually this state of affairs spreads to pore solutions, underground ice, and other forms of water in the interior of the Earth. We would like to emphasize that the increasing industrial load on the subsurface hydrosphere in the absence of compensating factors leads to irreversible consequences and the natural regime is replaced by an unbalanced one. The latter is exactly regulated by the sum of the industrial load which becomes more perceptible each year. In the foreseeable future an unbalanced regime will predominate.

Thus, geologists are faced with the very complex but interesting problem of the *control of the groundwater regime*. In essence the question is about a problem which is difficult and new in principle: we are faced with finding an optimum solution to the interrelationships of the subsurface hydrosphere with the load placed on it by industry. An excellent formulation of the problem is to say that what is required is a quantitative prognosis of the possible consequences and the devising of measures for the regulation of the regime of the subsurface hydrosphere. This problem, while it is of a particularly applied nature because in the final analysis it is expected to yield practical recommendations and the basis for any technical measures, rests upon a comprehensive theoretical foundation. After all, interference in the life of the subsurface hydrosphere demands a profound knowledge of the hydrogeological processes and the hydrogeological laws.

Of course, the problem above is just one of many problems which are worrying geologists. But those we have touched upon indicate an attractive future for hydrogeology. There are so many problems which are as yet unmentioned! It is sufficient to cite the study of water deep inside the planet, and the erection on this base of a new science — that of hydroplanetology.

Now, when hydrogeology is proceeding from the description of circumstances to the explanation of hydrogeological processes, to the revealing of the laws inherent in the subject and the devising of methods of scientific prognosis, its further development as a science demands a courageous synthesis of the 'descriptive', 'explanatory', and 'prognostic' stages. Hydrogeology is in need not only of an explanation of processes and scientific foresight, but also of a new set of facts. At times such information means something more than far-fetched concepts. And the theories, because of the absence of convincing foundations, in a number of cases, do not come up to the level required for answering unequivocally the various questions which arise from practical experience.

Being a complex science, hydrogeology is developing along many lines. Along with the quantitative and qualitative study of groundwater, other components of the subsurface hydrosphere, such as the history of the water in the interior of the Earth and its geological activity, are the subjects of constant interest. The period of accumulation of information about the subsurface hydrosphere has already been replaced by the summarizing of the processes and their laws and the formation of new concepts. Furthermore, it seems, we find ourselves on the road to the discovery of new knowledge.

Index

Printed by Printforce, United Kingdom